MW01230199

GRAPH ALGEBRAS AND AUTOMATA

MONOGRAPHS AND TEXTBOOKS IN PURE AND APPLIED MATHEMATICS

1. *K. Yano,* Integral Formulas in Riemannian Geometry (1970)
2. *S. Kobayashi,* Hyperbolic Manifolds and Holomorphic Mappings (1970)
3. *V. S. Vladimirov,* Equations of Mathematical Physics (A. Jeffrey, ed.; A. Littlewood, trans.) (1970)
4. *B. N. Pshenichnyi,* Necessary Conditions for an Extremum (L. Neustadt, translation ed.; K. Makowski, trans.) (1971)
5. *L. Narici et al.,* Functional Analysis and Valuation Theory (1971)
6. *S. S. Passman,* Infinite Group Rings (1971)
7. *L. Dornhoff,* Group Representation Theory. Part A: Ordinary Representation Theory. Part B: Modular Representation Theory (1971, 1972)
8. *W. Boothby and G. L. Weiss, eds.,* Symmetric Spaces (1972)
9. *Y. Matsushima,* Differentiable Manifolds (E. T. Kobayashi, trans.) (1972)
10. *L. E. Ward, Jr.,* Topology (1972)
11. *A. Babakhanian,* Cohomological Methods in Group Theory (1972)
12. *R. Gilmer,* Multiplicative Ideal Theory (1972)
13. *J. Yeh,* Stochastic Processes and the Wiener Integral (1973)
14. *J. Barros-Neto,* Introduction to the Theory of Distributions (1973)
15. *R. Larsen,* Functional Analysis (1973)
16. *K. Yano and S. Ishihara,* Tangent and Cotangent Bundles (1973)
17. *C. Procesi,* Rings with Polynomial Identities (1973)
18. *R. Hermann,* Geometry, Physics, and Systems (1973)
19. *N. R. Wallach,* Harmonic Analysis on Homogeneous Spaces (1973)
20. *J. Dieudonné,* Introduction to the Theory of Formal Groups (1973)
21. *I. Vaisman,* Cohomology and Differential Forms (1973)
22. *B.-Y. Chen,* Geometry of Submanifolds (1973)
23. *M. Marcus,* Finite Dimensional Multilinear Algebra (in two parts) (1973, 1975)
24. *R. Larsen,* Banach Algebras (1973)
25. *R. O. Kujala and A. L. Vitter, eds.,* Value Distribution Theory: Part A; Part B: Deficit and Bezout Estimates by Wilhelm Stoll (1973)
26. *K. B. Stolarsky,* Algebraic Numbers and Diophantine Approximation (1974)
27. *A. R. Magid,* The Separable Galois Theory of Commutative Rings (1974)
28. *B. R. McDonald,* Finite Rings with Identity (1974)
29. *J. Satake,* Linear Algebra (S. Koh et al., trans.) (1975)
30. *J. S. Golan,* Localization of Noncommutative Rings (1975)
31. *G. Klambauer,* Mathematical Analysis (1975)
32. *M. K. Agoston,* Algebraic Topology (1976)
33. *K. R. Goodearl,* Ring Theory (1976)
34. *L. E. Mansfield,* Linear Algebra with Geometric Applications (1976)
35. *N. J. Pullman,* Matrix Theory and Its Applications (1976)
36. *B. R. McDonald,* Geometric Algebra Over Local Rings (1976)
37. *C. W. Groetsch,* Generalized Inverses of Linear Operators (1977)
38. *J. E. Kuczkowski and J. L. Gersting,* Abstract Algebra (1977)
39. *C. O. Christenson and W. L. Voxman,* Aspects of Topology (1977)
40. *M. Nagata,* Field Theory (1977)
41. *R. L. Long,* Algebraic Number Theory (1977)
42. *W. F. Pfeffer,* Integrals and Measures (1977)
43. *R. L. Wheeden and A. Zygmund,* Measure and Integral (1977)
44. *J. H. Curtiss,* Introduction to Functions of a Complex Variable (1978)
45. *K. Hrbacek and T. Jech,* Introduction to Set Theory (1978)
46. *W. S. Massey,* Homology and Cohomology Theory (1978)
47. *M. Marcus,* Introduction to Modern Algebra (1978)
48. *E. C. Young,* Vector and Tensor Analysis (1978)
49. *S. B. Nadler, Jr.,* Hyperspaces of Sets (1978)
50. *S. K. Segal,* Topics in Group Kings (1978)
51. *A. C. M. van Rooij,* Non-Archimedean Functional Analysis (1978)
52. *L. Corwin and R. Szczarba,* Calculus in Vector Spaces (1979)
53. *C. Sadosky,* Interpolation of Operators and Singular Integrals (1979)
54. *J. Cronin,* Differential Equations (1980)
55. *C. W. Groetsch,* Elements of Applicable Functional Analysis (1980)

56. *I. Vaisman,* Foundations of Three-Dimensional Euclidean Geometry (1980)
57. *H. I. Freedan,* Deterministic Mathematical Models in Population Ecology (1980)
58. *S. B. Chae,* Lebesgue Integration (1980)
59. *C. S. Rees et al.,* Theory and Applications of Fourier Analysis (1981)
60. *L. Nachbin,* Introduction to Functional Analysis (R. M. Aron, trans.) (1981)
61. *G. Orzech and M. Orzech,* Plane Algebraic Curves (1981)
62. *R. Johnsonbaugh and W. E. Pfaffenberger,* Foundations of Mathematical Analysis (1981)
63. *W. L. Voxman and R. H. Goetschel,* Advanced Calculus (1981)
64. *L. J. Corwin and R. H. Szczarba,* Multivariable Calculus (1982)
65. *V. I. Istrățescu,* Introduction to Linear Operator Theory (1981)
66. *R. D. Järvinen,* Finite and Infinite Dimensional Linear Spaces (1981)
67. *J. K. Beem and P. E. Ehrlich,* Global Lorentzian Geometry (1981)
68. *D. L. Armacost,* The Structure of Locally Compact Abelian Groups (1981)
69. *J. W. Brewer and M. K. Smith, eds.,* Emmy Noether: A Tribute (1981)
70. *K. H. Kim,* Boolean Matrix Theory and Applications (1982)
71. *T. W. Wieting,* The Mathematical Theory of Chromatic Plane Ornaments (1982)
72. *D. B.Gauld,* Differential Topology (1982)
73. *R. L. Faber,* Foundations of Euclidean and Non-Euclidean Geometry (1983)
74. *M. Carmeli,* Statistical Theory and Random Matrices (1983)
75. *J. H. Carruth et al.,* The Theory of Topological Semigroups (1983)
76. *R. L. Faber,* Differential Geometry and Relativity Theory (1983)
77. *S. Barnett,* Polynomials and Linear Control Systems (1983)
78. *G. Karpilovsky,* Commutative Group Algebras (1983)
79. *F. Van Oystaeyen and A. Verschoren,* Relative Invariants of Rings (1983)
80. *I. Vaisman,* A First Course in Differential Geometry (1984)
81. *G. W. Swan,* Applications of Optimal Control Theory in Biomedicine (1984)
82. *T. Petrie and J. D. Randall,* Transformation Groups on Manifolds (1984)
83. *K. Goebel and S. Reich,* Uniform Convexity, Hyperbolic Geometry, and Nonexpansive Mappings (1984)
84. *T. Albu and C. Năstăsescu,* Relative Finiteness in Module Theory (1984)
85. *K. Hrbacek and T. Jech,* Introduction to Set Theory: Second Edition (1984)
86. *F. Van Oystaeyen and A. Verschoren,* Relative Invariants of Rings (1984)
87. *B. R. McDonald,* Linear Algebra Over Commutative Rings (1984)
88. *M. Namba,* Geometry of Projective Algebraic Curves (1984)
89. *G. F. Webb,* Theory of Nonlinear Age-Dependent Population Dynamics (1985)
90. *M. R. Bremner et al.,* Tables of Dominant Weight Multiplicities for Representations of Simple Lie Algebras (1985)
91. *A. E. Fekete,* Real Linear Algebra (1985)
92. *S. B. Chae,* Holomorphy and Calculus in Normed Spaces (1985)
93. *A. J. Jerri,* Introduction to Integral Equations with Applications (1985)
94. *G. Karpilovsky,* Projective Representations of Finite Groups (1985)
95. *L. Narici and E. Beckenstein,* Topological Vector Spaces (1985)
96. *J. Weeks,* The Shape of Space (1985)
97. *P. R. Gribik and K. O. Kortanek,* Extremal Methods of Operations Research (1985)
98. *J.-A. Chao and W. A. Woyczynski, eds.,* Probability Theory and Harmonic Analysis (1986)
99. *G. D. Crown et al.,* Abstract Algebra (1986)
100. *J. H. Carruth et al.,* The Theory of Topological Semigroups, Volume 2 (1986)
101. *R. S. Doran and V. A. Belfi,* Characterizations of C*-Algebras (1986)
102. *M. W. Jeter,* Mathematical Programming (1986)
103. *M. Altman,* A Unified Theory of Nonlinear Operator and Evolution Equations with Applications (1986)
104. *A. Verschoren,* Relative Invariants of Sheaves (1987)
105. *R. A. Usmani,* Applied Linear Algebra (1987)
106. *P. Blass and J. Lang,* Zariski Surfaces and Differential Equations in Characteristic $p > 0$ (1987)
107. *J. A. Reneke et al.,* Structured Hereditary Systems (1987)
108. *H. Busemann and B. B. Phadke,* Spaces with Distinguished Geodesics (1987)
109. *R. Harte,* Invertibility and Singularity for Bounded Linear Operators (1988)
110. *G. S. Ladde et al.,* Oscillation Theory of Differential Equations with Deviating Arguments (1987)
111. *L. Dudkin et al.,* Iterative Aggregation Theory (1987)
112. *T. Okubo,* Differential Geometry (1987)

113. *D. L. Stancl and M. L. Stancl,* Real Analysis with Point-Set Topology (1987)
114. *T. C. Gard,* Introduction to Stochastic Differential Equations (1988)
115. *S. S. Abhyankar,* Enumerative Combinatorics of Young Tableaux (1988)
116. *H. Strade and R. Farnsteiner,* Modular Lie Algebras and Their Representations (1988)
117. *J. A. Huckaba,* Commutative Rings with Zero Divisors (1988)
118. *W. D. Wallis,* Combinatorial Designs (1988)
119. *W. Więsław,* Topological Fields (1988)
120. *G. Karpilovsky,* Field Theory (1988)
121. *S. Caenepeel and F. Van Oystaeyen,* Brauer Groups and the Cohomology of Graded Rings (1989)
122. *W. Kozlowski,* Modular Function Spaces (1988)
123. *E. Lowen-Colebunders,* Function Classes of Cauchy Continuous Maps (1989)
124. *M. Pavel,* Fundamentals of Pattern Recognition (1989)
125. *V. Lakshmikantham et al.,* Stability Analysis of Nonlinear Systems (1989)
126. *R. Sivaramakrishnan,* The Classical Theory of Arithmetic Functions (1989)
127. *N. A. Watson,* Parabolic Equations on an Infinite Strip (1989)
128. *K. J. Hastings,* Introduction to the Mathematics of Operations Research (1989)
129. *B. Fine,* Algebraic Theory of the Bianchi Groups (1989)
130. *D. N. Dikranjan et al.,* Topological Groups (1989)
131. *J. C. Morgan II,* Point Set Theory (1990)
132. *P. Biler and A. Witkowski,* Problems in Mathematical Analysis (1990)
133. *H. J. Sussmann,* Nonlinear Controllability and Optimal Control (1990)
134. *J.-P. Florens et al.,* Elements of Bayesian Statistics (1990)
135. *N. Shell,* Topological Fields and Near Valuations (1990)
136. *B. F. Doolin and C. F. Martin,* Introduction to Differential Geometry for Engineers (1990)
137. *S. S. Holland, Jr.,* Applied Analysis by the Hilbert Space Method (1990)
138. *J. Okniński,* Semigroup Algebras (1990)
139. *K. Zhu,* Operator Theory in Function Spaces (1990)
140. *G. B. Price,* An Introduction to Multicomplex Spaces and Functions (1991)
141. *R. B. Darst,* Introduction to Linear Programming (1991)
142. *P. L. Sachdev,* Nonlinear Ordinary Differential Equations and Their Applications (1991)
143. *T. Husain,* Orthogonal Schauder Bases (1991)
144. *J. Foran,* Fundamentals of Real Analysis (1991)
145. *W. C. Brown,* Matrices and Vector Spaces (1991)
146. *M. M. Rao and Z. D. Ren,* Theory of Orlicz Spaces (1991)
147. *J. S. Golan and T. Head,* Modules and the Structures of Rings (1991)
148. *C. Small,* Arithmetic of Finite Fields (1991)
149. *K. Yang,* Complex Algebraic Geometry (1991)
150. *D. G. Hoffman et al.,* Coding Theory (1991)
151. *M. O. González,* Classical Complex Analysis (1992)
152. *M. O. González,* Complex Analysis (1992)
153. *L. W. Baggett,* Functional Analysis (1992)
154. *M. Sniedovich,* Dynamic Programming (1992)
155. *R. P. Agarwal,* Difference Equations and Inequalities (1992)
156. *C. Brezinski,* Biorthogonality and Its Applications to Numerical Analysis (1992)
157. *C. Swartz,* An Introduction to Functional Analysis (1992)
158. *S. B. Nadler, Jr.,* Continuum Theory (1992)
159. *M. A. Al-Gwaiz,* Theory of Distributions (1992)
160. *E. Perry,* Geometry: Axiomatic Developments with Problem Solving (1992)
161. *E. Castillo and M. R. Ruiz-Cobo,* Functional Equations and Modelling in Science and Engineering (1992)
162. *A. J. Jerri,* Integral and Discrete Transforms with Applications and Error Analysis (1992)
163. *A. Charlier et al.,* Tensors and the Clifford Algebra (1992)
164. *P. Biler and T. Nadzieja,* Problems and Examples in Differential Equations (1992)
165. *E. Hansen,* Global Optimization Using Interval Analysis (1992)
166. *S. Guerre-Delabrière,* Classical Sequences in Banach Spaces (1992)
167. *Y. C. Wong,* Introductory Theory of Topological Vector Spaces (1992)
168. *S. H. Kulkarni and B. V. Limaye,* Real Function Algebras (1992)
169. *W. C. Brown,* Matrices Over Commutative Rings (1993)
170. *J. Loustau and M. Dillon,* Linear Geometry with Computer Graphics (1993)
171. *W. V. Petryshyn,* Approximation-Solvability of Nonlinear Functional and Differential Equations (1993)

172. *E. C. Young*, Vector and Tensor Analysis: Second Edition (1993)
173. *T. A. Bick*, Elementary Boundary Value Problems (1993)
174. *M. Pavel*, Fundamentals of Pattern Recognition: Second Edition (1993)
175. *S. A. Albeverio et al.*, Noncommutative Distributions (1993)
176. *W. Fulks*, Complex Variables (1993)
177. *M. M. Rao*, Conditional Measures and Applications (1993)
178. *A. Janicki and A. Weron*, Simulation and Chaotic Behavior of α-Stable Stochastic Processes (1994)
179. *P. Neittaanmäki and D. Tiba*, Optimal Control of Nonlinear Parabolic Systems (1994)
180. *J. Cronin*, Differential Equations: Introduction and Qualitative Theory, Second Edition (1994)
181. *S. Heikkilä and V. Lakshmikantham*, Monotone Iterative Techniques for Discontinuous Nonlinear Differential Equations (1994)
182. *X. Mao*, Exponential Stability of Stochastic Differential Equations (1994)
183. *B. S. Thomson*, Symmetric Properties of Real Functions (1994)
184. *J. E. Rubio*, Optimization and Nonstandard Analysis (1994)
185. *J. L. Bueso et al.*, Compatibility, Stability, and Sheaves (1995)
186. *A. N. Michel and K. Wang*, Qualitative Theory of Dynamical Systems (1995)
187. *M. R. Darnel*, Theory of Lattice-Ordered Groups (1995)
188. *Z. Naniewicz and P. D. Panagiotopoulos*, Mathematical Theory of Hemivariational Inequalities and Applications (1995)
189. *L. J. Corwin and R. H. Szczarba*, Calculus in Vector Spaces: Second Edition (1995)
190. *L. H. Erbe et al.*, Oscillation Theory for Functional Differential Equations (1995)
191. *S. Agaian et al.*, Binary Polynomial Transforms and Nonlinear Digital Filters (1995)
192. *M. I. Gil'*, Norm Estimations for Operation-Valued Functions and Applications (1995)
193. *P. A. Grillet*, Semigroups: An Introduction to the Structure Theory (1995)
194. *S. Kichenassamy*, Nonlinear Wave Equations (1996)
195. *V. F. Krotov*, Global Methods in Optimal Control Theory (1996)
196. *K. I. Beidar et al.*, Rings with Generalized Identities (1996)
197. *V. I. Arnautov et al.*, Introduction to the Theory of Topological Rings and Modules (1996)
198. *G. Sierksma*, Linear and Integer Programming (1996)
199. *R. Lasser*, Introduction to Fourier Series (1996)
200. *V. Sima*, Algorithms for Linear-Quadratic Optimization (1996)
201. *D. Redmond*, Number Theory (1996)
202. *J. K. Beem et al.*, Global Lorentzian Geometry: Second Edition (1996)
203. *M. Fontana et al.*, Prüfer Domains (1997)
204. *H. Tanabe*, Functional Analytic Methods for Partial Differential Equations (1997)
205. *C. Q. Zhang*, Integer Flows and Cycle Covers of Graphs (1997)
206. *E. Spiegel and C. J. O'Donnell*, Incidence Algebras (1997)
207. *B. Jakubczyk and W. Respondek*, Geometry of Feedback and Optimal Control (1998)
208. *T. W. Haynes et al.*, Fundamentals of Domination in Graphs (1998)
209. *T. W. Haynes et al., eds.*, Domination in Graphs: Advanced Topics (1998)
210. *L. A. D'Alotto et al.*, A Unified Signal Algebra Approach to Two-Dimensional Parallel Digital Signal Processing (1998)
211. *F. Halter-Koch*, Ideal Systems (1998)
212. *N. K. Govil et al., eds.*, Approximation Theory (1998)
213. *R. Cross*, Multivalued Linear Operators (1998)
214. *A. A. Martynyuk*, Stability by Liapunov's Matrix Function Method with Applications (1998)
215. *A. Favini and A. Yagi*, Degenerate Differential Equations in Banach Spaces (1999)
216. *A. Illanes and S. Nadler, Jr.*, Hyperspaces: Fundamentals and Recent Advances (1999)
217. *G. Kato and D. Struppa*, Fundamentals of Algebraic Microlocal Analysis (1999)
218. *G. X.-Z. Yuan*, KKM Theory and Applications in Nonlinear Analysis (1999)
219. *D. Motreanu and N. H. Pavel*, Tangency, Flow Invariance for Differential Equations, and Optimization Problems (1999)
220. *K. Hrbacek and T. Jech*, Introduction to Set Theory, Third Edition (1999)
221. *G. E. Kolosov*, Optimal Design of Control Systems (1999)
222. *N. L. Johnson*, Subplane Covered Nets (2000)
223. *B. Fine and G. Rosenberger*, Algebraic Generalizations of Discrete Groups (1999)
224. *M. Väth*, Volterra and Integral Equations of Vector Functions (2000)
225. *S. S. Miller and P. T. Mocanu*, Differential Subordinations (2000)

226. *R. Li et al.*, Generalized Difference Methods for Differential Equations: Numerical Analysis of Finite Volume Methods (2000)
227. *H. Li and F. Van Oystaeyen*, A Primer of Algebraic Geometry (2000)
228. *R. P. Agarwal*, Difference Equations and Inequalities: Theory, Methods, and Applications, Second Edition (2000)
229. *A. B. Kharazishvili*, Strange Functions in Real Analysis (2000)
230. *J. M. Appell et al.*, Partial Integral Operators and Integro-Differential Equations (2000)
231. *A. I. Prilepko et al.*, Methods for Solving Inverse Problems in Mathematical Physics (2000)
232. *F. Van Oystaeyen*, Algebraic Geometry for Associative Algebras (2000)
233. *D. L. Jagerman*, Difference Equations with Applications to Queues (2000)
234. *D. R. Hankerson et al.*, Coding Theory and Cryptography: The Essentials, Second Edition, Revised and Expanded (2000)
235. *S. Dăscălescu et al.*, Hopf Algebras: An Introduction (2001)
236. *R. Hagen et al.*, C*-Algebras and Numerical Analysis (2001)
237. *Y. Talpaert*, Differential Geometry: With Applications to Mechanics and Physics (2001)
238. *R. H. Villarreal*, Monomial Algebras (2001)
239. *A. N. Michel et al.*, Qualitative Theory of Dynamical Systems: Second Edition (2001)
240. *A. A. Samarskii*, The Theory of Difference Schemes (2001)
241. *J. Knopfmacher and W.-B. Zhang*, Number Theory Arising from Finite Fields (2001)
242. *S. Leader*, The Kurzweil-Henstock Integral and Its Differentials (2001)
243. *M. Biliotti et al.*, Foundations of Translation Planes (2001)
244. *A. N. Kochubei*, Pseudo-Differential Equations and Stochastics over Non-Archimedean Fields (2001)
245. *G. Sierksma*, Linear and Integer Programming: Second Edition (2002)
246. *A. A. Martynyuk*, Qualitative Methods in Nonlinear Dynamics: Novel Approaches to Liapunov's Matrix Functions (2002)
247. *B. G. Pachpatte*, Inequalities for Finite Difference Equations (2002)
248. *A. N. Michel and D. Liu*, Qualitative Analysis and Synthesis of Recurrent Neural Networks (2002)
249. *J. R. Weeks*, The Shape of Space: Second Edition (2002)
250. *M. M. Rao and Z. D. Ren*, Applications of Orlicz Spaces (2002)
251. *V. Lakshmikantham and D. Trigiante*, Theory of Difference Equations: Numerical Methods and Applications, Second Edition (2002)
252. *T. Albu*, Cogalois Theory (2003)
253. *A. Bezdek*, Discrete Geometry (2003)
254. *M. J. Corless and A. E. Frazho*, Linear Systems and Control: An Operator Perspective (2003)
255. *I. Graham and G. Kohr*, Geometric Function Theory in One and Higher Dimensions (2003)
256. *G. V. Demidenko and S. V. Uspenskii*, Partial Differential Equations and Systems Not Solvable with Respect to the Highest-Order Derivative (2003)
257. *A. Kelarev*, Graph Algebras and Automata (2003)

Additional Volumes in Preparation

GRAPH ALGEBRAS AND AUTOMATA

Andrei Kelarev
University of Tasmania
Hobart, Tasmania, Austrailia

Marcel Dekker, Inc. New York • Basel

Library of Congress Cataloging-in-Publication Data
A catalog record for this book is available from the Library of Congress.

ISBN: 0-8247-4708-9

This book is printed on acid-free paper.

Headquarters
Marcel Dekker, Inc., 270 Madison Avenue, New York, NY 10016, U.S.A.
tel: 212-696-9000; fax: 212-685-4540

Distribution and Customer Service
Marcel Dekker, Inc., Cimarron Road, Monticello, New York 12701, U.S.A.
tel: 800-228-1160; fax: 845-796-1772

Eastern Hemisphere Distribution
Marcel Dekker AG, Hutgasse 4, Postfach 812, CH-4001 Basel, Switzerland
tel: 41-61-260-6300; fax: 41-61-260-6333

World Wide Web
http://www.dekker.com

The publisher offers discounts on this book when ordered in bulk quantities. For more information, write to Special Sales/Professional Marketing at the headquarters address above.

Current printing (last digit):

10 9 8 7 6 5 4 3 2 1

PRINTED IN THE UNITED STATES OF AMERICA

To my mother and my children

Preface

The concept of a graph algebra was introduced by G.F. McNulty and C.R. Shallon in 1983. Let $D = (V, E)$ be a directed graph. The *graph algebra* associated with D is the set $V \cup \{0\}$ equipped with multiplication defined by the rule

$$xy = \begin{cases} x & \text{if } x, y \in V \text{ and } (x, y) \in E, \\ 0 & \text{otherwise.} \end{cases}$$

Since then graph algebras have been actively investigated. It has become clear that they are related to various other concepts and play key roles in solutions to several problems in computer science, combinatorics, graph theory, operations research and universal algebra. In all of these areas graph algebras have been used to establish nontrivial relations among different notions, provide valuable insights into their essential properties, and apply methods of one of these theories in order to answer questions motivated by other branches. Thus, this topic lies in the intersection of large and active research areas: computer science, combinatorics, graph theory, operations research and universal algebra.

Courses in all of these directions can enjoy the benefits of using graph algebras as a convenient source of examples and exercises for students. In this way the students become familiar with many essential concepts and acquire practical skills by actively using them. Interesting examples provided by graph algebras help to learn the similarities and relations among various basic notions as a result of active work close to real research currently carried out in the literature.

This book has been written for undergraduate and graduate students as readers. It has been used in teaching the courses Applied Algebra, Automata Theory, Discrete Modelling, Mathematics for Computer Science, and Operations Research at the University of Tasmania. We have been using Excel, Java, Java Script, R, Visual Basic and computer algebra systems GAP, Magma, Mathematica, Maple, Matlab in teaching. Our exposition concentrates on introducing sufficient background knowledge, explaining all concepts for undergraduate students and preparing graduate students for future research work on this topic with abundant open problems and opportunities for new contributions. The monograph can be recommended for students in various related undergraduate courses.

Our book is the first one devoted to graph algebras. Many interesting results have appeared in publications that address advanced and established specialists. However, there did not exist any text collecting all of these results in a unified fashion with required preliminaries included for convenience of the readers, and making this topic accessible for students.

The investigation of graph algebras has started fairly recently. There are many open problems recorded in the literature, as well as many natural interesting questions that have not been answered yet. This creates excellent conditions for developing skills of gifted and talented students at an early stage of their education via involvement in research work that may lead to exciting new discoveries.

Our book summarizes numerous contributions collecting all known facts on graph algebras in a convenient form with preliminaries included for convenience of the reader. The author hopes that a complete text containing all results on graph algebras will be also useful for established researchers in computer science, combinatorics, graph theory, operations research and universal algebra. A separate chapter contains open problems.

This monograph is based on a series of publications by the author, which have been discussed with co-authors, colleagues, and editors organizing anonymous refereeing. The author is grateful to Olga Sokratova, who introduced him to the topics of automata theory during her appointment as a Visiting Research Associate at the University of Tasmania for the first semester of 2000; to his Ph.D. students Jilyana Cazaran, Allison Plant, Steve Quinn, and to our Honours students Brian Heazlewood, Christina Taylor, Venta Terauds and Melanie Webb for helpful discussions. The author would also like to express sincere appreciation to all mathematicians who have contributed to his research, in particular, to Professors L.M. Batten, A.D. Bell, S.N. Bespamyatnikh, S. Dăscălescu, D. Easdown, M.R. Fellows, K.R. Fuller, B.J. Gardner, S. Goberstein, T.E. Hall, K.J. Horadam, J.M. Howie, M. Ito, E. Jespers, J. Justin, J. Karhumaki, R. Lidl, J. Meakin, M. Miller, W.D. Munn, R. Oehmke, J. Okniński, F. Otto, D.S. Passman, F. Pastijn, G. Paun, G. Pirillo, C.E. Praeger, G. Rozenberg, A. Salagean, A. Salomaa, P. Schultz, L.N. Shevrin, I. Shparlinski, P. Shumyatsky, P. Solé, A. Solomon, M. Steinby, T. Stokes, E.J. Taft, M.L. Teply, P.G. Trotter, M.V. Volkov, A.P.J. van der Walt, R. Wiegandt, and to the Editors of Marcel Dekker, Inc., especially Maria Allegra.

Andrei Kelarev

Contents

Chapter 1

Preliminaries

In order to make this book self-contained and accessible for students of various backgrounds, we have included complete preliminaries for the convenience of the reader. This chapter can be used for instruction in class providing succinct notes if the lecturer supplies additional explanations, illustrating examples, and exercises. The students already familiar with the concepts summarized here will be able to quickly refresh their memory by referring back to the text. Sections of this chapter are arranged in the order convenient for learning new concepts: we begin with easier concepts illustrated by various examples, and only then introduce more general notions.

1.1 Sets

A *set* is a collection of objects, called *elements* or *members*. If a is an element of a set X, then we write $a \in X$ and say that a *belongs* to X. We can also say that a is in X, or is contained in X, or lies in X, or X has the element a. If a is not an element of the set X, then we write $a \notin X$.

If two sets X and Y have the same elements, then we say that they are *equal* and write $X = Y$. If two sets X and Y are not equal, then we write $X \neq Y$.

A set can be defined by describing its elements. For example, the following sentences define sets $\mathbb{N}$, $\mathbb{N}_0$, $\mathbb{Q}$, and $\mathbb{R}$:

$$\begin{array}{cl} \mathbb{N} & \text{is the set of all positive integers,} \\ \mathbb{N}_0 & \text{is the set of all nonnegative integers,} \\ \mathbb{Q} & \text{is the set of all rational numbers,} \\ \mathbb{R} & \text{is the set of all real numbers.} \end{array}$$

Sets can also be defined using the *set-builder notation*:

$$\{a \in X \mid P_1(a), \ldots, P_n(a)\}$$

stands for the set of all elements a of X with the properties $P_1(a), \ldots, P_n(a)$. For example, the same sets as above can be defined by

$$\begin{array}{rcl} \mathbb{N} & = & \{m \mid m \text{ is a positive integer}\}, \\ \mathbb{N}_0 & & \{n \mid n\text{is a nonnegative integer}\}, \\ \mathbb{Q} & = & \{q \mid q \text{ is a rational number}\}, \\ \mathbb{R} & = & \{r \mid r \text{ is a real number}\}. \end{array}$$

Some sets can be defined by listing their elements in braces:

$$X = \{a, b, c, d\},$$

$$Y = \{a, b, c, \ldots, z\},$$

$$Z = \{1, 2, 3, \ldots, 100\}.$$

In these examples we can guess what elements are hidden behind dots "..." if we look at the other members of the set. This method allows us to define infinite sets too:

$$\mathbb{N} = \{1, 2, 3, \ldots\}.$$

The order of elements in the *list notation* is irrelevant. For example, the following two sets are equal:

$$\{1, 2, 3\} = \{3, 2, 1\}.$$

The number of times an element occurs in the set is also ignorred, and the following two sets are equal too:

$$\{a, c, f\} = \{a, f, a, f, c\}.$$

If we want to take into account the number of occurrences of an element, then the concept of a *multiset* has to be used, see Section 1.3.

If every element of a set X belongs to a set Y, then we write $X \subseteq Y$ or $Y \supseteq X$, and say that X is a *subset* of Y, or Y is a *superset* of X, or X is *contained* in Y, or X is *included* in Y. Two sets X and Y are equal if and only if $X \subseteq Y$ are $X \supseteq Y$. This is often used in proofs: to verify an equality of sets one has to establish that both inclusions hold.

If $X \subseteq Y$ and $X \neq Y$, then we say that X is *properly contained* in Y, and write $X \subset Y$ or $Y \supset X$.

The *empty set* has no elements and is denoted by $\emptyset$. For every set X, the empty set $\emptyset$ and the whole set X are subsets of X. A subset Y of the set X is said to be *proper* if it is distinct from the sets $\emptyset$ and X. The set of all subsets of a set X is called the *power set* of X and is denoted by $\mathcal{P}(X)$ or 2^X.

The number of elements in a finite set X is called the *cardinality* of X and is denoted by $|X|$.

If X and Y are sets, then the *union* of these sets is denoted by $X \cup Y$ and is defined as the set of all elements which belong to at least one of the sets X and Y:

$$X \cup Y = \{a \mid a \in X \text{ or } a \in Y\}.$$

As usual, 'or' is understood in its inclusive sense, that is the definition above includes also all elements which belong to both sets X and Y simultaneously. The union of a family $(X_i)_{i \in I}$ of subsets of some set is denoted by $\cup_{i \in I} X_i$ and is defined as the set of all elements which belong to at least one of the subsets of the family. For example, the union of sets $X_1, X_2, \ldots, X_n$

is recorded as

$$\bigcup_{i=1}^{n} X_i = X_1 \cup X_2 \cup \cdots \cup X_n.$$

If X and Y are sets, then the *intersection* of these sets is the set denoted by $X \cap Y$ and defined by

$$X \cap Y = \{a \mid a \in X \text{ and } a \in Y\}.$$

The intersection of a family $(X_i)_{i \in I}$ of sets (indexed by the elements of I) is written as $\cap_{i \in I} X_i$ and is the set of all common elements of all sets in the family. In particular, the intersection of sets $X_1, X_2, \ldots, X_n$ can be written as

$$\bigcap_{i=1}^{n} X_i = X_1 \cap X_2 \cap \cdots \cap X_n.$$

A collection $\mathcal{K}$ of nonempty subsets of a nonempty set X is called a *partition* of X if X is the union of all sets in $\mathcal{K}$ and distinct sets of $\mathcal{K}$ do not intersect.

If S is a set and m is a positive integer, then the set of all m-element subsets of S is denoted by $\mathcal{P}_m(S)$. The general Ramsey Theorem can be recorded in several special cases of interest (see [73]).

Theorem 1.1. (Ramsey Theorem) ([50], Theorem 1.1) *Let m, n, k be positive integers where $m \leq n$. Then there exists a positive integer $R(m, n, k)$ such that, for each finite set E with $|E| \geq R(m, n, k)$ and for every partition of $\mathcal{P}_m(E)$ in k classes, there exists a subset T of E such that $|T| = n$ and all sets of $\mathcal{P}_m(T)$ lie in one class of the partition.*

Considering sets we usually fix a large set consisting of all elements we are interested in and called the *universal set* or *universe.* Then we look at various subsets of the universal set. For example, plane geometry investigates sets of points in the plane and in this case the universal set is the whole plane. Let U be the universal set. The *complement* of a set $X \subseteq U$ is the set of all elements of U that do not belong to X. It is denoted by X^c or $\overline{X}$:

$$X^c = \overline{X} = \{a \in U \mid a \notin X\}.$$

Exercise 1.1. Find the following sets:

(a) $\{1, 3, 4, 6, 7, 8, 9\} \cap \{2, 3, 5, 6, 8\}$;

(b) $\{3, 5, 6, 9\} \cup \{1, 3, 5\}$;

(c) $\{1, 4, 6, 7, 8, 9\} \cap \{1, 3, 7, 8\}$;

(d) $\{3, 5, 6, 9\} \cup \{1, 5, 8, 9\}$;

(e) $\{x \in \mathbb{N} \mid x \leq 20\}$;

(f) $\{x \mid x \leq 20, x \in \mathbb{N}\} \cap \{11, 14, 15, 21, 23, 37\}$;

(g) $\{x \mid x \geq 20, x \in \mathbb{N}\} \cap \{9, 12, 15, 22, 31, 79\}$.

Exercise 1.2. How many elements do the following sets contain?

(a) $\mathcal{P}(\{a, b\})$;

(b) $2^{\{1,2,3\}}$;

(c) $\mathcal{P}(\{x, y, z, t\})$;

(d) $2^{\{1,2,3,4,5,6\}}$;

(e) $\{x \mid x \leq 20, x \in \mathbb{N}\}$;

(a) $\{a, c, d, e, f, h\} \cap \{b, c, e, g, h\}$;

(b) $\{c, e, e, h\} \cup \{a, c, e\}$;

(c) $\{a, d, e, f, g, h\} \cap \{a, c, f, g\}$;

(d) $\{c, e, e, h\} \cup \{a, e, g, h\}$;

(e) $\{x \in \mathbb{N} \mid x \leq 20\}$;

(f) $\{x \mid x \leq 30, x \in \mathbb{N}\} \cap \{12, 16, 22, 25, 32, 38\}$;

(g) $\{x \mid x \geq 30, x \in \mathbb{N}\} \cap \{13, 15, 17, 21, 23, 33, 37\}$.

Familiar Venn diagrams represent arbitrary sets as sets of points in the plane and can be used to illustrate properties of sets and laws of set operations. Here we include an improved version of Venn diagrams, known as Karnaugh maps. They are just modifications of Venn diagrams representing sets as sets of squares in a rectangular table. To indicate which of the unit squares or boxes belong to a set, we write 1 in the box, and we may write 0 in a box to show that it does not belong to the set being illustrated.

Karnaugh map for two sets:

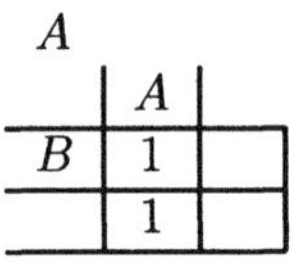

A

	A	
B	1	
	1	

Karnaugh map for three sets:

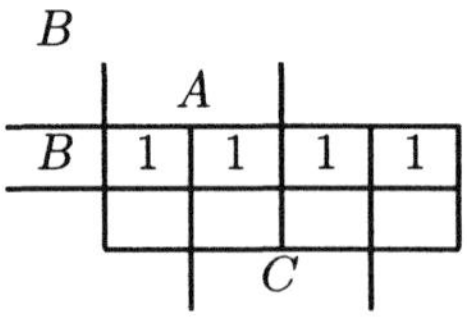

B

	A			
B	1	1	1	1
		C		

Karnaugh map for four sets:

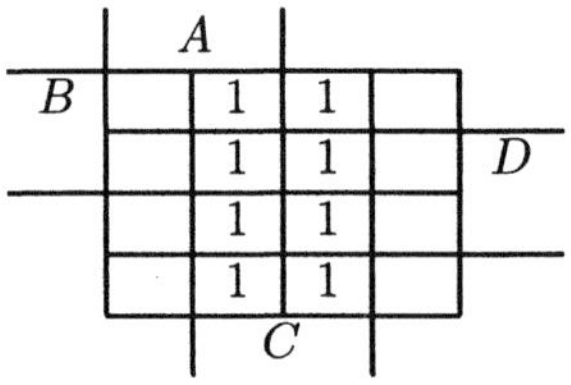

	A				
B		1	1		
		1	1		D
		1	1		
		1	1		
		C			

Karnaugh map for five sets

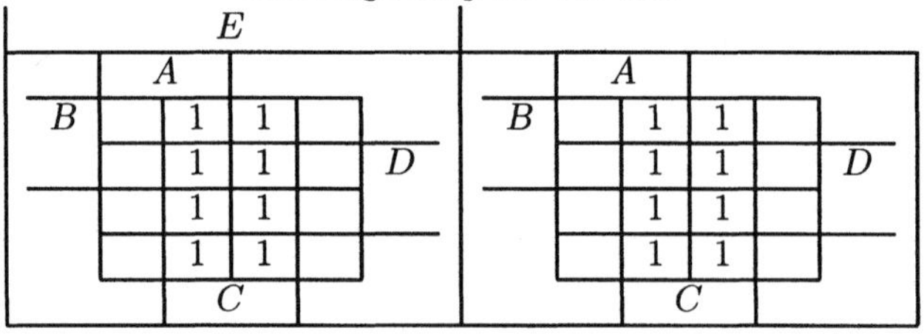

E

	A						A				
B		1	1			B		1	1		
		1	1		D			1	1		D
		1	1					1	1		
		1	1					1	1		
		C						C			

Karnaugh map for six sets

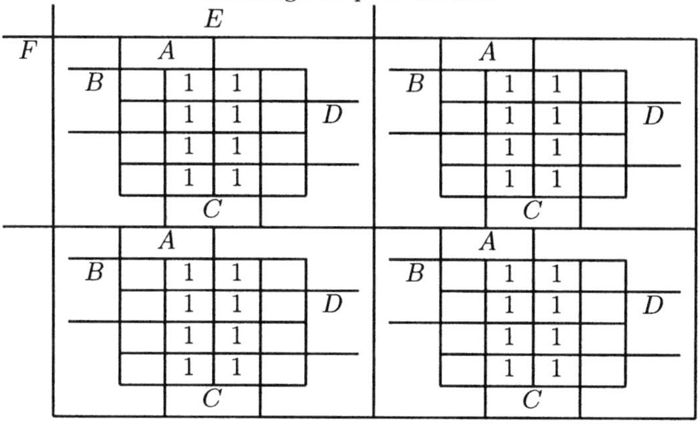

	E	E	E	E				
		A	A			A	A	
F, B		1	1			1	1	
F, B, D		1	1			1	1	
F, D		1	1			1	1	
F		1	1			1	1	
			C	C			C	C
		A	A			A	A	
B		1	1			1	1	
B, D		1	1			1	1	
D		1	1			1	1	
		1	1			1	1	
			C	C			C	C

Exercise 1.3. Use Karnaugh maps to illustrate the following de Morgan's laws

$$\overline{A \cap B} = \overline{A} \cup \overline{B},$$

$$\overline{A \cup B} = \overline{A} \cap \overline{B},$$

that hold for all sets A and B.

Exercise 1.4. Use Karnaugh maps to illustrate the absorption laws

$$A \cup (A \cap B) = A,$$

$$A \cap (A \cup B) = A,$$

that hold for all sets A, B and C.

Exercise 1.5. Use Karnaugh maps to illustrate the associative laws

$$A \cap (B \cap C) = (A \cap B) \cap C,$$

$$A \cup (B \cup C) = (A \cup B) \cup C,$$

that hold for all sets A, B and C.

1.2 Sequences

In a *pair* or *ordered pair* (a, b) the order of elements is essential: the first element is a and the second one is b. Two ordered pairs (a_1, b_1) and (a_2, b_2)

are *equal* if and only if $a_1 = a_2$ and $b_1 = b_2$. The *Cartesian product* or *direct product* of sets X and Y is the set of all ordered pairs (a, b) such that $a \in X$ and $b \in Y$. It is denoted by $X \times Y$:

$$X \times Y = \{(a, b) \mid a \in X \text{ and } b \in Y\}.$$

Let n be a positive integer. Ordered sequences of n elements are also called *n-tuples.* Two n-tuples $(a_1, a_2, \ldots, a_n)$ and $(b_1, b_2, \ldots, b_n)$ are *equal* if and only if the corresponding elements of these sequences coincide: $a_1 = b_1$, $a_2 = b_2$, ..., and $a_n = b_n$. The *Cartesian product* or *direct product* of sets $X_1, X_2, \ldots, X_n$ is the set of all n-tuples $(a_1, \ldots, a_n)$ such that $a_1 \in X_1$, ..., $a_n \in X_n$. It is denoted by

$$X_1 \times X_2 \times \cdots \times X_n = \{(a_1, \ldots, a_n) \mid a_1 \in X_1, \ldots, a_n \in X_n\}.$$

The direct product of n copies of the same set X is called the *n-th Cartesian power* or *direct power* of X, and is denoted by X^n. For finite sets $X_1, X_2, \ldots, X_n$, the cardinality of direct product satisfies the following equality:

$$|X_1 \times X_2 \times \cdots \times X_n| = |X_1| \times |X_2| \times \cdots \times |X_n|.$$

It can be used to count ordered sequences of elements and is referred to as the *product rule.*

Exercise 1.6. How many elements are there in the following sets?

(a) $\mathcal{P}(\{a, b\})$;

(b) $2^{\{1,2,3\}}$;

(c) $2^{\{1,2,3\}} \times 2^{\{a,b\}}$;

1.3 Functions or Mappings

A *function* or *mapping* from a set X to a set Y is a rule that assigns to each element of X one element of Y. More formally, a *function* from X to Y can be defined as a subset of the direct product $X \times Y$ such that every element a of X is the first member of precisely one pair in the subset. Notation $f : X \to Y$ is used to show that f is a mapping from X to Y.

The set X is called the *domain* of f, and the set Y is called the *codomain* (or *range*) of f. The set of all functions $f : X \to Y$ is denoted by Y^X.

If $f : X \to Y$ and $x \in X$, then the unique element of Y assigned to x is denoted by $f(x)$ and is called the *image* of x under f. In this case we write $x \mapsto f(x)$. For some functions it may be convenient to use notation x^f instead of $f(x)$. If V is a subset of X, then the set of all images of elements of V under f is called the *image* of V and is denoted by $f(V)$, that is

$$f(V) = \{f(x) \mid x \in V\}.$$

Exercise 1.7. How many elements are there in the following sets?

(a) $\{x, y\}^{\{1,2,3\}}$;

(b) $\{1, 2, 3\}^{\{a,b\}}$;

(c) $\{x, y, z\}^{\{1,2,3\}} \times 2^{\{a,b\}}$;

(d) $\mathcal{P}(\{1, 2\}) \times 2^{\{a,b,c\}} \times \{x, y, z\}^{\{t,u,v,w\}}$.

For example, if S is a set, then a *multiset* is a function $F : S \to \mathbb{N}_0$, where the nonnegative integer $F(x)$ indicates the number of occurrences of x in the multiset F.

If $f : X \to Y$ and $g : Y \to U$ are mappings, then the *composition* of f and g is the mapping $f \circ g : X \to U$ defined by

$$(f \circ g)(x) = f(g(x)).$$

In particular, the image $(f \circ g)(x)$ is undefined whenever $g(x)$ does not belong to the domain of the function f. We can also write fg for $f \circ g$.

The identical mapping from X to X is denoted by 1, or 1_X is defined by $1(x) = x$ for all $x \in X$. A function $f : X \to Y$ is said to be *invertible* if there exists a function $g : Y \to X$ such that $gf = 1_X$.

If $f : X \to Y$ and $f(X) = Y$, then the mapping f is said to be *onto*. The mapping f is said to be *one-to-one* if distinct elements of X have unequal images, that is $x_1 \neq x_2$ implies $f(x_1) \neq f(x_2)$, whenever $x_1, x_2 \in X$. A one-to-one mapping is sometimes called an *injection*, an onto mapping is called a *surjection*. A mapping that is both one-to-one and onto is called a *bijection* or a *one-to-one correspondence*.

Theorem 1.2. ([55], Theorem 2.2, [201], 1.10) *A mapping $f : X \to X$ is invertible if and only if it is a bijection.*

For arbitrary (not necessarily finite) sets A and B, we say that they have the same *cardinality* if there exists a bijection from A onto B. A set is said to be *countable* if it has the same cardinality as the set of all positive integers.

The set $\{1, \ldots, n\}$ is also often denoted by $[n]$ or $[1 : n]$. Let $X = \{1, \ldots, n\}$. Each mapping $f : X \to X$ can be defined with the table of all values

$$f = \begin{pmatrix} 1 & 2 & \ldots & n \\ f(1) & f(2) & \ldots & f(n) \end{pmatrix}.$$

A mapping $f : X \to X$ is called a *permutation* if it is a bijection, that is if

$$X = \{f(1), f(2), \ldots, f(n)\}.$$

The composition of permutations f and g is more often called the *product* of f and g. It is equal to the permutation

$$fg = \begin{pmatrix} 1 & 2 & \ldots & n \\ f(g(1)) & f(g(2)) & \ldots & f(g(n)) \end{pmatrix}.$$

The set of all permutations on $\{1, \ldots, n\}$ is called the *symmetric group of degree n* and is denoted by S_n. The *order* of the symmetric group of degree n, that is the number of permutations in it, is given by $|S_n| = n! = 1 \times 2 \times \cdots \times n$.

Properties of permutations were used by a Norwegian mathematician N.H. Abel and a French mathematician Évariste Galois to prove that there does not exist an exact formula for roots of polynomial equations of degree greater than 5, to find examples of polynomial equations of degree 5 with roots which cannot be expressed in terms of integers, arithmetical operations and radicals, and to give an exact criterion of how to determine whether the roots of a given equation can be expressed in this way.

Properties of permutations are used in computer algorithms for finding best arrangements of elements in solutions to various practical problems, as well as in chemistry and crystallography for counting elements in certain sets.

As an example of application relying on permutations let us discuss an excellent permutation cipher that uses a rotating square grid.[1] We take a small 4×4 square for the purpose of this example. To decide which of the boxes can be cut out, we number them as follows

1	2	3	1
3	4	4	2
2	4	4	3
1	3	2	1

and then cut out one box with each number:

1	2	3	1
3		4	
2	4	4	
	3	2	1

This gives us the following rotating square grid:

X	X	X	X
X		X	
X	X	X	
	X	X	X

[1]This system was popular in my primary school and is reproduced here from memory.

Let's encipher the text

SURJECTION IS ONTO.

The square grid rearranges plaintext in four steps. First, we put the square on paper and start writing the text in the empty boxes:

X	X	X	X
X		X	
X	X	X	
	X	X	X

	S		U
			R
J			

Second, we turn the square clockwise by 90° and put it on the same place. Now new boxes are open for letters of the text, and we can write down the next part.

	X	X	X
X	X		X
X	X	X	X
X			X

E			
	S	**C**	U
			R
J	**T**	**I**	

Third, we rotate the grid again and write down the next 4 letters.

X	X	X	
	X	X	X
	X		X
X	X	X	X

E			**O**
N	S	C	U
I		**S**	R
J	T	I	

Finally, the remaining boxes can be filled in even without the grid:

E	**O**	**N**	O
N	S	C	U
I	**T**	S	R
J	T	I	**O**

Exercise 1.8. Construct a 6×6 square grid and use it to encipher the message

BIJECTION IS A ONE-TO-ONE AND ONTO MAPPING.

Let $X = \{1, \ldots, n\}$, and let $i_1, i_2, \ldots, i_k$ be pairwise distinct elements of X. Then the *cycle*

$$(i_1, i_2, \ldots, i_k)$$

is the permutation f such that

$$\begin{aligned}
f(i_1) &= i_2, \\
f(i_2) &= i_3, \\
&\ldots \\
f(i_{k-1}) &= i_k, \\
f(i_k) &= i_1, \\
f(x) &= x \text{ for all } x \notin \{i_1, \ldots, i_k\}.
\end{aligned}$$

It is also called a k-*cycle*, or a *cycle of order* k. The *order* of a permutation f is the smallest positive integer m such that $f^m = 1$, where 1 stands for the identical permutation. Every k-cycle has order k.

Two cycles $(i_1, i_2, \ldots, i_k)$ and $(j_1, j_2, \ldots, j_\ell)$ are said to be *disjoint* if they have no common entries, that is

$$\{i_1, i_2, \ldots, i_k\} \cap \{j_1, j_2, \ldots, j_\ell\} = \emptyset.$$

Theorem 1.3. ([55], §6) *Each permutation is equal to a product of disjoint cycles. The product is unique up to rewriting cycles in another order.*

The product of disjoint cycles equal to a permutation is also called the *disjoint cycle notation* for the permutation.

Example 1.1. Find a product of disjoint cycles equal to the permutation

$$f = \begin{pmatrix} 1 & 2 & 3 & 4 & 5 & 6 & 7 \\ 4 & 7 & 5 & 2 & 3 & 6 & 1 \end{pmatrix}$$

Solution. $f = (1, 4, 2, 7)(3, 5)(6)$. □

Exercise 1.9. Find a product of disjoint cycles equal to the following permutations

(a) $f = \begin{pmatrix} 1 & 2 & 3 & 4 & 5 & 6 & 7 & 8 & 9 \\ 1 & 8 & 7 & 9 & 3 & 6 & 4 & 2 & 5 \end{pmatrix}$,

(b) $f = \begin{pmatrix} 1 & 2 & 3 & 4 & 5 & 6 & 7 & 8 & 9 \\ 8 & 9 & 1 & 6 & 5 & 4 & 7 & 3 & 2 \end{pmatrix}$.

A 2-cycle is called a *transposition.* A permutation is said to be *even* (resp., *odd*) if it is equal to a product of an even (resp., odd) number of transpositions.

Theorem 1.4. ([55], Problem 6.9) *Every permutation can be written as a product of transpositions. For each permutation, the products or transpositions equal to it either all have even numbers of transpositions, or all have odd numbers of transpositions.*

Exercise 1.10. Find some products of transpositions equal to the permutations listed in Exercise 1.9.

Theorem 1.5. *Every symmetric group is partitioned into two disjoint subsets: the set of all even permutations, and the set of all odd permutations.*

Exercise 1.11. Make the list of all permutations of the set $\{1, 2, 3\}$, and partition it into the class of all even permutations and the class of all odd permutations.

The *sign* of a permutation f is defined by

$$\operatorname{sign}(f) = \begin{cases} 1 & \text{if } f \text{ is even,} \\ -1 & \text{if } f \text{ is odd.} \end{cases}$$

Theorem 1.6. *For all $f, g \in S_n$, the following sign equality holds:*

$$sign(fg) = sign(f)\, sign(g). \tag{1.1}$$

The following facts follow from the sign equality:

(a) the product of two even permutations is even;

(b) the product of two odd permutations is even;

(c) the product of an odd permutation and an even permutation is odd;

(d) a permutation is even if and only if it is equal to a product of an even number of transpositions;

(e) a permutation is odd if and only if it is equal to a product of an odd number of transpositions.

Signs of permutations are used in matrix theory to define determinants that help to write down various important formulas in almost all directions of applied science.

1.4 Operations

A *binary operation* on a set X is a mapping from the direct product $X \times X$ to X. If $*$ is a binary operation on X and $x, y \in X$, then the image of the pair (x, y) under $*$ is usually denoted using the *infix* notation: $x * y$.

Example 1.2. (i) Subtraction is a binary operation on the set of integers:

$$(x, y) \mapsto x - y.$$

(ii) The mapping $(x, y) \mapsto x^y$ is a binary operation on the set of positive integers.

The values of a binary operation can be listed in a *Cayley table.* The rows and columns of a Cayley table are labeled by elements of the set, and entries contain values of the operation. For example, the field of order two $\mathbb{F}_2 = GF(2)$ is the set $\{0, 1\}$ with two operations $+, \cdot$ defined by the following Cayley tables:

+	0	1
0	0	1
1	1	0

·	0	1
0	0	0
1	0	1

Let n be a positive integer. An *n-ary operation* on a set X is a mapping from the Cartesian power X^n to X. If

$$f : X \times X \times \cdots \times X \to X$$

is an operation, then the image of the sequence $(x_1, x_2, \ldots, x_n)$ is denoted by $f(x_1, x_2, \ldots, x_n)$. In particular, a *unary* operation on X is a mapping from X to X. A *nullary* operation on X is a mapping from $X^0 = \emptyset$ to X. This means that a nullary operation only chooses a certain special element of X, and properties of the element can be recorded by referring to this operation.

1.5 Modular Arithmetic

Next, we review the most essential properties of integers.

Definition 1.1. (Division with Remainder) ([55], §10) Let n be a nonzero integer. Then, for each integer m, there exist unique integers q, r such that

$$m = qn + r \text{ and } 0 \leq r \leq n - 1.$$

These numbers q and r are called the *quotient* and *remainder* of m on division by n, respectively. If $r = 0$, then we write $n|m$, and say that n *divides* m or that n is a *divisor* of m. The remainder on division of m by n is also often denoted by

$$m \mod n.$$

We can find q and r by long division with remainder, or using computers. Hasse diagrams can be used to draw the sets of divisors of integers, as the following example shows.

Example 1.3. Let $B = \{1, 2, 3, 5, 6, 10, 15, 30\}$ be the set of positive divisors of 30. We can illustrate the *set of divisors* or the *divisibility relation* | with a *Hasse diagram* (see Figure 1.1). All integers are represented by points. The point of each integer is connected to the points of its divisors with edges or with sequences of edges going in one direction. The points of divisors of each integer lie lower in the diagram.

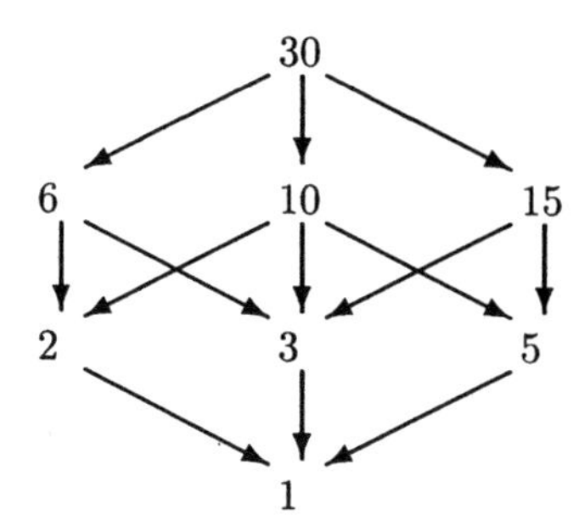

Figure 1.1: Hasse diagram of the set of divisors of 30

Exercise 1.12. Draw the Hasse diagrams for the sets of divisors of the following integers:

(a) 12; (b) 18; (c) 24;

(d) 32; (e) 54.

Definition 1.2. If x and y have the same remainder on division by n, then we say that x is *congruent* to y *modulo* n, and write

$$x \equiv y \mod n.$$

The number n is called the *modulus* of the congruence.

This notation is convenient, because we can use congruences like ordinary equalities of numbers. The only difference is that when we divide both sides of a congruence by a divisor of a modulus, then the modulus also has to be divided by it. We can add and multiply congruences.

Theorem 1.7. ([201], 2.112) *If* $a \equiv b \mod n$ *and* $c \equiv d \mod n$, *then*

$$\begin{aligned} a + c &\equiv b + d \mod n, \\ a - c &\equiv b - d \mod n, \text{ and} \\ a \cdot c &\equiv b \cdot d \mod n. \end{aligned}$$

We can always divide both sides of a congruence together with the modulus by a common divisor of all three integers.

Lemma 1.1. *Let a, b, c, n be integers, where $c \neq 0$. Then*

$$a \equiv b \mod n \iff ca \equiv cb \mod cn \tag{1.2}$$

Proof. By definition, the congruence $a \equiv b \mod n$ means that $a - b = qn$ for some integer q. Since $c > 0$, this is equivalent to $c(a-b) = cqn$, i.e., $ca \equiv cb \mod cn$. □

In other words, if d is a common divisor of a, b, n, and if $a \equiv b \mod n$, then $a/d \equiv b/d \mod n/d$.

Definition 1.3. Two integers a and b are *coprime* or *relatively prime* if they do not have common divisors greater than 1.

Note that 1 is coprime with all integers. We can divide a congruence by a number coprime with the modulus n.

Theorem 1.8. ([201], 2.116, 2.117) *If d is a common divisor of a and b which is coprime with n, and $a \equiv b \mod n$, then $a/d \equiv b/d \mod n$.*

Example 1.4. If we do not divide the modulus as indicated in Lemma 1.1, this may lead to errors. Indeed,

$$2 \equiv 6 \mod 4, \text{ but}$$
$$1 = \frac{2}{2} \not\equiv \frac{6}{2} \mod 4.$$

Theorem 1.8 and Lemma 1.1 allow us to divide a congruence by any integer x if we first divide it, together with the modulus n, by the greatest common divisor $d = \gcd(x, n)$, and then divide the left and right hand sides only by the number $\frac{x}{d}$ coprime with the new modulus $\frac{n}{d}$.

A positive integer p is called a *prime* if it has exactly two divisors, 1 and p. This means that 1 is not a prime number.

Theorem 1.9. (Fundamental Theorem of Arithmetic) ([55], §13, [201], 2.97) *For each integer m greater than 1 there exist pairwise distinct primes $p_1, \ldots, p_k$ and positive integers $a_1, \ldots, a_k$ such that*

$$m = p_1^{a_1} \cdots p_k^{a_k}.$$

This factorization is unique up to the order of primes in the product.

1.6 Extended Euclidean Algorithm

The *greatest common divisor* of integers $a_1, a_2, \ldots, a_n$ is denoted by

$$\gcd(a_1, a_2, \ldots, a_n).$$

It's easy to find it if prime factorizations of integers are known.

Theorem 1.10. ([55], §13, [201], 2.98) *Let* $m = p_1^{a_1} \cdots p_k^{a_k}$ *and* $n = p_1^{b_1} \cdots p_k^{b_k}$ *be prime power factorizations of* m *and* n, *i.e.,* $p_1, \ldots, p_k$ *are pairwise distinct primes and* $a_1, \ldots, a_k, b_1, \ldots, b_k$ *are positive integers. Then*

$$m \textit{ divides } n \iff a_1 \leq b_1, \ldots, a_k \leq b_k, \tag{1.3}$$

and the greatest common divisor of m *and* n *is given by*

$$gcd(m, n) = p_1^{\min\{a_1, b_1\}} \cdots p_k^{\min\{a_k, b_k\}}. \tag{1.4}$$

Lemma 1.2. *All common divisors of* a *and* b *divide the greatest common divisor* $gcd(a, b)$.

Lemma 1.3. *If* $a = bq_1 + r_1$, *then* $\gcd(a, b) = \gcd(b, r_1)$.

Exercise 1.13. Prove Lemmas 1.2 and 1.3.

Extended Euclidean Algorithm finds $d = \gcd(a, b)$ and x, y such that $d = ax + by$. Suppose that $b < a$ are positive integers. To start steps of the algorithm, fill in the first two rows of the following table.

Remainders	Coefficients		Row operations
$r_1 = a$	$x_1 = 1$	$y_1 = 0$	R_1
$r_2 = b$	$x_2 = 0$	$y_2 = 1$	R_2
$r_3 = a - q_3 b$	$x_3 = 1$	$y_3 = -q_3$	$R_3 = R_1 - q_3 R_2$
$r_4 = b - q_4 r_3$	$x_4 = -q_4$	$y_4 = 1 + q_4 q_3$	$R_4 = R_2 - q_4 R_3$
$\vdots$	$\vdots$	$\vdots$	$\vdots$
$r_{n-2} = r_{n-4} - r_{n-3} q_{n-2}$	x_{n-2}	y_{n-2}	$R_{n-2} = R_{n-4} - q_{n-2} R_{n-3}$
$r_{n-1} = r_{n-3} - r_{n-2} q_{n-1}$	x_{n-1}	y_{n-1}	$R_{n-1} = R_{n-3} - q_{n-1} R_{n-2}$
$r_n = 0$			R_n

To complete the remaining rows, divide $r_1 = a$ by $r_2 = b$ with remainder, get $a = q_3 b + r_3$, subtract the second row q_3 times from the first row, and write down the row $R_3 = R_1 - q_3 R_2$ in the third row. Now do the same with rows R_2 and R_3, that is divide r_2 by r_3 with remainder, get $r_2 = q_4 r_3 + r_4$, and write down the row $R_4 = R_2 - q_4 R_3$ in the fourth row. Repeat the process until the first zero remainder $r_n = 0$ is obtained. Then the last nonzero remainder $r_{n-1} \neq 0$ is equal to the greatest common divisor $d = \gcd(a, b)$.

Since the first two rows satisfy the equations $r_1 = x_1 a + y_1 b$ and $r_2 = x_2 a + y_2 b$, it follows that the third row satisfies $r_3 = x_3 a + y_3 b$, and so on. Finally, we have $d = r_{n-1} = x_{n-1} a + y_{n-1} b$, as required.

In other words, the same recursive operation $s_k = s_{k-2} - s_{k-1} q_{k-1}$ is applied to each of the first three columns. It follows that the entries r_i, x_i, y_i in every row satisfy $r_i = ax_i + by_i$. Finally, we get $d = r_{n-1} = ax_{n-1} + by_{n-1}$.

Example 1.5. Use the extended Euclidean algorithm to find $d = \gcd(23, 5)$ and x, y such that $23x + 5y = d$.

Solution.

$$\begin{array}{rrrl} 23 & 1 & 0 & \\ 5 & 0 & 1 & \\ 3 & 1 & -4 & R_1 - 4R_2 \\ 2 & -1 & 5 & R_2 - R_3 \\ 1 & 2 & -9 & R_3 - R_4 \\ 0 & & & \end{array} \qquad \text{Here} \quad \begin{array}{rcl} 23 & = & 5 \cdot 4 + 3 \\ 5 & = & 3 \cdot 1 + 2 \\ 3 & = & 2 \cdot 1 + 1 \\ 2 & = & 1 \cdot 2 \end{array}$$

Therefore $\gcd(23, 5) = 1 = 23 \times 2 + 5 \times (-9)$. □

It is not necessary to divide r_{i-1} by r_i every time. The greatest common divisor will be found by just subtracting r_i from r_{i-1} a few times until the same division is accomplished.

Example 1.6. Use extended Euclidean algorithm to find $d = \gcd(203, 112)$ and x, y satisfying $203x + 112y = d = (203, 112)$.

Solution.

$$\begin{aligned} 203 &= 112 \cdot 1 + 91 \\ 112 &= 91 \cdot 1 + 21 \\ 91 &= 21 \cdot 4 + 7 \\ 21 &= 7 \cdot 3 \end{aligned}$$

In this example the last column lists all quotients.

203	1	0	
112	0	1	*1
91	1	−1	*1
21	−1	2	*4
7	5	−9	*3
0			

Therefore $d = 7 = (203, 112)$ and $(203, 112) = 7 = 203 \cdot 5 + 112 \cdot (-9)$. □

The *least common multiple* of integers $a_1, a_2, \ldots, a_n$ is the smallest positive integer divisible by all of these numbers. It is denoted by

$$\operatorname{lcm}(a_1, a_2, \ldots, a_n).$$

Theorem 1.11. ([201], 2.89, 2.98) *Let m and n be integers. Then*

$$lcm(m,n) = \frac{mn}{gcd(m,n)}.$$

If prime power factorizations of m and n are $m = p_1^{a_1} \cdots p_k^{a_k}$ and $n = p_1^{b_1} \cdots p_k^{b_k}$, then

$$lcm(m,n) = p_1^{\max\{a_1,b_1\}} \cdots p_k^{\max\{a_k,b_k\}}. \tag{1.5}$$

Theorem 1.12. (Linear Diophantine Equation Theorem) ([252], Chapter 4) *The linear Diophantine equation $ax + by = c$ has integer solutions (x, y) if and only if $d = gcd(a,b)$ divides c. If $ax_0 + by_0 = d$, then all solutions are given by the following formulas, where t runs over $\mathbb{Z}$:*

$$\begin{aligned} x &= \frac{c}{d} \cdot x_0 - \frac{b}{d} \cdot t, \\ y &= \frac{c}{d} \cdot y_0 + \frac{a}{d} \cdot t. \end{aligned}$$

Exercise 1.14. Prove Theorem 1.12 by referring to extended Euclidean algorithm.

For example, in order to solve the Diophantine equation $4x+6y = 14$, we find $\gcd(4,6) = 2$ using the extended Euclidean algorithm, and get $-1 \times 4 + 1 \times 6 = 2$. Therefore formulas for all integer solutions of the equation are

$$\begin{aligned} x &= -\frac{14}{2} + 3 \cdot t, \\ y &= \frac{14}{2} - 2 \cdot t. \end{aligned}$$

Similarly, the Euclidean algorithm finds that $\gcd(23,5) = 1$ and $2 \times 23 - 9 \times 5 = 1$. Therefore all integer solutions to $23x + 5y = 7$ are given by

$$\begin{aligned} x &= 7 \times 2 + (-5)t, \\ y &= 7 \times (-9) + 23t. \end{aligned}$$

Theorem 1.13. (Linear Congruence Equation Theorem) ([201], 2.119) *The linear congruence $ax \equiv b \mod n$ has integer solutions x if and only* *if $d = gcd(a, n)$ divides b. A particular solution x_0 can be found with Euclidean algorithm, and then all solutions are given by*

$$x = x_0 + nt, \qquad \text{where } t \in \mathbb{Z}.$$

1.7 Euler's Theorem and RSA Cryptosystem

Definition 1.4. *Euler's totient function* $\phi(n)$ is defined for each positive integer n as the number of positive integers coprime with n and less than n. It is also called the *Euler's phi function.*

Theorem 1.14. (Formula for the Euler's phi Function) ([190], 13.15, [201], 2.101) *If $n = p_1^{a_1} \cdots p_k^{a_k}$, where $p_1, \ldots, p_k$ are pairwise distinct primes, then the Euler's totient function is given by the formula*

$$\phi(n) = n\left(1 - \frac{1}{p_1}\right) \cdots \left(1 - \frac{1}{p_k}\right). \tag{1.6}$$

Lemma 1.4. *If a and b are coprime, then*

$$\phi(ab) = \phi(a)\phi(b). \tag{1.7}$$

If p is a prime and r is a positive integer, then

$$\phi(p^r) = p^r - p^{r-1}. \tag{1.8}$$

Equality (1.6) immediately implies (1.7) and (1.8).

Example 1.7. If $n = p_1 p_2$, then

$$\phi(n) = (p_1 - 1)(p_2 - 1).$$

For example, if $n = 6$, then $\phi(6) = (2 - 1)(3 - 1) = 2$.

Exercise 1.15. Prove the formula for the Euler's totient function

(a) in the case where $n = p$ is a prime number (Fermat's Little Theorem);

(b) in the case where $n = p_1 p_2$ is a product of two prime numbers;

(c) in the case where $n = p^a$ is a power of a prime;

(d) in the general case.

Example 1.8. The prime power factorization of 24 is $24 = 2^3 3$. By (1.6) we get

$$\phi(24) = 24(1 - \frac{1}{2})(1 - \frac{1}{3}) = 24(\frac{1}{2} \times \frac{2}{3}) = 8.$$

Alternatively, (1.7) implies $\phi(24) = \phi(2^3)\phi(3) = 4 \times 2 = 8$, because $\phi(2^3) = 2^3 - 2^2 = 4$ by (1.8).

The following Euler's Theorem is used in the RSA cryptosystem.

Theorem 1.15. (Euler's Theorem) ([190], §22, [201], 2.126(i)) *If x is an integer coprime with n, then*

$$x^{\phi(n)} \equiv 1 \mod n. \tag{1.9}$$

Proof. Denote by S the set of all positive integers less than n and coprime with n. By the definition of $\phi(n)$, we get $|S| = \phi(n)$. Fix x coprime to n, and consider the function $f(z)$ defined by

$$f(z) = \text{the remainder of } xz \text{ on division by } n.$$

Suppose that there exist z_1, z_2 such that $f(z_1) = f(z_2)$. This means that $xz_1 \equiv xz_2 \mod n$. Since x is coprime with n, we can divide the congruence by x, and get $z_1 \equiv z_2 \mod n$. This means that f is one-to-one.

For each $z \in S$, the product xz is coprime with n, and so $f(x)$ lies in S. Thus f is a one-to-one function from S to S. It follows that $S = \{z_1, \ldots, z_{\phi(n)}\}$ is the same as the set $\{f(z_1), \ldots, f(z_{\phi(n)})\}$. Therefore

$$\begin{aligned} z_1 \cdots z_{\phi(n)} &= f(z_1) \cdots f(z_{\phi(n)}) \\ &\equiv (xz_1) \cdots (xz_{\phi(n)}) \mod n \\ &\equiv x^{\phi(n)}(z_1 \cdots z_{\phi(n)}) \mod n. \end{aligned}$$

Cancelling $z_1 \cdots z_{\phi(n)}$ in this congruence gives us $x^{\phi(n)} \equiv 1 \mod n$, as required. □

Since $\phi(p) = p-1$ for each prime p, the following corollary is a special case of Euler's Theorem.

Theorem 1.16. (Fermat's Little Theorem) ([190], §13, Exercise 10, [201], 2.127(i)) *If p is a prime, then*

$$x^p \equiv x \mod p$$

holds for all integers x.

If the modulus n has no square divisors > 1, then it is not necessary to require that x be coprime with n. The following theorem has been independently discovered by students in our Operations Research course.[2]

Theorem 1.17. (Euler's Theorem for square-free modulus) ([201], 2.126(ii)) *Let x and n be positive integers such that n has no square divisors > 1. Then*

$$x^{\phi(n)+1} \equiv x \mod n \qquad \text{(Euler's equality)} \tag{1.10}$$

Proof of Theorem 1.17. We use induction on the number k of common prime divisors of x and n. The induction base $k = 0$ is the Euler's theorem for coprime x and n which we have proved above.

Suppose that a prime p divides x and n. It follows from the formula (1.6) for Euler's phi function that

$$\phi(n) = (p-1)\phi\left(\frac{n}{p}\right). \tag{1.11}$$

[2]It is interesting to note that Theorem 1.17 has been independently discovered by my students with the use of computer algebra system GAP. Verifying Euler's formula for RSA cryptosystem with GAP examples, students found out that Euler's equality may remain valid in the case where the integer and modulus are not coprime. Experiments suggested that this is always true if the modulus is square-free, and then we worked out our complete proof. This fact is stated, for example, in 2.126(ii), A.J. Menezes, P.C. van Oorschot and S.A. Vanstone, "Handbook of Applied Cryptography", 1996.

Since p^2 does not divide n, we see that p is coprime with $\frac{n}{p}$, and therefore

$$\begin{aligned}
1 &\equiv p^{\phi(\frac{n}{p})} \mod \frac{n}{p} \\
&\equiv 1 \cdot p^{\phi(\frac{n}{p})} \mod \frac{n}{p} \\
&\equiv p^{2\phi(\frac{n}{p})} \mod \frac{n}{p} \\
&\equiv \dots \\
&\equiv p^{(p-1)\phi(\frac{n}{p})} \mod \frac{n}{p} \\
&\equiv p^{\phi(n)} \mod \frac{n}{p}
\end{aligned}$$

By Lemma 1.1, we can multiply both sides of a congruence and the modulus by p, and get

$$p \equiv p^{\phi(n)+1} \mod n \tag{1.12}$$

Besides, $\frac{x}{p}$ and $\frac{n}{p}$ have fewer common prime divisors, and the induction hypothesis implies

$$\left(\frac{x}{p}\right) \equiv \left(\frac{x}{p}\right)^{\phi(\frac{n}{p})+1} \mod \frac{n}{p} \tag{1.13}$$

Therefore

$$\begin{aligned}
\left(\frac{x}{p}\right)^{\phi(\frac{n}{p})+1} &\equiv \left(\frac{x}{p}\right)^{\phi(\frac{n}{p})} \times \left(\frac{x}{p}\right) \mod \frac{n}{p} \\
&\equiv \left(\frac{x}{p}\right)^{\phi(\frac{n}{p})} \times \left(\frac{x}{p}\right)^{\phi(\frac{n}{p})+1} \mod \frac{n}{p} \\
&\equiv \left(\frac{x}{p}\right)^{2\phi(\frac{n}{p})+1} \mod \frac{n}{p}
\end{aligned} \tag{1.14}$$

Hence we get

$$\begin{aligned}
\left(\frac{x}{p}\right) &\equiv \left(\frac{x}{p}\right)^{\phi(\frac{n}{p})+1} \mod \frac{n}{p} \\
&\equiv \left(\frac{x}{p}\right)^{2\phi(\frac{n}{p})+1} \mod \frac{n}{p} \quad \text{(by (1.13) and (1.14))} \\
&\equiv \dots \\
&\equiv \left(\frac{x}{p}\right)^{(p-1)\phi(\frac{n}{p})+1} \mod \frac{n}{p} \\
&\equiv \left(\frac{x}{p}\right)^{\phi(n)+1} \mod \frac{n}{p}
\end{aligned}$$

Multiplying both sides of the congruence and the modulus by p gives us

$$x \equiv p\left(\frac{x}{p}\right)^{\phi(n)+1} \mod n. \tag{1.15}$$

We can replace p in (1.15) by its value from (1.12) as follows:

$$\begin{aligned}
x &\equiv p^{\phi(n)+1}\left(\frac{x}{p}\right)^{\phi(n)+1} \mod n \\
&\equiv \left(p \cdot \frac{x}{p}\right)^{\phi(n)+1} \mod n \\
&\equiv x^{\phi(n)+1} \mod n.
\end{aligned}$$

This completes our proof. □

Easy examples show that the main equality of Euler's theorem does not hold for numbers x if both x and the modulus n are divisible by a common square of a prime.

The RSA Cryptosystem. Two large primes p_1, p_2 and the decoding number d are chosen and are kept secret. An encoding number e is found such that

$$ed \equiv 1 \mod (p_1 - 1)(p_2 - 1). \tag{1.16}$$

The encoding number and the modulus $m = p_1 p_2$ are made public. Each number $s \leq m$ is encoded as r, where

$$s^e \equiv r \mod m. \tag{1.17}$$

If the number r is received, then it is deciphered with

$$r^d \equiv s \mod m. \tag{1.18}$$

The following two easy examples are included just to illustrate these calculations with small prime numbers.

Example 1.9. If $p_1 = 11$ and $p_2 = 13$, then $m = p_1 p_2 = 143$ and $\phi(m) = (p_1 - 1)(p_2 - 1) = 120$. We can take $e = 7$ and use Euclidean algorithm to find d. We get $1 = \gcd(7, 120) = -17 \times 7 + 1 \times 120$. Hence $d = 120 - 17 = 103$ gives us the product ed congruent to 1 modulo 120. Now if we take the message "code" and replace its letters with their numbers $3, 15, 4, 5$ in the alphabet, then in order to encipher the message we raise them to the power $e = 7$.

$$\begin{aligned}
3^e &\equiv 42 \mod m; \\
15^e &\equiv 115 \mod m; \\
4^e &\equiv 82 \mod m; \\
5^e &\equiv 47 \mod m.
\end{aligned}$$

Thus the ciphertext is "42,115,82,47". Since we know the deciphering key $d = 103$, we can decipher the message as follows.

$$\begin{aligned}
42^d &\equiv 3 \mod m; \\
115^d &\equiv 15 \mod m; \\
82^d &\equiv 4 \mod m; \\
47^d &\equiv 5 \mod m.
\end{aligned}$$

Everyone can encode messages, but only the person who sets up the RSA cryptosystem can decode messages, because it is very difficult to find decoding key d given e and m only. Everyone can find d if the prime factorization $m = p_1 \times p_2$ becomes known. However, there do not exist efficient factorization algorithms for arbitrary large integers, and a lot of research has been devoted to this topic.

The following well-known example of calculating with very large numbers is worth looking at.

Exercise 1.16. Find the last two digits of the integer $9^{(9^9)}$.

Solution. We could use the Euler's Theorem to solve this exercise. Since $\phi(100) = 40$, the Euler's equation becomes

$$x^{40} \equiv 1 \mod 100.$$

Thus our example shows that some numbers are better than all numbers and have even more convenient equations than the Euler's equation that holds for all numbers coprime with the modulus.

$$9^{10} \equiv 1 \mod 100; \tag{1.19}$$

$$9^{11} \equiv 9 \mod 100;$$

$$9^{12} \equiv 81 \mod 100.$$

We see that the same numbers $9, 81, \ldots$ occur again, and so the sequence is periodic. If we ever get an equality

$$9^m \equiv r \mod 100,$$

we can multiply it with (1.19) and get

$$9^{m+10} \equiv r \mod 100.$$

Therefore we can subtract 10 from the exponent without changing the last two digits of a power of 9. Hence the last two digits of a power of 9 depend only on the last digit of the exponent. It follows that $9^{(9^9)}$ has the same last two digits as the number 9^m, where m is the last digit of 9^9. We have already found it in the equality (1.19). Thus we get

$$9^{(9^9)} \equiv 9^9 \equiv 89 \mod 100.$$

□

Exercise 1.17. Find the last two digits of the following integers

(a) $7^{(7^7)}$; (b) $2^{(2^2)}$; (c) $2^{\left(2^{(2^2)}\right)}$;

(d) $3^{(3^3)}$; (e) $3^{\left(3^{(3^3)}\right)}$.

Theorem 1.18. (Chinese Remainder Theorem) ([201], 2.120) *Let $m_1, \ldots,$ m_k be pairwise coprime positive integers, and let $c_1, \ldots, c_k$ be arbitrary integers. Then the system of congruences*

$$x \equiv c_i \mod m_i, \text{ where } i = 1, \ldots, k, \tag{1.20}$$

has a solution that is unique modulo $m_1 \cdots m_k$.

Theorem 1.19. (Chinese Remainder Theorem: general case) ([190], Theorem 15.1) *Let $m_1, \ldots, m_k$ be positive integers, and let $c_1, \ldots, c_k$ be arbitrary integers. Then the system of congruences (1.20) has a solution if and only if all residues $c_1, \ldots, c_k$ are equal to each other modulo the greatest common divisor $d = gcd(m_1, \ldots, m_k)$. Solution is unique modulo the least common multiple $lcm(m_1, \ldots, m_k)$.*

This theorem can be used for fast addition and multiplication of large numbers, if these operations are replaced by parallel simultaneous additions or multiplications of small numbers.

Exercise 1.18. Use extended Euclidean algorithm and prove the Chinese Remainder Theorem

(i) for $k = 1$;

(ii) for $k = 2$;

(iii) for $k > 2$.

Exercise 1.19. Use extended Euclidean algorithm to find minimum positive solutions to the following systems of congruences

(i)
$$x \equiv 1 \mod 2,$$
$$x \equiv 2 \mod 3.$$

(ii)
$$x \equiv 0 \mod 2,$$
$$x \equiv 2 \mod 3,$$
$$x \equiv 1 \mod 5.$$

(iii)
$$x \equiv 1 \mod 2,$$
$$x \equiv 1 \mod 3,$$
$$x \equiv 3 \mod 5.$$

1.8 Relations

Let X and Y be sets. A *relation* ϱ from X to Y is a subset of the Cartesian product $X \times Y$. If a pair (a, b) belongs to a relation ϱ, then we may also use the *infix notation* $a\varrho b$ and say that "a is in relation ϱ to b". Relations from a set X to the same set X are called *relations on* X.

If $\varphi \subseteq X \times Y$ and $\psi \subseteq Y \times Z$ are relations, then the *product* or *composition* of φ and ψ is the relation

$$\varphi \circ \psi = \{(a, c) \mid (b, c) \in \varphi, (a, b) \in \psi \text{ for some } b \in Y\}.$$

The *converse* of the relation φ is

$$\varphi^{-1} = \{(b, a) \mid (a, b) \in \varphi\}.$$

Every set S has the *equality relation* denoted by $=$, or $=_S$, or ι, or ι_S consisting of all pairs (a, a), where $a \in S$. The *complete relation* is the whole direct product $S \times S$. A relation is said to be *proper* if it is distinct from ι and $S \times S$.

A relation ϱ on a set X is said to be *reflexive* if $a\varrho a$, for all $a \in X$. A relation is reflexive if and only if it contains the *equality relation*. A relation

ϱ on X is said to be *symmetric* if $a\varrho b$ implies $b\varrho a$ for all $a, b \in X$. A relation ϱ on X is said to be *transitive* if $a\varrho b$ and $b\varrho c$ imply $a\varrho c$, for all $a, b, c \in X$.

A relation on a set is called an *equivalence relation* if it is reflexive, symmetric and transitive. Let ϱ be an equivalence relation on a set X, and let $a \in X$. The *equivalence class* of a relative to ϱ is the set of all elements b such that $b\varrho a$. Every element of an equivalence class is called a *representative* of the class. The equivalence class of ϱ containing the element a is denoted by $[a]$, or a/ϱ, or a^ϱ. The set of all equivalence classes of ϱ is denoted by X/ϱ. The *index* of an equivalence relation ϱ on Q is the number of classes in the quotient set Q/ϱ.

Theorem 1.20. (Partitions and Equivalence Relations) ([55], Theorem 9.1) *Let X be a set, and let ϱ be an equivalence relation on X. Then the set X/ϱ of all equivalence classes of ϱ forms a partition of X. The relation ϱ coincides with the set of all pairs (x, y) such that x and y are in the same equivalence class of this partition.*

Let ρ be a relation on X. The *reflexive closure* of ρ is the relation $\rho^r = \rho \cup 1_X$. For a positive integer n, the *power* ρ^n consists of all pairs (x, y) such that there exist $(x, y_1), (y_1, y_2), \ldots, (y_{n-1}, y) \in \rho$. The *transitive closure* of ρ is defined by

$$\rho^\infty = \bigcup\{\rho^n \mid n \geq 1\}.$$

The reflexive and transitive closure of the relation ρ is denoted by ρ^*. For an arbitrary relation ρ on X, the smallest *equivalence generated* by ρ is given by

$$\rho^e = [\rho \cup \rho^{-1} \cup 1_X]^\infty.$$

A relation ϱ on a set X is said to be *antisymmetric* if $x\varrho y$ and $y\varrho x$ imply $x = y$, whenever $x, y \in X$. A reflexive, antisymmetric, and transitive relation is called a *partial order*. A set with a partial order is called a *partially ordered set* or a *poset*.

A partial order $\leq$ on X is called a *linear order* if, for each $x, y \in X$, either $x \leq y$ or $y \leq x$. A set with a linear order is called a *linearly ordered set* or a *chain*.

Let $(X, \leq)$ be a partially ordered set. An element x of X is called a *largest element* of X if $y \leq x$ for all $y \in X$. An element y of X is called a *smallest element least element* of X if $y \leq x$ for all $x \in X$. An element z of X is said to be *maximal* in X if there are no larger elements in X, that is, if $z \leq x$ implies $x = z$ for all $x \in X$. An element t of X is said to be *minimal* in X if there are no smaller elements in X, that is, if $x \leq t$ implies $x = t$ for all $x \in X$. Clearly, a set may have at most one largest element and at most one smallest element.

Evidently, every largest element is maximal, and every smallest element is minimal. The following example shows that a set may have infinitely many maximal elements, and that a maximal element does not have to be the largest element in a set.

Example 1.10. Let $\mathbb{N}$ be the set of all positive integers, and let S be the set consisting of the empty set and all one-element subsets of $\mathbb{N}$ ordered by the ordinary inclusion of sets. Then the empty set is the smallest element of S, all singletons are maximal elements of S, but S does not have a largest element.

Let $(X, \leq)$ be a partially ordered set with a subset Y. An element $x \in X$ is called an *upper bound* of Y if $y \leq x$ for all $y \in Y$. An element $x \in X$ is called a *lower bound* of Y if $x \leq y$ for all $y \in Y$. If the subset X has the largest lower bound, then it is also called the *infimum* of X (or the *exact lower bound*). If the subset X has the smallest upper bound, then it is also called the *supremum* of X (or the *exact upper bound*).

The following lemma is equivalent to the Axiom of Choice, which is one of the axioms that can be used in proofs.

Lemma 1.5. (Zorn's Lemma) ([190], 1.4) *Let $(X, \leq)$ be a partially ordered set. If every ascending chain of elements of X has an upper bound in X, then X has at least one maximal element. If every descending chain of* *elements of X has a lower bound in X, then X has at least one minimal element.*

Lemma 1.6. (Least Integer Principle) ([55], §10) *Each nonempty set of positive integers has a smallest element.*

A partially ordered set $(X, \leq)$ is said to be *lattice ordered* if every pair of elements of X has a supremum and infimum.

1.9 Graphs

A *graph* is a pair $D = (V, E)$, where V is a set of elements called *vertices* or *nodes*, and E is a set of pairs of vertices, called *edges* or *arcs*. The set of all vertices (edges) of the graph D is denoted by $V(D)$ (resp., $E(D)$). The number of vertices is called the *order* of the graph. A *loop* is an edge of the form (v, v), where $v \in V$.

A *walk* of length n in a graph $D = (V, E)$ is a sequence of vertices $v_0, v_1, \ldots, v_n$ such that $(v_i, v_{i+1}) \in E$ for $i = 0, 1, \ldots, n-1$. A *trail* is a walk where all edges are distinct. A *path* is a walk where all vertices are distinct.

A graph $D = (V, E)$ is said to be *strongly connected* if, for every two distinct vertices $u, v \in V$, there exists a walk from u to v.

The *in-degree* (*out-degree*) of a vertex v of the graph $D = (V, E)$ is the number of edges $(u, v) \in E$ (resp., $(v, u) \in E$), where $u \in V$. The *in-neighbourhood* and the *out-neighbourhood* of a vertex u of a graph $D = (V, E)$ are the sets

$$\text{In}(u) = \{w \in V \mid (w, u) \in E\} \quad \text{and} \quad \text{Out}(u) = \{w \in V \mid (u, w) \in E\}.$$

Clearly, the in-degree (out-degree) of a vertex $v \in V$ is the cardinality of $\text{In}(v)$ (resp., $\text{Out}(v)$):

$$\text{indeg}(v) = |\,\text{In}(v)| \qquad \text{outdeg}(v) = |\,\text{In}(v)|.$$

A graph is said to be *undirected* if $(u, v) \in E$ implies $(v, u) \in E$ for all $u, v \in V$. The *underlying undirected graph* of a graph $D = (V, E)$ is the graph $D' = (V, E')$ where

$$E' = E \cup \{(u, v) \mid (v, u) \in E\}.$$

The *neighbourhood* of a vertex v in an undirected graph $D = (V, E)$ is the set $N(v) = \{w \in V \mid (v, w) \in E\}$.

A graph $D_1 = (V_1, E_1)$ is called a *subgraph* of the graph $D = (V, E)$ if $V_1 \subseteq V$ and $E_1 \subseteq E$. A subgraph D_1 of D is said to be *induced* if $E_1 = E \cap (V_1 \times V_1)$.

A graph D is said to be *connected* if its underlying undirected graph is strongly connected. A maximal connected subgraph of a graph D is called a *connected component* of D.

Given a graph $D = (V, E)$ and a vertex $x \in V$, denote by $\text{Out}^*(x) = \text{Out}^*_D(x)$ the set of all vertices $y \in V$ such that there exists a path from x to y in D. Now, a graph $D = (V, E)$ is strongly connected if $V = \text{Out}^*(x)$ for all $x \in V$.

An *undirected walk* of length n in a graph $D = (V, E)$ is a walk in the underlying undirected graph of D, i.e., a sequence of vertices $v_0, v_1, \ldots, v_n$ such that $(v_i, v_{i+1}) \in E$ or $(v_{i+1}, v_i) \in E$ for $i = 0, 1, \ldots, n-1$. An *undirected trail* is an undirected walk where all edges are distinct. An *undirected path* is an undirected walk where all vertices are distinct.

The *sum* $D_1 + D_2$ of graphs $D_1 = (V_1, E_1)$ and $D_2 = (V_2, E_2)$ is the graph with vertex set $V_1 \cup V_2$ and edge set $E_1 \cup E_2 \cup E_{1,2}$, where

$$E_{1,2} = \{(x, y) \mid x \in V_1, y \in V_2\}.$$

If D_1 and D_2 have no common vertices, then we say that the sum is *direct*, and we denote it by $D_1 \oplus D_2$. A *null graph* is a graph without edges. The null graph of order m is denoted by N_m. We assume that *complete graphs* contain all edges including loops. The complete graph of order n is denoted by K_n. The direct sum of these graphs will be denoted by

$$K_n^m = N_m \oplus K_n.$$

If $D_i = (V_i, E_i)$ is a family of graph indexed by the elements $i \in I$, then the *direct product* $D = \prod_{i \in I} D_i$ is has the set $V = \prod_{i \in V}$ of vertices and the set

$$E = \{(x, y) \mid (x(i), y(i)) \in E_i \text{ for all } i \in I\}$$

of edges. In this notation the $x(i)$ stands for the i-th component of the sequence $x = (x_i)_{i \in I}$, i.e., the value of the function x at i.

1.10 Exercises

1 **(a)** Compute the product of permutations

$$\begin{pmatrix} 1 & 2 & 3 & 4 \\ 2 & 4 & 1 & 3 \end{pmatrix} \begin{pmatrix} 1 & 2 & 3 & 4 \\ 4 & 1 & 2 & 3 \end{pmatrix}$$

(b) Find disjoint cycle notation for the permutation

$$\begin{pmatrix} 1 & 2 & 3 & 4 & 5 \\ 5 & 1 & 4 & 3 & 2 \end{pmatrix}.$$

2 **(a)** Find the image $\sigma(3)$, where $\sigma = (5, 7, 4, 1)(2, 6, 3)$.

(b) Calculate the order of the permutation $(2, 6, 3)(5, 7, 4, 1)$.

(c) Calculate the order of the permutation $(2, 6)(9, 10, 11, 12, 4, 1)(5, 7, 8, 3)$.

3 **(a)** Compute the product of permutations

$$\begin{pmatrix} 1 & 2 & 3 & 4 & 5 \\ 2 & 4 & 1 & 5 & 3 \end{pmatrix} \begin{pmatrix} 1 & 2 & 3 & 4 & 5 \\ 4 & 5 & 2 & 3 & 1 \end{pmatrix}$$

(b) Find disjoint cycle notation for the permutation

$$\begin{pmatrix} 1 & 2 & 3 & 4 & 5 & 6 & 7 \\ 5 & 1 & 7 & 6 & 2 & 3 & 4 \end{pmatrix}$$

4 **(a)** Find the image $\sigma(4)$, where $\sigma = (6,7,4,1,10)(5,9)(8,2,3)$.

(b) Calculate the order of the permutation $(6,3,8)(2,5,7,10,1)(9,4)$.

(c) Calculate the order of the permutation $(7,10,6)(9,2,5,1,4)(3,8)$.

5 Find a solution to the following system of congruences:

$$x \equiv 0 \mod 2$$
$$x \equiv 1 \mod 3$$
$$x \equiv 4 \mod 5$$

6 **(a)** Use extended Euclidean algorithm to find $d = gcd(12, 30)$ and x, y satisfying $12x + 30y = d$.

(b) Use the results of (a) to find formulas for all solution to the Diophantine equation $12x + 30y = 6$.

(c) Find formulas for all solutions to the Diophantine equation $12x + 30y = 42$.

7 **(a)** Use extended Euclidean algorithm to show that 15 and 26 are co-prime, and to find x, y satisfying $15x + 26y = 1$.

(b) Write down formulas for all solutions to the Diophantine equation $15x + 26y = 1$.

(c) Use the results of (b) to find a particular solution to the congruence

$$15x \equiv 5 \mod 26.$$

(d) State formulas for all solutions to the congruence given in (c).

8 **(a)** Use extended Euclidean algorithm to find $d = \gcd(308, 231)$ and integers x, y such that $d = 308x + 231y$.

(b) Write down formulas for all solutions of the Diophantine equation $308x + 231y = 154$.

(c) Use Karnaugh maps to verify that the following generalized de Morgan's law holds for all sets A, B, C, D :

$$\overline{A \cap B \cap C \cap D} = \overline{A} \cup \overline{B} \cup \overline{C} \cup \overline{D}.$$

9 **(a)** Use extended Euclidean algorithm to find $d = \gcd(390, 455)$ and integers x, y such that $d = 390x + 455y$.

(b) Write down formulas for all solutions of the Diophantine equation $390x + 455y = 455$.

(c) Use Karnaugh maps to verify that the following generalized de Morgan's law holds for all sets A, B, C, D :

$$\overline{A \cup B \cup C \cup D} = \overline{A} \cap \overline{B} \cap \overline{C} \cap \overline{D}.$$

10 **(a)** Find the last two digits of $7^{(7^7)}$.

(b) Find the last two digits of

$$3^{\left(3^{\left(3^3\right)}\right)}.$$

(c) Let a, b, c be integers. Refer only to extended Euclidean algorithm and prove that if $a|bc$ and $\gcd(a, b) = 1$, then $a|c$.

Chapter 2

Algebraic Structures

2.1 Words and Free Monoids

An *alphabet* X is any finite set. The elements of X are called *letters* or *symbols*. A *word* or *string* over the alphabet X is a finite sequence of letters from X. A word $(x_1, x_2, \ldots, x_n)$ can be written also as $x_1 x_2 \cdots x_n$. The *length* of the word $w = x_1 x_2 \cdots x_n$, where $x_i \in X$, is the number n of letters in w. It is denoted by $|w|$.

If $u = (x_1, \ldots, x_n)$ and $v = (y_1, \ldots, y_m)$ are words, then the *catenation* or *concatenation* of u and v is obtained by *juxtaposing* them, i.e., by writing them one immediately after another

$$uv = (x_1, \ldots, x_n, y_1, \ldots, y_m)$$

and is denoted by uv. For example, the concatenation of words $xxzy$ and yx is $(xxzy)(yx) = xxzyyx$.

It is often convenient to include in consideration a special word of length 0, called the *empty word* and denoted by 1. It satisfies $w1 = w = 1w$ for all words w. Two other symbols are also often used for the empty word: λ and ε. They help when it makes sense to distinguish the empty word from other meanings of the symbol 1. The length of the empty word is 0.

Let X be an alphabet. The set of all finite words over X with concatenation is called the *free monoid* generated by X, and is denoted by X^*. The set of all nonempty words over an alphabet X is denoted by X^+. It is called the *free semigroup* generated by X. Thus $X^* = X^+ \cup \{1\}$. A *language* over X is a set of word over X, that is a subset of the free monoid X^*. A language on X is *trivial* if it is equal to $\emptyset$ or X^*. The set X can be identified with the set of words of length 1, and so we write $X \subset X^*$.

Each letter may occur several times in a word. For a subset Y of the alphabet X, the number of letters of w that belong to Y is denoted by $|w|_Y$. If $x \in X$, then the number of times the letter x occurs in w is denoted by $|w|_x$. Thus $w_x = |w|_{\{x\}}$. The set of all letters that occur in a word w is denoted by $\mathrm{alph}(w)$.

A word v is a *subsequence* of the word w if letters of v form a subsequence of the sequence of letters of w. For example, the word $bdeghj$ is a subsequence of the word $abcdefghijk$.

A word v is called a *segment* or a *factor* of the word w if there exist words $u_1, u_2 \in X^*$ such that $w = u_1 v u_2$. A segment v of w is said to be *proper* if $v \neq w$. For example, the word $cdefg$ is a segment of the word $abcdefghijk$.

If $w = uv$ for some $u, v, w \in X^*$, then u is called a *prefix* (or a *left factor*) of w and v is called a *suffix* (or *right factor*) of w. For example, the word $abcd$ is a prefix of the word $abcdefghijk$, and the word $ghijk$ is a suffix of the word $abcdefghijk$.

Let $s = s_1, s_2, \ldots$ be a finite or infinite sequence of words in X^+. An *m-factorization* of s is a sequence $t_1, t_2, \ldots, t_m$, where all t_k are consecutive segments of s, that is $t_k = s_{i_k} \ldots s_{i_{k+1}-1}$ for $k = 1, 2, \ldots, m$ and some $1 \leq i_1 < i_2 < \ldots < i_{m+1}$. Note that an m-factorization may let aside some prefix and suffix of s.

Let X be an alphabet, and $Y \subseteq X^*$. If, for all $n, m \geq 1$ and all

$x_1, \ldots, x_n, x'_1, \ldots, x'_m$ in Y, the equality

$$x_1 x_2 \ldots x_n = x'_1 x'_2 \ldots x'_m$$

implies

$$n = m \text{ and } x_i = x'_i \quad \text{for } i = 1, \ldots, n,$$

then Y is called a *code* or a *variable length code* over X. In other words, Y is a code if each element of Y^+ is equal to a unique product of codewords. A code cannot contain the empty word 1, since $1 \cdot 1 = 1$. Each subset of a code is a code too. For any alphabet X, the set $Y = X^n$ of sequences of symbols of length n is a code, called the *uniform code* of length n.

Let Y be a subset of X^*. Then Y is called a *prefix code* if no element in Y is a prefix of any other element of Y. This means that, for all $y, y' \in Y$, $yw = y'$ implies $w = 1$. A set Y is called a *suffix code* if $wy = y'$ implies $w = 1$. A set Y is called a *biprefix code* if it is both prefix and suffix. Each prefix code or suffix code is a variable length code.

Example 2.1. The code $\{b, ba, baa, baaa\}$ is suffix, but not prefix. All uniform codes are biprefix. The sets $Y = \{b, ab, aab, aaab\}$ and

$$Z = \{a^n b^n \mid n \geq 1\}$$

are prefix; Z is also suffix, but Y is not. The variable length code $\{a, ab, bba\}$ is neither prefix, nor suffix.

Theorem 2.1. (Defect Theorem) ([27], Theorem 4.1) *If a finite set $Y \subseteq X^*$ is not a code, then there exists a code Z such that*

$$|Z| \leq |Y| - 1 \text{ and } Y \subseteq Z^*.$$

Corollary 2.1. *If u, v are words which are not powers of a single word, then $\{u, v\}$ is a code.*

A word is said to be *primitive* if it is not a power of another word.

Theorem 2.2. (Primitive Word Theorem) ([27], Theorem 2.1) *For each word w there exists a unique primitive word v such that w is a power of v. In particular, if $u^a = v^b$, then there exists a primitive word x such that $u, v \in x^*$.*

A positive integer k is called a *period* of the word $w = w_1 \dots w_n$ if $w_i = w_{i+k}$ for all i with $1 \leq i, i+k \leq n$.

Theorem 2.3. (Fine & Wilf's Theorem) ([51], Theorem 1.4.1) *If w is a word with periods p and q and $|w| \geq p+q-\ gcd(p,q)$, then $gcd(p,q)$ is also a period of w.*

Theorem 2.4. ([27], Theorem 6.1) *If two powers u^a and v^b of words u and v have a common prefix of length at least $|u|+|v|-\ gcd(|u|,|v|)$, then u and v are powers of the same word.*

Theorem 2.5. ([51], Proposition 1.6.1) *Let w be a word with k letters and let $f_w(n)$ be the number of subwords of length n in w. Then the following conditions are equivalent:*

(i) *there exist positive integers n, p such that $w_i = w_{i+p}$ for all $i \geq n$;*

(ii) *there exists a constant c such that $f_w(n) < c$ for all $n \geq 0$;*

(iii) *there exists a positive integer n such that $f_w(n) \leq n+k-2$;*

(iv) *there exists an integer n such that $f_w(n) = f_w(n+1)$.*

The function f_w is called the *subword complexity* of w. A word w satisfying condition (i) of Theorem 2.5 is said to be *ultimately periodic*.

Theorem 2.6. (Lyndon's Theorem) ([195], Theorems 2.1, 2.2) (i) *If $xy = yz$, for some $x, y, z \in X^*$, $x \neq 1$, then there exist words u, v and a nonnegative integer k such that $x = uv$, $y = (uv)^k u$, $z = vu$.*

(ii) *Two words commute iff they are powers of the same word.*

(iii) *The set of all words commuting with a word w coincides with v^*, where v is the unique primitive word such that w is a power of v.*

Definition 2.1. Let B be a finite set called the *source alphabet*, $A = \{0,1\}$ the set of code symbols, and let $\pi : B \to \mathbb{R}_+$ be a function. For any mapping of the form $\alpha : B \to A^*$, define $\lambda : (A^*)^B \to \mathbb{R}^+$ by

$$\lambda(\alpha) = \sum_{b \in B} |\alpha(b)| \cdot \pi(b)$$

A *binary Huffman encoding* of B with respect to π is a mapping $\alpha : B \to A^*$ for which $\alpha(B)$ is a prefix code and $\lambda(\alpha)$ is minimal.

If π is interpreted as a probability or frequency distribution of elements in B, then $\lambda(\alpha)$ is the expected value of the length of words encoding elements of B. Hence a Huffman code gives the shortest binary encoding of elements from B.

Since every Huffman code is a prefix code, it is easy to break a stream of code symbols up into the corresponding codewords. No codeword has any other codeword as a prefix, so the decoder can look at the stream one symbol at a time. As soon as a valid codeword is seen, it can decode that word and start looking for the next word. If a code is not prefix, then it will require looking at more symbols to decide whether a given sequence should be interpreted as a codeword in its own right, or as the beginning of a larger codeword.

Here is a method for finding a binary Huffman code for a given probability distribution π.

Step 1. List all elements of B in decreasing order of probability.

Step 2. Choose the last two symbols (those with the smallest probability of occurrence), and delete them from the table. Create a new column in the table which contains all remaining symbols and a new letter with probability equal to the sum of probabilities of the deleted letters. The entries of the new column should be ordered by their probabilities again.

Repeat combining last symbols and adding columns until the last column with only two entries is added to the table.

Step 3. Label the entries in the last column with 0 and 1. Use the column with ready codewords to fill in all labels in the preceding column. Move back from the last column to the first original column in the table allocating codewords to the alphabets in each column. To this end rewrite all codewords of the rightmost column for the same symbols in the preceding column. In order to distinguish the codewords of two symbols that have been combined, make two longer codewords by adding 0 and 1 to the codeword of their new combined symbol. Use the column with ready codewords to fill in all labels in the preceding column.

Example 2.2. Suppose that we want to encode the string "this is test" ignoring the blank space in text. Let B be the set of symbols in the phrase, $B = \{t, h, i, e, s\}$. The probabilities associated with the elements of B are given in Figure 2.1.

t	.3
h	.1
i	.2
e	.1
s	.3

Figure 2.1: The frequency function π

First, we sort the table in decreasing order of frequencies. Then replace the last two entries with a combined entry, maintaining the list in decreasing frequency order. The first few steps in the process are given in Figure 2.2. When there are only two entries left, we stop.

t	.3	t	.3	(he)i	.4	ts	.6
s	.3	s	.3	t	.3	i(he)	.4
i	.2	he	.2	s	.3		
h	.1	i	.2				
e	.1						

Figure 2.2: Steps of the process

After that we move from right to left and assign binary codewords to all symbols in the columns as shown in the following table (Figure 2.3). Let us agree to write 0 above 1. Note that in order to find the codewords in the column to the left of a column that has been already filled in, we rewrite all codeword of the same letters, and append 0 and 1 to the codewords of the two new symbols that have been combined in one symbol in the right column.

t	.3	00	t	.3	00	(he)i	.4	1	ts	.6	0
s	.3	01	s	.3	01	t	.3	00	(he)i	.4	1
i	.2	10	he	.2	10	s	.3	01			
h	.1	100	i	.2	11						
e	.1	101									

Figure 2.3: Huffman codewords

The final Huffman code is given in Figure 2.4.

t	00
s	01
i	10
h	100
e	101

Figure 2.4: The final Huffman code

Note how the symbols with the highest frequency are assigned the shortest codes. To gain the maximum reduction in length from this encoding method, the frequencies assigned to each symbol should roughly match the actual number of occurrences of each symbol.

Now we can replace each symbol with its codeword and rewrite "this is test" into "00 100 10 01 10 01 00 101 01 00". The standard encoding of letters in binary digits (called ASCII) uses 7 digits per letter. The Huffman encoding required less than half this amount. This means that the text could be transmitted over a computer network in half the time, or can be stored on a disk using half the space that ASCII would use. In order to decode the data the table of codewords can be sent or stored together with longer texts. If a text is very short, we may use standard tables of frequencies of letters to avoid adding the table of codewords.

It is possible that when we combine entries from the table, the result may have equal frequencies with one or more existing entries. Should we put the new entry before or after the existing ones? Although it makes no difference to the average length of the code, it may make a difference to the variance in code length. It is usually better to have a code that gives more consistent lengths, so we should reduce the variance. To do this, whenever a combined entry has equal frequency with an existing entry, we should put it *above* the existing entries.

Example 2.3. Let the source alphabet be $B = \{a, b, c, d\}$ with frequencies given in Figure 2.5.

a	$4/11$
b	$3/11$
c	$2/11$
d	$2/11$

Figure 2.5: Frequencies for B

The first step involves combining c with d, giving a combined frequency of $4/11$. This is equal to the frequency of a. To reduce variance, we put the combined entry cd above a, as shown in the second column of Figure 2.6.

a	$4/11$	cd	$4/11$	ab	$7/11$
b	$3/11$	a	$4/11$	cd	$4/11$
c	$2/11$	b	$3/11$		
d	$2/11$				

Figure 2.6: Reducing the variance

The final code constructed according to the rule for reducing variance is given in Figure 2.7.

a	00
b	01
c	10
d	11

Figure 2.7: Reduced-variance code

This code actually has zero variance, since every codeword has the same length. If we had placed the *cd* entry below the *a* entry instead, we would have constructed a different code, given in Figure 2.8.

a	1
b	01
c	000
d	001

Figure 2.8: Code resulting from a different ordering choice

The alternative code clearly has greater variance – a is encoded by one symbol, while c and d are encoded by three. Both codes have an average length of 2 symbols per source letter, but the first code is much more consistent in the length of its codewords.

In some applications it may be useful to allow more code symbols. It is possible to construct Huffman codes over larger alphabets, and such codes share many of the desirable properties of binary Huffman codes. As in the binary case, we combine the least-frequent letters together into a new symbol. The number of letters combined in the first step depends on the total number of letters we are encoding.

If we have to encode m letters using n-letter alphabet, then we choose n' to satisfy $2 \leq n' \leq m$ and $n' = m \mod n - 1$. In the first step, we combine the least frequent n' letters. All further steps combine n letters, and we stop when we have n or less letters left in the table.

If $n' < n$, then the resulting Huffman code is not maximal – there is at least one unused codeword. The above process ensures that no unused codewords are shorter than any used codewords. In concept, it is equivalent to adding letters to the source alphabet ('padding') and assigning them frequencies of 0 to show that they are unused.

After that we allocate codewords to the symbols in each column moving from right to left across the table as in the binary case. At each move we add n symbols of our alphabet to the codeword of the combined symbol in order to obtain n different longer codewords for the older symbols that had been combined.

Example 2.4. Suppose we wish to encode the text "this test is not too long". However, now our encoding alphabet has four symbols: $\{0, 1, 2, 3\}$. As before, we construct the table of frequencies for each letter in Figure 2.9.

t	.5
h	.1
i	.2
s	.3
e	.1
n	.2
o	.4
l	.1
g	.1

Figure 2.9: The frequency function π

After sorting the table in decreasing order of frequency, we work out how many entries to combine in the first step. There are 9 rows for 4 code symbols. Since $9 \equiv 3 \mod (4-1)$, we combine 3 entries in the first step, and 4 in all subsequent steps, as in Figure 2.10.

t	.25	t	.25	sinh	.4
o	.2	o	.2	t	.25
s	.15	elg	.15	o	.2
i	.1	s	.15	elg	.15
n	.1	i	.1		
h	.05	n	.1		
e	.05	h	.05		
l	.05				
g	.05				

Figure 2.10: Steps of the process

The process terminates more rapidly, since at each step we are replacing n entries with 1. This is also why the first step may combine a different number of entries – we wish the final stage to have exactly n entries.

Next, we move back from right to left and assign codewords to all symbols in each column as shown in Figure 2.11.

t	.25	1	t	.25	1	sinh	.4	0
o	.2	2	o	.2	2	t	.25	1
s	.15	00	elg	.15	3	o	.2	2
i	.1	01	s	.15	00	elg	.15	3
n	.1	02	i	.1	01			
h	.05	03	n	.1	02			
e	.05	30	h	.05	03			
l	.05	31						
g	.05	32						

Figure 2.11: Steps in the process

The final table is given in Figure 2.12.

t	1
o	2
s	00
i	01
n	02
h	03
e	30
l	31
g	32

Figure 2.12: The final quaternary Huffman code

2.2 Groupoids

A set with a binary operation is called a *groupoid.* Given a binary operation on a set we can use *multiplicative notation* referring to this operation as a *product* and denoting it by $\cdot$. Sometimes it is more convenient to use *additive notation* where the operation is called *addition* and is recorded as $+$. If A and B are subsets of a groupoid G, then we define

$$AB = \{ab \mid a \in A, b \in B\}.$$

Groupoids can be used to encode trees and graphs. Conversely, we can use graphs to define the following examples of groupoids. Let $D = (V, E)$ be a graph. The *graph algebra* $\mathrm{Alg}(D)$ associated with D is the set $V \cup \{0\}$ equipped with multiplication defined by the rule

$$xy = \begin{cases} x & \text{if } x, y \in V \text{ and } (x, y) \in E, \\ 0 & \text{otherwise.} \end{cases}$$

An equivalence relation ϱ on a groupoid G is called a *congruence* if it is *compatible* with the operation $\cdot$ of G, that is if $(x, y) \in \varrho$ implies $(xz, yz) \in \varrho$ and $(zx, zy) \in \varrho$ for all $x, y, z \in G$. A groupoid G is said to be *simple* if it has at most two congruences: the equality relation and $G \times G$.

If ϱ is a congruence on a groupoid G, then we can define a binary operation on the set G/ϱ of all equivalence classes of ϱ by the rule $[x][y] = [xy]$. This means that in order to find the product of two classes one chooses an element x in the first class and an element y is the second class, computes their product xy and finds the equivalence class $[xy]$. This definition is correct in the sense that no ambiguity arises, because it is possible to show that the resulting class $[xy]$ depends only on the whole classes $[x]$ and $[y]$, but does not depend on the choice of particular representatives x and y. Thus, the set G/ϱ is also a groupoid. It is called the *quotient groupoid* of G modulo ϱ.

Let G be a groupoid. An element 1 of G is called a *neutral element* of the groupoid G if $1g = g1 = g$ for all $g \in G$. In multiplicative notation, where we the operation is called multiplication, the neutral element is also called the *identity element* of the groupoid G.

Lemma 2.1. *If a groupoid has an identity element, then it is unique.*

Proof. Suppose that there exist two identity elements e_1 and e_2. By definition this means that the following equalities hold for all x:

$$e_1 x = x e_1 = x, \tag{2.1}$$

$$e_2 x = x e_2 = x. \tag{2.2}$$

If we substitute e_2 for x in (2.1), then we get $e_1e_2 = e_2$. Similarly, substituting e_1 for x in (2.2) gives us $e_1e_2 = e_1$. Thus $e_1 = e_1e_2 = e_2$, which completes the proof. □

An element 0 is called an *absorbing element* of the groupoid G if $0g = g0 = 0$, for every $g \in G$. In multiplicative notation an absorbing element is also called a *zero* of the groupoid G if $0g = g0 = 0$, for every $g \in G$.

Denote by $G^1 = G \cup \{1\}$ (and $G^0 = G \cup \{0\}$) the groupoid G with identity (resp., zero) adjoined. If G has an identity element (or zero), then we assume that $G^1 = G$ (resp., $G^0 = G$). Otherwise, $G^1 = G \cup \{1\}$ and $G \cup \{0\}$ where $1 \notin G$ and $0 \notin G$, respectively. If M is a subset of a groupoid G and $x \in G$, then $xM^1 = xM \cup \{x\}$ and $M^1x = Mx \cup \{x\}$.

If $X \subseteq G$, then the *subgroupoid* generated by X in G is usually denoted by $\langle X \rangle$ or X^+. It consists of all products of elements of X. The *ideal* (*left ideal*, *right ideal*) generated by X in G is the smallest subgroupoid I of G containing X such that $GI \cup IG \subseteq I$ (resp., $GI \subseteq I$, $IG \subseteq I$). An ideal generated by one element is called a *principal ideal.*

The cardinality of the set G is called the *order* of the groupoid G and is denoted by $|G|$. The *order* of the element g is the order of the subgroupoid $\langle g \rangle$ it generates. It is denoted by $|g|$.

An element g of a groupoid G is said to be *periodic* if the subgroupoid $\langle g \rangle$ is finite. A subset T of S is *periodic* if every element of T is periodic. An element e of G is called an *idempotent* if $e = e^2$. The set of all idempotents of G is denoted by $E(G)$.

Let G, H be groupoids. A mapping $f : G \to H$ is called a *morphism* or a *homomorphism* if it preserves the operation, i.e., if

$$f(xy) = f(x)f(y)$$

for all $x, y \in G$. An *endomorphism* is a homomorphism of a groupoid onto itself. The homomorphism f is called an *isomorphism* if it is one-to-one and onto. An *automorphism* is an isomorphism from a groupoid onto itself.

It is easy to verify that every isomorphism is an invertible mapping where the inverse function is an isomorphism too.

Informally speaking, two groupoids G, H are *isomorphic* if they are the same up to notation used for their elements. This means that we can relabel all elements of G to obtain an exact copy of H. More formally, G and H are *isomorphic* if there exists an isomorphism from G to H (the relabeling function is the isomorphism).

A language L in X^* is said to be *recognized* by a groupoid S, if there exists a morphism $f\colon X^* \to G$ and a subset T of G such that $L = f^{-1}(T)$.

The *direct product* $S \times T$ of groupoids S and T is the set

$$\{(s,t) \mid s \in S, t \in T\}$$

with multiplication defined by $(s,t)(s',t') = (ss', tt')$.

A groupoid S with zero 0 is said to be a 0-*direct union* of its subgroupoids S_i, where $i \in I$, if and only if $S = \cup_{i \in I} S_i$ and $S_i S_j = S_i \cap S_j = 0$ for all $i \neq j$.

2.3 Semigroups and Monoids

A binary operation on a set G is *associative* if it satisfies the following associative law: $x(yz) = (xy)z$ for all $x, y, z \in G$. A *semigroup* $(S, \cdot)$ is a set S equipped with an associative binary operation "$\cdot$". The abbreviation S can be used instead of $(S, \cdot)$. For example, the set of natural numbers or positive integers is a semigroup with respect to addition, and at the same time it is a semigroup with respect to multiplication. If we want to distinguish between these two semigroups on the same set, we can use the following notation

$$(\mathbb{N}, +) \qquad (\mathbb{N}, \cdot)$$

where the symbol of operation is indicated explicitly.

Let S be a semigroup, and let $s = s_1, s_2, \ldots$ be a finite or infinite sequence of elements of S. A finite subsequence $s_i, s_{i+1}, \ldots, s_{i+k}$ is called a *segment* of s. The product $s_i s_{i+1} \ldots s_{i+k}$ is called the *value* of the segment. An *m-factorization* of s is a sequence $t_1, t_2, \ldots, t_m$, where all t_k are the values of consecutive segments of s, that is $t_k = s_{i_k} \cdots s_{i_{k+1}-1}$ for $k = 1, 2, \ldots, m$ and some $1 \le i_1 < i_2 < \ldots < i_{m+1}$. Note that an m-factorization may let aside some left part and some right part of s. Similarly, when s is infinite, an *ω-factorization* of s is an infinite sequence $t = t_1, t_2, \ldots$, where all t_k are the values of consecutive segments of s.

Theorem 2.7. (Ramsey Theorem) ([50], Theorem 1.2) *Each infinite sequence $s = s_1, s_2, \ldots$ of elements of every finite semigroup S has an ω-factorization of the form $e, e, \ldots$, where e is an idempotent of S.*

Let m, n be positive integers. The finite *monogenic semigroup* or *cyclic semigroup* with generator x is denoted by

$$\langle x \mid x^{m+n} = x^m \rangle$$

and is defined as the set $S = \{x, x^2, \ldots, x^{m+n-1}\}$ with multiplication defined by the following *exponent law*

$$x^a x^b = x^{a+b}, \tag{2.3}$$

for all positive integers a, b, and the equality

$$x^{m+n} = x^m. \tag{2.4}$$

This means that in order to compute the product $x^a x^b$ of two elements $x^a, x^b \in S$, we have to find $a + b$, and if it is $\geq m + n$, then we subtract n until the resulting integer c gets below $m + n$. The number m and n are called the *index* and *period* of S. The element x is called the *generator* of S. The monogenic semigroup of index 0 and period n is also called the *cyclic group* of order n and is denoted by $\mathbb{Z}_n$.

The *Cayley table* of a semigroup lists all values of the binary operation.

Exercise 2.1. Complete the Cayley tables of the monogenic semigroups with

(i) index 3 and period 2;

(ii) index 2 and period 3;

(iii) index 3 and period 3;

(iv) index 4 and period 6.

Two semigroups S and H are *isomorphic* if they are isomorphic as groupoids. This again means that they are the same up to notation used for their elements. In other words. we can relabel all elements of G to obtain an exact copy of H. More formally, G and H are *isomorphic* if there exists an isomorphism from G to H (the relabeling function is the isomorphism).

A nonempty subset T of a semigroup S is called a *subsemigroup* if it is closed under multiplication, i.e., if $x, y \in T \Rightarrow xy \in T$.

Exercise 2.2. (a) Prove that the set of all even positive integers is a subsemigroup of the multiplicative semigroup of all positive integers.

(b) Prove that the set of all odd positive integers is a subsemigroup of the multiplicative semigroup of all positive integers.

(c) Prove that, for every positive integer k, the set of all integers divisible by k is a subsemigroup of the multiplicative semigroup of all integers.

(d) Show that the set of all odd integers is not a subsemigroup of the additive semigroup of all integers.

Exercise 2.3. Let S be a semigroup with subsemigroups A and B.

(a) Prove that the intersection $A \cap B$ is a subsemigroup of S.

(b) Give an example to show that the union $A \cup B$ may fail to be a subsemigroup of S.

Theorem 2.8. ([234]) *Every infinite nil semigroup contains an infinite nilpotent subsemigroup.*

Let S be a semigroup with a subset T. The smallest subsemigroup of S containing T of S is called the *subsemigroup generated* by T. It is denoted by $\langle T \rangle$ or T^+. The set T is called a *generating set* of the subsemigroup $\langle T \rangle$.

Lemma 2.2. *For all subsets T and G of S, the following conditions are equivalent:*

(i) *G is the smallest subsemigroup of S containing T;*

(ii) *G is equal to the intersection of all subsemigroups of S containing T;*

(iii) *G is the set of all products of the form $t_1 t_2 \cdots t_n$, where n is a positive integer and $t_1, \ldots, t_n \in T$.*

Exercise 2.4. Find all subsemigroups of the monogenic semigroup with

(i) index 3 and period 2;

(ii) index 2 and period 3;

(iii) index 3 and period 3;

(iv) index 4 and period 6.

A semigroup with identity element is called a *monoid.* Let S be a monoid with identity element 1. A subsemigroup G of S is called a *submonoid* if it contains 1. The intersection of all submonoids of S containing a subset T of S is the smallest submonoid of S containing T. It is called the *submonoid generated* by T, and it is denoted by T^*. The set T is called a *generating set* of the submonoid T^*. The operation of generating a submonoid is called the *Kleene's $*$-operation.*

Let S be a semigroup. A subsemigroup I of S is called an *ideal* of S if $IS \cup SI \subseteq I$. The smallest ideal of S containing T of S is called the *ideal* generated by T. It is denoted by S^1TS^1, or $\mathrm{id}(T)$, or (T). It always exists, as the following lemma shows. The set T is called a *generating set* of the ideal.

Lemma 2.3. *For all subsets T and I of S, the following conditions are equivalent:*

(i) *I is the smallest ideal of S containing T;*

(ii) *G is equal to the intersection of all ideals of S containing T;*

(iii) *I is the set S^1TS^1 of all products xty, where $x, y \in S^1$, $t \in T$.*

Exercise 2.5. (a) Prove that the set of all even positive integers is an ideal of the multiplicative semigroup of all positive integers.

(b) Prove that, for every positive integer k, the set of all integers divisible by k is an ideal of the multiplicative semigroup of all integers.

(c) Is it true that the set of all odd integers is an ideal of the multiplicative semigroup of integers? Explain your answer.

Exercise 2.6. Let S be a semigroup with ideals A and B.

(a) Prove that the intersection $A \cap B$ is an ideal of S.

(b) Prove that the union $A \cup B$ is an ideal of S.

Exercise 2.7. Find all ideals of the monogenic semigroup with

(i) index 3 and period 2;

(ii) index 2 and period 3;

(iii) index 3 and period 3;

(iv) index 4 and period 6.

A nonempty subset I of a semigroup S is a *left* (resp., *right*, *two-sided*) *ideal* of S if $SI \subseteq I$ (resp., $IS \subseteq I$, both $SI \subseteq I$ and $IS \subseteq I$). Two-sided ideals are also called *ideals*. If T is a subset of a semigroup S, then the smallest ideal (left ideal, right ideal) of S containing T of S is called the *ideal* (*left ideal*, *right ideal*) generated by T, and the set T is called a *generating set* of the ideal. The ideal (left ideal, right ideal) generated by T in G coincides with the set $G^1TG^1 = \{gth \mid g, h \in G^1,\ t \in T\}$ (resp., $G^1T = \{gt \mid g \in G^1,\ t \in T\}$, $TG^1 = \{tg \mid g \in G^1,\ t \in T)$.

Exercise 2.8. *Prove that, for all subsets T and I of every semigroup S, the following conditions are equivalent:*

(i) *I is the smallest left ideal of S containing T;*

(ii) *G is equal to the intersection of all left ideals of S containing T;*

(iii) *I is the set S^1T of all products xt, where $x \in S^1$, $t \in T$.*

Exercise 2.9. Let S be a semigroup with left ideals A and B.

(a) Prove that the intersection $A \cap B$ is a left ideal of S.

(b) Prove that the union $A \cup B$ is a left ideal of S.

Exercise 2.10. *Prove that, for all subsets T and I of every semigroup S, the following conditions are equivalent:*

(i) *I is the smallest right ideal of S containing T;*

(ii) *G is equal to the intersection of all right ideals of S containing T;*

(iii) *I is the set TS^1 of all products tx, where $x \in S^1$, $t \in T$.*

Exercise 2.11. Let S be a semigroup with right ideals A and B.

(a) Prove that the intersection $A \cap B$ is a right ideal of S.

(b) Prove that the union $A \cup B$ is a right ideal of S.

Descriptions of left, right and two-sided ideals of semigroups are summarized in the following theorem.

Theorem 2.9. ([75], §*II*.1) *If G is a semigroup, then the ideal (left ideal, right ideal) generated by T in G coincides with the set G^1TG^1 (resp., G^1T, TG^1).*

Let S be a monoid with identity element 1, and let $x \in S$. An element y is called an *inverse* of the element x if $xy = yx = 1$.

Lemma 2.4. (Uniqueness of Inverses) *Let S be a monoid. If an element $x \in S$ has an inverse, then the inverse is unique.*

Proof. Let 1 be the identity element of S. Suppose that y and z are two inverse elements of x. This means that $xy = yx = 1$ and $xz = zx = 1$. Hence we get $y = y1 = y(xz) = (yx)z = 1z = z$. Thus $y = z$. □

This is why it is always possible to denote the inverse of x by x^{-1}.

Lemma 2.5. (Inverse of Product) *Let S be a monoid with identity 1, and let $x, y \in S$ be such that the inverses x^{-1}, y^{-1}, $(xy)^{-1}$ exist. Then*

$$(xy)^{-1} = y^{-1}x^{-1}.$$

Proof. The following two products are both equal to 1:

$$xy(y^{-1}x^{-1}) = x(yy^{-1})x^{-1} = x1x^{-1} = xx^{-1} = 1,$$

$$(y^{-1}x^{-1})xy = y^{-1}(x^{-1}x)y = y^{-1}y = 1.$$

Therefore $y^{-1}x^{-1}$ is the inverse of xy. It follows from Lemma 2.4 that it is equal to $(xy)^{-1}$. □

An element s of a semigroup S is *periodic* if and only if there exist positive integers m, n such that $s^{m+n} = s^m$. A semigroup S is *periodic* if all elements of S are periodic.

A semigroup is said to be *aperiodic* or *combinatorial* if it has no subgroups of order greater than one. A semigroup S is said to be *torsion-free* if $s^n = t^n$ implies $s = t$ for all $s, t \in S$ and any positive integer n.

Let S be a semigroup with zero 0. Then S is called a *null* semigroup, or a semigroup with *zero multiplication,* if the product of any two elements is zero, i.e., $S^2 = 0$. An element x of S is *nil* or *nilpotent* if $x^n = 0$ for a positive integer n. The whole S is *nilpotent* if $S^n = 0$ for some n. A semigroup is said to be *nil* if it entirely consists of nilpotent elements.

The following simple fact has been known for a long time (see, for example, a remark in [235], 1.4.3.

Theorem 2.10. *Every finite nil semigroup is nilpotent.*

Recall that an equivalence relation ρ on a semigroup is a *congruence* if and only if, for all $s, t, s', t' \in S$,

$$(s,t),(s',t') \in \rho \Longrightarrow (ss',tt') \in \rho.$$

For an arbitrary binary relation ρ on a semigroup S, $\rho^{\#}$ denotes the smallest congruence on S containing ρ. It is given by $\rho^{\#} = (\rho^c)^e$, where

$$\rho^c = \{(xay, xby) \mid x, y \in S^1, (a,b) \in \rho\}.$$

If $c, d \in S$ are such that $c = xay, d = xby$, for some $x, y \in S^1$, where either (a,b) or (b,a) belongs to a relation ρ, then c is connected to d by an *elementary ρ-transition*. This gives us the following lemma.

Lemma 2.6. *Let ρ be a relation on a semigroup S, and let $a, b \in S$. Then $(a,b) \in \rho^{\#}$ if and only if either $a = b$ or, for some $n \in \mathbb{Z}^+$, there is a sequence*

$$a = z_1 \to z_2 \to \cdots \to z_n = b$$

of elementary ρ-transitions connecting a to b.

The *quotient semigroup* S/ρ of a semigroup S by a congruence ρ is the set of all equivalence classes of ρ with a binary operation $[a][b] = [ab]$. For a congruence ρ, the natural map $\rho^\natural : S \to S/\rho$ is defined by $\rho^\natural(s) = s/\rho$, for all $s \in S$.

A *finitely presented semigroup* (resp., *finitely presented monoid*) is a quotient of a free semigroup (resp., free monoid) modulo a finitely generated congruence. If $X = \{x_1, \ldots, x_n\}$ and $r_1, \ldots, r_k \in X^+ \times X^+$, then the notation

$$\langle x_1, \ldots, x_n \mid r_1, \ldots, r_k \rangle,$$

stands for the quotient of the free semigroup X^+ modulo the congruence generated by $r_1, \ldots, r_k$.

Exercise 2.12. Find the Cayley table of the semigroup

$$\langle x, y \mid x^2 = x, y^2 = y, xy = yx \rangle.$$

Exercise 2.13. Find the Cayley table of the semigroup

$$\langle x, y, z \mid x^2 = x, y^2 = y, z^2 = z, xy = yx, xz = zx, yz = zy \rangle.$$

Exercise 2.14. Find the Cayley table of the semigroup

$$\langle x, y \mid x^2 = x, y^2 = y, (xy)^2 = xy, (yx)^2 = yx, (xyx)^2 = xyx, (yxy)^2 = yxy \rangle.$$

2.4 Cayley Graphs

Let G be a semigroup and let T be a subset of G. The *Cayley graph* $\mathrm{Cay}(G,T)$ of G relative to T is defined as the graph with vertex set G and edge set $E(T)$ consisting of those ordered pairs (x, y) such that $xs = y$ for some $s \in T$.

For example, if $G = \mathbb{Z}_3 = \{e, x, x^2\}$ and $T = \{x\}$, then the Cayley graph $\mathrm{Cay}(G,T)$ is displayed in Figure 2.13. If $G = \mathbb{Z}_6 = \{e, x, x^2, x^3, x^4, x^5\}$ and $T = \{x, x^2\}$, then Figure 2.14 sketches the Cayley graph $\mathrm{Cay}(G,T)$.

Exercise 2.15. Let S be the monogenic semigroup with index m, period n and generator x. Draw the Cayley graph $\mathrm{Cay}(S,T)$ for

(i) $m = 3, n = 2$ and $T = \{x\}$;

(ii) $m = 3, n = 2$ and $T = \{x^2\}$;

(iii) $m = 3, n = 2$ and $T = \{x^2, x^3, x^4\}$;

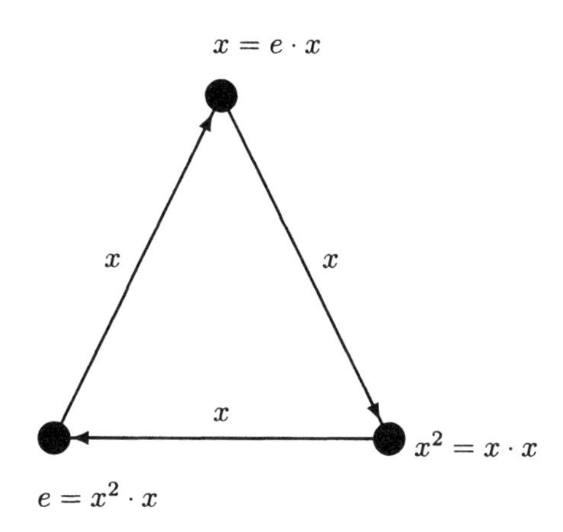

Figure 2.13: $\mathrm{Cay}(\mathbb{Z}_3, \{x\})$

(iv) $m = 3, n = 6$ and $T = \{x\}$;

(v) $m = 3, n = 6$ and $T = \{x, x^2\}$;

(vi) $m = 3, n = 6$ and $T = \{x^3\}$;

(vii) $m = 3, n = 6$ and $T = \{x^3, x^4\}$;

(viii) $m = 3, n = 6$ and $T = S$.

Exercise 2.16. Given the Cayley table of a semigroup, draw some of its Cayley graphs.

Exercise 2.17. Let S be the semigroup with the following Cayley table

	e_0	g_0	h_0	e_1	g_1	h_1
e_0	e_0	g_0	g_0	e_0	g_0	g_0
g_0	g_0	e_0	e_0	g_0	e_0	e_0
h_0	g_0	e_0	e_0	g_0	e_0	e_0
e_1	e_0	g_0	g_0	e_1	g_1	g_1
g_1	g_0	e_0	e_0	g_1	e_1	e_1
h_1	g_0	e_0	e_0	g_1	e_1	e_1

Draw the Cayley graph $\mathrm{Cay}(S, T)$ for

(i) $T = \{e_0\}$;

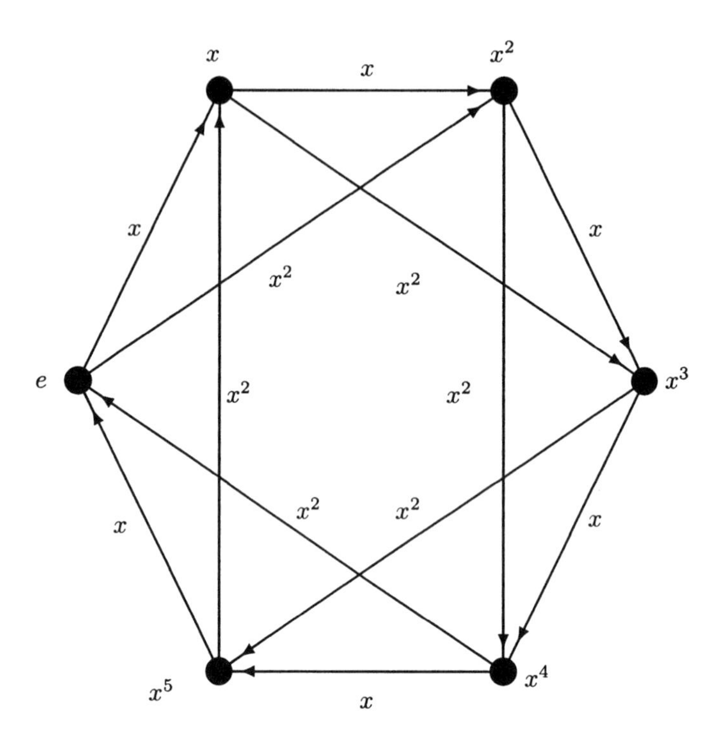

Figure 2.14: $\text{Cay}(\mathbb{Z}_6, \{x, x^2\})$

(ii) $T = \{g_0\}$;

(iii) $T = \{e_0, g_0\}$;

(iv) $T = \{e_1\}$;

(v) $T = \{g_1\}$;

(vi) $T = \{e_1, g_1\}$.

Exercise 2.18. Let G be the direct product $\mathbb{Z}_2 \times \mathbb{Z}_2$. Draw the Cayley graph $\text{Cay}(S, T)$ for

(i) $T = \{(0, 0)\}$;

(ii) $T = \{(1,1)\}$;

(iii) $T = \{(0,0),(1,1)\}$;

(iv) $T = \{(0,1)\}$;

(v) $T = \{(1,1)\}$;

(vi) $T = G$.

Exercise 2.19. Let G be the direct product $\mathbb{Z}_4 \times \mathbb{Z}_4$. Draw the Cayley graph $\mathrm{Cay}(S,T)$ for

(i) $T = \{(0,0)\}$;

(ii) $T = \{(1,1)\}$;

(iii) $T = \{(2,2)\}$;

(iv) $T = \{(0,2)\}$;

(v) $T = \{(2,0)\}$;

(vi) $T = G$.

Exercise 2.20. Draw the Cayley graph $\mathrm{Cay}(S,T)$, where $T = \{x\}$ and

$$S = \langle x, y \mid x^2 = x, y^2 = y, xy = yx \rangle.$$

Exercise 2.21. Draw the Cayley graph $\mathrm{Cay}(S,T)$, where $T = \{x\}$ and

$$S = \langle x, y, z \mid x^2 = x, y^2 = y, z^2 = z, xy = yx, xz = zx, yz = zy. \rangle$$

Exercise 2.22. Draw the Cayley graph $\mathrm{Cay}(S,T)$, where $T = \{x\}$ and

$$\langle x, y \mid x^2 = x, y^2 = y, (xy)^2 = xy, (yx)^2 = yx, (xyx)^2 = xyx, (yxy)^2 = yxy \rangle.$$

2.5 Groups

A semigroup G with identity element 1 is called a *group* if every element $g \in G$ has an inverse g^{-1} such that $gg^{-1} = g^{-1}g = 1$. For example, the set $\mathbb{R}_+$ of positive real numbers and the set $\mathbb{Q}_+$ of positive rational numbers are groups with respect to multiplication. The set $\mathbb{Z}$ of integers is a group with respect to addition. It is called the *infinite cyclic group.*

In general terminology, where the operation of the group is not necessarily called multiplication, the special element of the group is called a *neutral* element, exactly as it has been explained in the case of monoids.

A commutative group is said to be *abelian.* Abelian groups are often considered in the *additive notation* where the operation of the group $(G, +, 0)$ is denoted by $+$ and the neutral element is 0. For example, the set $\mathbb{Z}$ of all integers is an abelian group with respect to addition.

If n is a positive integer, then the *cyclic group* of order n is the set

$$\mathbb{Z}_n = \{0, 1, 2, \ldots, n-1\}$$

with respect to addition $+$ defined modulo n. This means that in order to find the product of two elements $x, y \in \mathbb{Z}$ we have to find the sum of x and y as ordinary integers and take its remainder on division by n. If in some discussion it is more convenient for us to refer to the operation of the cyclic group as a multiplication, then we can denote all elements of $\mathbb{Z}_n$ as follows

$$\mathbb{Z}_n = \{1, g, g^2, \ldots, g^{n-1}\}$$

and define the operation by the rules

$$g^0 = 1, \qquad g^a g^b = g^{(a+b \mod n)}.$$

A semigroup S is said to be *left* (*right*) *cancellative* if $zx = zy$ implies $x = y$ (resp., $xz = yz$ implies $x = y$) for all $x, y, z \in S$. A semigroup is *cancellative* if it is both left and right cancellative.

Exercise 2.23. Prove that every finite cancellative semigroup is a group.

Let G be a group. A *subgroup* H of G is a subset closed with respect to the multiplication of the group and the operation of taking the inverse element. If $T \subseteq G$, then the *subgroup generated by* T is the smallest subgroup containing T. It consisit of all products of all elements of T and their inverses. The set T is called a *generating set* of the subgroup. A subgroup is *finitely generated* if it has a finite generating set.

A *right coset* of the subgroup H is a set of the form

$$Hg = \{hg \mid h \in H\},$$

where $g \in G$. A *left coset* of the subgroup H is a set of the form

$$gH = \{gh \mid h \in H\}.$$

Lemma 2.7. (Coset Partition Lemma) ([227], 1.3.4) *The set of left cosets has the same cardinality as the set of right cosets. The left cosets of every group form a partition of the group, and the same can be said of the right cosets as well.*

The cardinality of the set of left (or right) cosets is denoted by

$$|G : H|$$

and is called the *index* of H in G.

If T is a nonempty subset of G and $g \in G$, then the *conjugate* of T by g is the set

$$T^g = g^{-1}Tg = \{g^{-1}tg \mid t \in T\}.$$

Exercise 2.24. Let G be a group, and let H be a subgroup of G. Prove that the following conditions are equivalent:

(i) the set $H^G = \{g^{-1}hg \mid h \in H, g \in G\}$ is equal to H;

(ii) for each $g \in G$, the set $H^g = \{g^{-1}hg \mid h \in H\}$ is equal to H;

(iii) $gH = Hg$, for each $g \in G$;

(iv) the set of left cosets is equal to the set of right cosets.

A subgroup H of G is said to be *normal* if it satisfies the conditions of Exercise 2.24. If N is a normal subgroup of G, then the *quotient group* G/N is the set of all right cosets of N in G with the following operations: $(Ng_1)(Ng_2) = Ng_1g_2$ and $(Ng)^{-1} = Ng^{-1}$ for all $g, g_1, g_2 \in G$.

Now we can give an alternative definition of a finite cyclic group. Let n be a positive integer. The set nZ of all integers divisible by n is a subgroup of the infinite cyclic group, and the quotient group $\mathbb{Z}_n = \mathbb{Z}/n\mathbb{Z}$ is called the *cyclic group of order n*.

For any real number x, denote by $\lfloor x \rfloor$ the *floor* or *integral part* of x, i.e., the largest integer not exceeding x. Let p be a prime. The *quasicyclic group* is the set

$$\mathbb{Z}_{p^\infty} = \left\{ \frac{m}{p^n} \mid m, n \in \mathbb{Z}, 0 \le \frac{m}{p^n} < 1 \right\}$$

with operation defined by

$$x + y = x + y - \lfloor x + y \rfloor.$$

Alternatively, we can define the quasicyclic group as the quotient group

$$\left\{ \frac{m}{p^n} \mid m, n \in \mathbb{Z}, 0 \le \frac{m}{p^n} < 1 \right\} / \mathbb{Z}.$$

It is also possible to use multiplicative notation for the elements of the quasicyclic group $\mathbb{Z}_{p^\infty}$ by assuming that

$$\mathbb{Z}_{p^\infty} = \left\{ g^{\frac{m}{p^n}} \mid m, n \in \mathbb{Z}, 0 \le \frac{m}{p^n} < 1 \right\}$$

and the multiplication is again defined by the rules

$$g^0 = 1, \qquad g^a g^b = g^{(a+b-\lfloor a+b \rfloor)}.$$

Let p be a prime. An element of a group is called a *p-element* if its order is a power of p. A group G is called a *p-group* if it entirely consists of p-elements. For every abelian group G, the set of all p-elements forms a subgroup called the *p-primary component* of G. Periodic groups are more often called *torsion* groups.

Two groups S and H are *isomorphic* if they are isomorphic as groupoids. This means that they are the same up to notation used for their elements. Again we can relabel all elements of G to obtain an exact copy of H. Formally speaking, G and H are *isomorphic* if there exists an isomorphism from G to H. It is easy to prove that every isomorphism from G to H maps the identity element of G to the identity element of H, and preserves inverse elements too.

Theorem 2.11. (Principal Theorem on Finite Abelian Groups) ([190], 10.25) *Every finite abelian group is isomorphic to the direct product of its primary components and is isomorphic to the direct product of a finite number of cyclic groups*

$$G \cong G_1 \times G_2 \times \cdots \times G_n$$

such that the order $|G_i|$ divides $|G_{i+1}|$ for $i = 1, \ldots, n-1$.

Exercise 2.25. Prove that the following groups are isomorphic:

(a) the cyclic group $\mathbb{Z}_6$ and the direct product $\mathbb{Z}_2 \times \mathbb{Z}_3$;

(b) $\mathbb{Z}_{12}$ and $\mathbb{Z}_3 \times \mathbb{Z}_4$;

(b) $\mathbb{Z}_{30}$ and $\mathbb{Z}_2 \times \mathbb{Z}_3 \times \mathbb{Z}_5$.

Theorem 2.12. (Primary Decomposition Theorem) ([227], 4.1.1) *Every torsion abelian group is isomorphic to the direct product of its primary components.*

Theorem 2.13. (Finitely Generated Abelian Groups) ([98], Theorem 8.1.1) *Every finitely generated abelian group is isomorphic to the direct product of a finite number of cyclic groups. Each subgroup of every finitely generated abelian group is finitely generated.*

Theorem 2.14. (Lagrange's Theorem) ([190], 10.17, [201], 2.171) *If G is a finite group, and H is a subgroup of G, then the order of H divides the* *order of G and, moreover,*

$$|G| = |H| \times |G : H|.$$

Theorem 2.15. (Sylow's Theorem) ([227], 1.6.16) *Let G be a finite group of order $|G| = p^a m$, where p is a prime, and m is a positive integer not divisible by p.*

(i) *Each p-subgroup of G is contained in a subgroup of order p^a.*

(ii) *If n_p is the number of subgroups of order p^a, then $n_p \equiv 1 \mod p$.*

(iii) *All subgroups of order p^a are conjugate in G.*

A series of subgroups of the group G

$$\{1\} = N_0 \subseteq N_1 \subseteq N_2 \subseteq \cdots \subseteq N_k = G \tag{2.5}$$

is said to be *normal* if N_{i-1} is a normal subgroup of N_i for every $i = 1, \ldots, k$. A group G is *solvable* if it has a finite normal series (2.5) such that all factors N_i/N_{i-1} are abelian for all $i = 1, \ldots, k$.

The *center* of a group G is the set of elements x such that $xy = yx$ for every $y \in G$. A group is *nilpotent* if it has a finite normal series (2.5) such that, for $i = 0, \ldots, k-1$, the N_i is a normal subgroup of the whole G and N_{i+1} is the center of G/N_i. A group is *polycyclic* if it has a finite normal series (2.5) such that all factors N_i/N_{i-1} are cyclic groups for $i = 1, \ldots, k$.

Theorem 2.16. (Finite Nilpotent Groups) ([227], 5.2.4) *A finite group is nilpotent if and only if it is a direct product of its primary components. In particular, for each prime p, every finite p-group is nilpotent.*

A group G is said to be *finite-by-abelian-by-finite* if it has subgroups

$$\{e\} \subseteq H \subseteq K \subseteq G$$

such that H is finite and normal in K, K/H is abelian, K is normal in G, and G/K is finite. A group G is *abelian-by-finite* if it has a normal abelian subgroup K such that G/K is finite.

2.6 Commutative Semigroups

The structure of all commutative semigroups can be described using semilattices and Archimedean semigroups. Let us first introduce these two classes of semigroups. A *semilattice* is a commutative semigroup entirely consisting of idempotents. A natural partial order $\leq$ can be defined on every semilattice by the rule

$$x \leq y \Leftrightarrow xy = x.$$

We can use Hasse diagrams to represent semilattices.

Example 2.5. Prove that the set S of all divisors of 12 is a semilattice with respect to the operation $xy = \gcd(x, y)$.

(a) Draw the Hasse diagram of S.

(b) Complete the Cayley table of S.

(c) Draw the Cayley graph $\text{Cay}(S, \{2\})$.

(d) Draw the Cayley graph $\text{Cay}(S, \{1, 2\})$.

Exercise 2.26. Given the Hasse diagram of a semilattice, draw some of its Cayley graphs.

Exercise 2.27. Given the Hasse diagram of a semilattice, write down its Cayley table.

A semigroup S is said to be *Archimedean* if and only if, for every $s, t \in S$, a power of s belongs to the ideal generated in S by t.

Exercise 2.28. Prove that all monogenic semigroups are Archimedean.

Proof. Consider a finite monogenic semigroup S with generator x. Take two elements s and t of S, where $s = x^a$, $t = x^b$. The ideal generated in I by t is the set

$$tS^1 = \{x^b, x^{b+1}, x^{b+2}, \dots\}.$$

Clearly, $y^b = x^{ab}$ belongs to tS^1, because $ab \geq b$. By definition, this means that S is Archimedean. $\square$

A semigroup S is said to be a *semilattice* Y *of its subsemigroups* S_y, where $y \in Y$, if $S = \cup_{y \in Y} S_y$ is a disjoint union of its subsemigroups S_y, and $S_x S_y \subseteq S_{xy}$ for all $x, y \in Y$.

Theorem 2.17. (Semilattice Decomposition of Commutative Semigroups) ([75], Theorem 2.2) *Every commutative semigroup is uniquely represented as a semilattice of Archimedean semigroups.*

If a commutative semigroup S is a semilattice Y of Archimedean semigroups S_y, then the semigroups S_y are called the *Archimedean components* of S. The largest subgroup of S_y will be denoted by G_y. If S_y has no idempotents, then $G_y = \emptyset$.

Let I and J be ideals of a semigroup S such that $J \subseteq I$. The *Rees quotient semigroup* I/J is the semigroup with zero obtained from I by identifying all elements of the ideal J with 0. If I has zero and $J = \{0\}$, then $I/J = I$. In the case where $J = \emptyset$, we assume that $I/J = I$. The quotient semigroup I/J is called a *factor* of S. We identify all elements of $I \setminus J$ with their images in I/J, and say that all elements of $I \setminus J$ belong to I/J. Take any element g in S, put $I = S^1 g S^1$ and denote by J the set of all elements which generate principal ideals properly contained in I. Then J is also an ideal of S, and I/J is called a *principal factor* of S.

A *Rees extension* or *ideal extension* of a semigroup S by a semigroup Q is a semigroup E such that S is an ideal of E and the Rees quotient E/S is isomorphic to Q. If an Archimedean component S_y has an idempotent, it will be denoted by e_y. Then the ideal $G_y = e_y S_y$ is a group, and the quotient semigroup S_y/G_y is nil.

Theorem 2.18. ([75], Proposition 2.3) *If a commutative Archimedean semigroup has an idempotent, then it is an ideal extension of a group by a nil semigroup.*

Corollary 2.2. *Every periodic commutative Archimedean semigroup is an* *ideal extension of a torsion group by a nil semigroup. Every finite commutative Archimedean semigroup is an ideal extension of a finite group by a nilpotent semigroup.*

For finite commutatiove semigroups Theorem 2.17 can be rewritten as the following proposition which is easier to use for solving exercises.

Proposition 2.1. *Let S be a finite commutative semigroup, Y the set of all idempotents of S, and let*

$$S_y = \{x \in S \mid x^n = y \text{ for some } n\},$$

$$G_y = \{x \in S_y \mid x^{n+1} = x \text{ for some } n\}.$$

Then Y is a semilattice, every set S_y is an Archimedean semigroup, each set G_y is the largest subgroup of S_y and is the smallest ideal of S_y, and

$$S = \bigcup_{y \in Y} S_y$$

is the unique semilattice decomposition of S.

Exercise 2.29. Prove Proposition 2.1.

Exercise 2.30. Find the number of Archimedean components of the commutative semigroup with the following Cayley table

	e_0	g_0	e_1	g_1
e_0	e_0	g_0	e_0	g_0
g_0	g_0	e_0	g_0	e_0
e_1	e_0	g_0	e_1	g_1
g_1	g_0	e_0	g_1	e_1

Exercise 2.31. Find the number of Archimedean components of the commutative semigroup with the following Cayley table

	e_0	g_0	e_1	g_1	e_2	g_2
e_0	e_0	g_0	e_0	g_0	e_0	g_0
g_0	g_0	e_0	g_0	e_0	g_0	e_0
e_1	e_0	g_0	e_1	g_1	e_0	g_0
g_1	g_0	e_0	g_1	e_1	g_0	e_0
e_2	e_0	g_0	e_0	g_0	e_2	g_2
g_2	g_0	e_0	g_0	e_0	g_2	e_2

Exercise 2.32. Find the semilattice decomposition of the commutative semigroup with the Cayley table

	e_0	g_0	e_1	g_1
e_0	e_0	g_0	e_0	g_0
g_0	g_0	e_0	g_0	e_0
e_1	e_0	g_0	e_1	g_1
g_1	g_0	e_0	g_1	e_1

Exercise 2.33. Find the semilattice decomposition of the commutative semigroup with the Cayley table

	e_0	g_0	e_1	g_1	e_2	g_2
e_0	e_0	g_0	e_0	g_0	e_0	g_0
g_0	g_0	e_0	g_0	e_0	g_0	e_0
e_1	e_0	g_0	e_1	g_1	e_0	g_0
g_1	g_0	e_0	g_1	e_1	g_0	e_0
e_2	e_0	g_0	e_0	g_0	e_2	g_2
g_2	g_0	e_0	g_0	e_0	g_2	e_2

Exercise 2.34. Find the semilattice decomposition of

(i) the multiplicative semigroup $(\mathbb{N}, \cdot)$ of all positive integers.

(ii) the multiplicative semigroup $(\mathbb{N}_0, \cdot)$ of all nonnegative integers.

(iii) the additive semigroup $(\mathbb{N}, +)$ of all positive integers.

(ii) the additive semigroup $(\mathbb{N}_0, +)$ of all nonnegative integers.

Exercise 2.35. Find the number of Archimedean components of the commutative semigroup with the following Cayley table

	e_0	g_0	h_0	e_1	g_1	h_1
e_0	e_0	g_0	g_0	e_0	g_0	g_0
g_0	g_0	e_0	e_0	g_0	e_0	e_0
h_0	g_0	e_0	e_0	g_0	e_0	e_0
e_1	e_0	g_0	g_0	e_1	g_1	g_1
g_1	g_0	e_0	e_0	g_1	e_1	e_1
h_1	g_0	e_0	e_0	g_1	e_1	e_1

Exercise 2.36. Find the semilattice decomposition of the commutative semigroup with the Cayley table

	e_0	g_0	h_0	e_1	g_1	h_1
e_0	e_0	g_0	g_0	e_0	g_0	g_0
g_0	g_0	e_0	e_0	g_0	e_0	e_0
h_0	g_0	e_0	e_0	g_0	e_0	e_0
e_1	e_0	g_0	g_0	e_1	g_1	g_1
g_1	g_0	e_0	e_0	g_1	e_1	e_1
h_1	g_0	e_0	e_0	g_1	e_1	e_1

A *band* is a semigroup entirely consisting of idempotents. A band is called a *semilattice* (*rectangular band, left zero band, right zero band, left regular band, right regular band*) if it satisfies the identity $xy = yx$ (resp., $xyx = x$, $xy = x$, $xy = y$, $xyx = xy$, $xyx = yx$). A partial order $\leq$ is defined on the set of idempotents $E(S)$ by $e \leq f \Leftrightarrow ef = fe = e$. A semilattice is called a *chain* if it is linearly ordered. By the *height* of a semilattice Y we mean the supremum of the cardinalities of all chains contained in Y.

A semigroup S is said to be *inverse* if, for each $s \in S$, there exists a unique $t \in S$ such that $sts = s$ and $tst = t$. If S is an inverse semigroup, then every two idempotents of S commute, and so the set $E(S)$ of idempotents of S is a semilattice.

2.7 Rees Matrix Semigroups

Exactly as in the case of groupoids, a semigroup S with zero 0 adjoined is denoted by S^0 and is defined by

$$S^0 = \begin{cases} S & \text{if } S \text{ has a zero,} \\ S \cup \{0\} & \text{otherwise.} \end{cases}$$

By analogy with the case of S^0, the *monoid* S^1 is defined by

$$S^1 = \begin{cases} S & \text{if } S \text{ has an identity element,} \\ S \cup \{1\} & \text{otherwise.} \end{cases}$$

An element x in a monoid S is called a *unit* if there exists $y \in S$ such that $xy = yx = 1$. By Lemma 2.5, the set of all units of S forms a group. This group is called the *group of units*.

Suppose that G is a group, I and Λ are nonempty sets, and $P = [p_{\lambda i}]$ is a $(\Lambda \times I)$-matrix with entries $p_{\lambda i} \in G$, for all $\lambda \in \Lambda$, $i \in I$. The *Rees matrix semigroup* $M(G; I, \Lambda; P)$ over G with *sandwich-matrix* P consists of all triples $(g; i, \lambda)$, where $i \in I$, $\lambda \in \Lambda$, and $g \in G$, endowed with multiplication defined by the rule

$$(g_1; i_1, \lambda_1)(g_2; i_2, \lambda_2) = (g_1 p_{\lambda_1 i_2} g_2; i_1, \lambda_2).$$

A semigroup S is called *simple* if S is the only ideal of S. A semigroup is said to be *completely simple* if it has no proper ideals and has a minimal idempotent with respect to the natural partial order. For example, all finite simple semigroups are completely simple. Similarly, a semigroup with zero is *completely* 0*-simple* if it has no proper nonzero ideals and has a minimal nonzero idempotent.

Theorem 2.19. (Rees Theorem) ([89], Theorem 3.2.3) *Every completely simple semigroup is isomorphic to a Rees matrix semigroup $M(G; I, \Lambda; P)$ over a group G. Every completely 0-simple semigroup is isomorphic to a Rees matrix semigroup $M^0(G; I, \Lambda; P)$ over a group G with zero adjoined. Conversely, every semigroup $M(G; I, \Lambda; P)$ is completely simple, and a semigroup $M^0(G; I, \Lambda; P)$ is completely 0-simple if and only if each row and every column of P contains at least one nonzero entry.*

Corollary 2.3. *For every periodic semigroup S, the following conditions are equivalent:*

(i) *S is simple;*

(ii) *S is completely simple;*

(iii) *S is isomorphic to a Rees matrix semigroup over a group.*

Corollary 2.4. *For every finite semigroup S, the following conditions are equivalent:*

(i) *S is* 0*-simple;*

(ii) *S is completely* 0*-simple;*

(iii) *S is isomorphic to a Rees matrix semigroup over a group with zero such that each row and every column of the sandwich matrix has a nonzero entry.*

A semigroup is *completely regular* if it is a union of groups.

Theorem 2.20. (Clifford's Theorem) ([89], Theorem 4.1.3) *For every semigroup S, the following conditions are equivalent:*

(i) *S is a union of groups;*

(ii) *S is a semilattice of Rees matrix semigroups over groups.*

Corollary 2.5. *Every band is a semilattice of rectangular bands.*

Theorem 2.21. (Finite Archimedean Semigroup Theorem) ([235], Proposition 1.5.1) *For every finite semigroup S, the following conditions are equivalent:*

(i) *S is Archimedean;*

(ii) *S does not contain the two-element semilattice;*

(iii) *has an ideal I which is a Rees matrix semigroup over a group, and some power S^n is contained in I.*

Lemma 2.8. *Let S be a simple semigroup which is not completely simple. Then S contains a subsemigroup isomorphic to the bicyclic monoid*

$$\langle a, b \mid ab = 1 \rangle.$$

Exercise 2.37. Complete the Cayley tables of the Rees matrix semigroup $M(G; I, \Lambda; P)$, where

(i) $I = \{1\}$, $\Lambda = \{1, 2\}$, $G = \{e, g\}$, $P = \begin{bmatrix} e \\ e \end{bmatrix}$;

(ii) $I = \{1\}$, $\Lambda = \{1,2\}$, $G = \{e,g\}$, $P = \begin{bmatrix} e \\ g \end{bmatrix}$;

(iii) $I = \{1,2\}$, $\Lambda = \{1\}$, $G = \{e,g\}$, $P = \begin{bmatrix} e & e \end{bmatrix}$;

(iv) $I = \{1,2\}$, $\Lambda = \{1,2\}$, $G = \{e,g\}$, $P = \begin{bmatrix} e & g \\ e & e \end{bmatrix}$;

(v) $I = \{1,2\}$, $\Lambda = \{1,2\}$, $G = \{e,g\}$, $P = \begin{bmatrix} e & g \\ g & g \end{bmatrix}$.

Exercise 2.38. Draw the Cayley graph $\mathrm{Cay}(S,T)$, where $S = M(G; I, \Lambda; P)$ is a Rees matrix semigroup, and

(i) $I = \{1\}$, $\Lambda = \{1,2\}$, $G = \{e,g\}$, $P = \begin{bmatrix} e \\ e \end{bmatrix}$, $T = \{(g;1,1)\}$;

(ii) $I = \{1\}$, $\Lambda = \{1,2\}$, $G = \{e,g\}$, $P = \begin{bmatrix} e \\ g \end{bmatrix}$, $T = \{(g;1,1)\}$;

(iii) $I = \{1,2\}$, $\Lambda = \{1\}$, $G = \{e,g\}$, $P = \begin{bmatrix} e & e \end{bmatrix}$, $T = \{(g;1,1)\}$;

(iv) $I = \{1,2\}$, $\Lambda = \{1,2\}$, $G = \{e,g\}$, $P = \begin{bmatrix} e & g \\ e & e \end{bmatrix}$, $T = \{(g;1,1)\}$;

(v) $I = \{1,2\}$, $\Lambda = \{1,2\}$, $G = \{e,g\}$, $P = \begin{bmatrix} e & g \\ g & g \end{bmatrix}$, $T = \{(g;1,1)\}$.

Let G be a group, $M = M(G; I, \Lambda; P)$, and let $i \in I$, $\lambda \in \Lambda$. Put

$$\begin{aligned} G_{*\lambda} &= \{(g;i,\lambda) \mid g \in G, i \in I\}, \\ G_{i*} &= \{(g;i,\lambda) \mid g \in G, \lambda \in \Lambda\}, \\ G_{i\lambda} &= \{(g;i,\lambda) \mid g \in G\}. \end{aligned}$$

In the case where $M = M^0(G; I, \Lambda; P)$ the zero is included in all of these sets. That is, we let

$$\begin{aligned} G_{*\lambda} &= \{0\} \cup \{(g;i,\lambda) \mid g \in G, i \in I\}, \\ G_{i*} &= \{0\} \cup \{(g;i,\lambda) \mid g \in G, \lambda \in \Lambda\}, \\ G_{i\lambda} &= \{0\} \cup \{(g;i,\lambda) \mid g \in G\}. \end{aligned}$$

Lemma 2.9. *Let G be a group, and let $M = M(G; I, \Lambda; P)$ be a completely simple semigroup or let $M = M^0(G; I, \Lambda; P)$ be a completely 0-simple semigroup. Then, for all $i, j \in I$, $\lambda, \mu \in \Lambda$,*

(i) *$G_{*\lambda}$ is an $\mathcal{L}$-class of M and a minimal nonzero left ideal of M;*

(ii) *G_{i*} is an $\mathcal{R}$-class of M and a minimal nonzero right ideal of M;*

(iii) *$G_{i\lambda}$ is an $\mathcal{H}$-class of M, a left ideal of G_{i*} and a right ideal of $G_{*\lambda}$;*

(iv) *$|G_{i\lambda}|=|G_{j\mu}|$;*

(v) *each maximal subgroup of M coincides with $G_{j\mu}$, for some $j \in I$, $\mu \in \Lambda$;*

(vi) *if $p_{\lambda i} \neq 0$, then $G_{i\lambda}$ is a maximal subgroup of M isomorphic to G;*

(vii) *if $p_{\lambda i} = 0$, then $G_{i\lambda}^2 = 0$;*

(viii) *every $\mathcal{L}$-class of M contains at least one maximal subgroup, $G_{j\mu}$;*

(ix) *every $\mathcal{R}$-class of M contains at least one maximal subgroup, $G_{j\mu}$.*

Exercise 2.39. Complete the Cayley tables of the Rees matrix semigroup $M(G^0; I, \Lambda; P)$ over a group G with zero adjoined, where

(i) $I = \{1\}$, $\Lambda = \{1, 2\}$, $G = \{e\}$, $P = \begin{bmatrix} 0 \\ e \end{bmatrix}$;

(ii) $I = \{1\}$, $\Lambda = \{1, 2\}$, $G = \{e\}$, $P = \begin{bmatrix} e \\ 0 \end{bmatrix}$;

(iii) $I = \{1, 2\}$, $\Lambda = \{1\}$, $G = \{e\}$, $P = \begin{bmatrix} e & e \end{bmatrix}$;

(iv) $I = \{1, 2\}$, $\Lambda = \{1, 2\}$, $G = \{e\}$, $P = \begin{bmatrix} e & 0 \\ e & e \end{bmatrix}$;

(v) $I = \{1, 2\}$, $\Lambda = \{1, 2\}$, $G = \{e\}$, $P = \begin{bmatrix} e & 0 \\ 0 & 0 \end{bmatrix}$.

Exercise 2.40. Draw the Cayley graph $\mathrm{Cay}(S, T)$, where $S = M(G^0; I, \Lambda; P)$ is a Rees matrix semigroup over a group G with zero adjoined, and

(i) $I = \{1\}$, $\Lambda = \{1, 2\}$, $G = \{e\}$, $P = \begin{bmatrix} e \\ e \end{bmatrix}$, $T = \{(g; 1, 1)\}$;

(ii) $I = \{1\}$, $\Lambda = \{1, 2\}$, $G = \{e\}$, $P = \begin{bmatrix} e \\ 0 \end{bmatrix}$, $T = \{(e; 1, 1)\}$;

(iii) $I = \{1, 2\}$, $\Lambda = \{1\}$, $G = \{e\}$, $P = \begin{bmatrix} e & e \end{bmatrix}$, $T = \{(e; 1, 1)\}$;

(iv) $I = \{1, 2\}$, $\Lambda = \{1, 2\}$, $G = \{e\}$, $P = \begin{bmatrix} e & 0 \\ e & e \end{bmatrix}$, $T = \{(e; 1, 1)\}$;

(v) $I = \{1, 2\}$, $\Lambda = \{1, 2\}$, $G = \{e\}$, $P = \begin{bmatrix} e & 0 \\ 0 & 0 \end{bmatrix}$, $T = \{(e; 1, 1)\}$.

A semigroup is *left* (*right*) *simple* if it has no proper left (resp., right) ideals.

Exercise 2.41. Prove that every finite left and right simple semigroup is a group.

If I is an ideal of a semigroup, then the *Rees quotient semigroup* is the semigroup with zero obtained by identifying with 0 all elements of the ideal I. If I, J are ideals of a semigroup S and $J \subset I$, then the Rees quotient semigroup I/J is called a *factor* of S. In the case where $J = \emptyset$, we put $I/J = I$.

A *finite ideal series* is a chain of ideals

$$\emptyset = S_0 \subset S_1 \subset \cdots \subset S_n = S.$$

A *principal series* of a semigroup S is a finite maximal chain

$$\emptyset = S_0 \subset S_1 \subset \cdots \subset S_n = S$$

of ideals of $S_i (i = 0, 1, \dots n)$. The *factors* of the principal series are the Rees quotient semigroups $S_i/S_{i-1} (i = 1, 2, \dots n)$.

Theorem 2.22. ([89], Propositions 3.1.4, 3.1.5) *Each principal factor of a semigroup is either simple, or 0-simple, or null.*

If S admits a principal series and each principal factor is either simple or 0-simple, then S is said to be *semisimple.*

A semigroup is *nilpotent* if $S^n = 0$ for some natural number n. An element s is *nil* if $s^n = 0$ for some n. If all elements of S are nil then S is called a *nilsemigroup.*

A *left zero semigroup* S satisfies the identity $ab = a$, for all $a, b \in S$.

2.8 Finite Semigroups

The well-known main theorem of this section is valid for all finite semigroups and, moreover, but for a larger class of semigroups, and we include it in this more general case. An *epigroup* is a semigroup S such that a power of each element belongs to some subgroup of S. All finite semigroups and periodic semigroups are epigroups.

Theorem 2.23. (Principal Ideal Chain Theorem) ([235], Proposition 11.1) *Every epigroup S with a finite number of idempotents has a finite ideal chain*

$$0 = S_0 \subseteq S_1 \subseteq \cdots \subseteq S_n = S^0 \tag{2.6}$$

such that, for each $i = 0, \ldots, n-1$, the factor S_{i+1}/S_i is either a nilsemigroup or a Rees matrix semigroup over a group with zero and finite sandwich matrix without zero rows and columns.

Theorem 2.24. *Every finite semigroup S has a finite ideal chain*

$$0 = S_0 \subseteq S_1 \subseteq \cdots \subseteq S_n = S^0 \tag{2.7}$$

such that, for each $i = 0, \ldots, n-1$, either S_{i+1}/S_i is the two-element null semigroup, or a Rees matrix semigroup over a group with zero and finite sandwich matrix without zero rows and columns.

The ideal chain (2.7) is called the *principal ideal chain.*

Exercise 2.42. Find the principal ideal chain for finite semigroup with the following Cayley table

	e_0	g_0	e_1	g_1	e_2	g_2
e_0	e_0	g_0	e_0	g_0	e_0	g_0
g_0	g_0	e_0	g_0	e_0	g_0	e_0
e_1	e_0	g_0	e_1	g_1	e_0	g_0
g_1	g_0	e_0	g_1	e_1	g_0	e_0
e_2	e_0	g_0	e_0	g_0	e_2	g_2
g_2	g_0	e_0	g_0	e_0	g_2	e_2

Exercise 2.43. Find the principal ideal chain for finite semigroup with the following Cayley table

	a	b	c	d	e	f
a	a	a	a	a	a	a
b	a	b	b	b	b	b
c	a	b	c	c	c	c
d	a	b	c	d	d	d
e	a	b	c	d	e	e
f	a	b	c	d	e	f

Exercise 2.44. Find the principal ideal chain for finite semigroup with the following Cayley table

	a	b	c	d	e	f
a	a	b	c	d	e	f
b	a	b	c	d	e	f
c	a	b	c	d	e	f
d	a	b	c	d	e	f
e	a	b	c	d	e	f
f	a	b	c	d	e	f

Exercise 2.45. Find the principal ideal chain for finite semigroup with the following Cayley table

	a	b	c	d	e	f
a	a	a	a	a	a	a
b	b	b	b	b	b	b
c	c	c	c	c	c	c
d	d	d	d	d	d	d
e	e	e	e	e	e	e
f	f	f	f	f	f	f

Exercise 2.46. Find the principal ideal chain for finite semigroup with the following Cayley table

	x_{11}	x_{12}	x_{13}	x_{21}	x_{22}	x_{23}
x_{11}	x_{11}	x_{12}	x_{13}	x_{11}	x_{12}	x_{13}
x_{12}	x_{11}	x_{12}	x_{13}	x_{11}	x_{12}	x_{13}
x_{13}	x_{11}	x_{12}	x_{13}	x_{11}	x_{12}	x_{13}
x_{21}	x_{21}	x_{22}	x_{23}	x_{21}	x_{22}	x_{23}
x_{22}	x_{21}	x_{22}	x_{23}	x_{21}	x_{22}	x_{23}
x_{23}	x_{21}	x_{22}	x_{23}	x_{21}	x_{22}	x_{23}

Exercise 2.47. Find the principal ideal chain for finite semigroup with the following Cayley table

	x_{11}	x_{12}	x_{21}	x_{22}	x_{31}	x_{32}
x_{11}	x_{11}	x_{12}	x_{11}	x_{12}	x_{11}	x_{12}
x_{12}	x_{11}	x_{12}	x_{11}	x_{12}	x_{11}	x_{12}
x_{21}	x_{11}	x_{12}	x_{21}	x_{22}	x_{21}	x_{22}
x_{22}	x_{11}	x_{12}	x_{21}	x_{22}	x_{21}	x_{22}
x_{31}	x_{11}	x_{12}	x_{21}	x_{22}	x_{31}	x_{32}
x_{32}	x_{11}	x_{12}	x_{21}	x_{22}	x_{31}	x_{32}

Theorem 2.25. ([89], Theorem 1.1.3) *Every rectangular band is isomorphic to a direct product of a left zero band and a right zero band.*

Let S be a semigroup. The *Green's equivalences* $\mathcal{L}$, $\mathcal{R}$, $\mathcal{J}$, $\mathcal{H}$, and $\mathcal{D}$ are defined by

$$x\mathcal{L}y \Leftrightarrow S^1x = S^1y,$$

$$x\mathcal{R}y \Leftrightarrow xS^1 = yS^1,$$

$$x\mathcal{J}y \Leftrightarrow S^1xS^1 = S^1yS^1,$$

$$\mathcal{H} = \mathcal{L} \cap \mathcal{R} \text{ and } \mathcal{D} = \mathcal{L}\mathcal{R} = \mathcal{R}\mathcal{L},$$

where $\mathcal{L}\mathcal{R}$ and $\mathcal{R}\mathcal{L}$ denote the usual composition of relations. Denote by J_x (R_x, L_x, H_x) the $\mathcal{J}$-class (resp., $\mathcal{R}$-class, $\mathcal{L}$-class, $\mathcal{H}$-class) of S containing x, i.e., the set of elements generating the same ideals (resp., right ideals, left ideals, left ideals and right ideals) as $x \in S$. Put $P_e = \{x \in eSe \mid xy = e \text{ for some } y \in eSe\}$. Then $H_e \subseteq P_e \subseteq R_e$ and $P_e = R_e \cap eSe$. For any two elements x, y in one $\mathcal{H}$-class, $H_x \cong H_y$.

Lemma 2.10. ([32]) *Let S be a semigroup with an idempotent e. Then H_e is a subgroup of S with identity element e. It contains all subgroups of S containing e.*

Exercise 2.48. Find the principal ideal chain of the semigroup

$$\langle x, y \mid x^2 = x, y^2 = y, xy = yx \rangle.$$

Exercise 2.49. Find the principal ideal chain of the semigroup

$$\langle x, y, z \mid x^2 = x, y^2 = y, z^2 = z, xy = yx, xz = zx, yz = zy \rangle.$$

Exercise 2.50. Find the principal ideal chain of the semigroup

$$\langle x, y \mid x^2 = x, y^2 = y, (xy)^2 = xy, (yx)^2 = yx, (xyx)^2 = xyx, (yxy)^2 = yxy \rangle.$$

2.9 Lattices

A *lattice* is a set L with two binary operations $\wedge$ and $\vee$ satisfying the following laws for all $x, y, z \in L$:

(L1) $x \wedge y = y \wedge x$ and $x \vee y = y \vee x$ (commutative laws);

(L2) $x \wedge (y \wedge z) = (x \wedge y) \wedge z$ and $x \vee (y \vee z) = (x \vee y) \vee z$ (associative laws);

(L3) $x \wedge (y \vee z) = x$ and $x \vee (y \wedge z) = x$ (absorption laws);

(L4) $x \wedge x = x$ and $x \vee x = x$ (idempotent laws).

In fact, the idempotent laws follow from the absorption laws: $x \vee x = x \vee (x \wedge (x \vee x)) = x$. The operations $\wedge$ and $\vee$ are called *meet* and *join* (or intersection and union, or product and sum).

Theorem 2.26. ([190], 1.8) *Every lattice ordered set $(X, \leq)$ is a lattice with operations $x \wedge y = \inf(x, y)$ and $x \vee y = \sup(x, y)$. Every lattice is a lattice ordered set with the partial order defined by $x \leq y \Leftrightarrow x \wedge y = x$.*

A lattice L is said to be *distributive* if the following two distributive laws hold, for all $x, y, z \in L$,

$$\begin{aligned} x \vee (y \wedge z) &= (x \vee y) \wedge (x \vee z), \\ x \wedge (y \vee z) &= (x \wedge y) \vee (x \wedge z). \end{aligned}$$

The smallest examples of nondistributive lattices are the *diamond* and the *pentagon* shown in Figure 2.15.

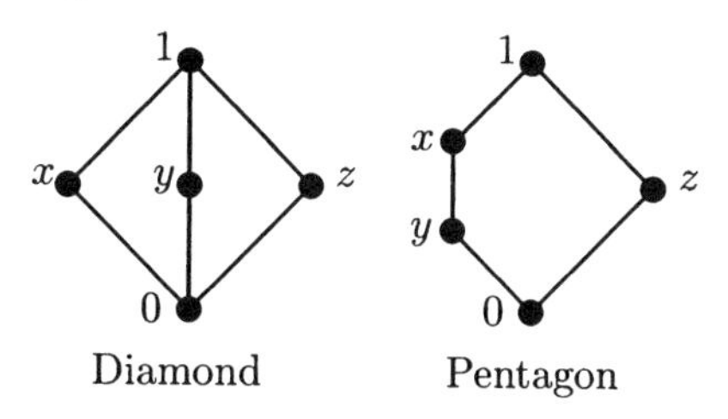

Figure 2.15: Diamond and Pentagon Lattices

Exercise 2.51. Prove that the diamond and pentagon lattices are not distributive.

A subset X of a lattice L is called a *sublattice* if it is closed under $\wedge$ and $\vee$.

Theorem 2.27. ([190], 2.3) *A lattice is distributive if and only if it does not contain any sublattices isomorphic to the diamond and the pentagon.*

Theorem 2.28. ([190], 2.6) *A lattice is distributive if and only if the following cancellation rule*

$$x \wedge y = x \wedge z,\ x \vee y = x \vee z \Longrightarrow y = z$$

holds for all $x, y, z \in L$.

The largest element of a lattice L is called the *identity element* of L and is denoted by 1. An element x of L is the identity element of L if and only if it $xy = yx = y$ for all $y \in L$. The smallest element of a lattice L is called the *zero* of L and is denoted by 0. An element x of L is zero if and only if it $xy = yx = x$ for all $y \in L$. The elements 0 and 1 of a lattice are also called the *universal bounds.* A lattice is said to be *bounded* if it has a zero and identity element.

Let L be a lattice L with 0 and 1, and let $x \in L$. An element y of L is called a *complement* of x if $x \wedge y = 0$ and $x \vee y = 1$. A lattice is said to be *complemented* if it has 0 and 1, and if every element has a complement. The pentagon lattice shows that one element may have several complements in a lattice which is not distributive (see Figure 2.15).

Theorem 2.29. ([190], 2.9) *An element of a distributive lattice may have at most one complement.*

Notation x' is used to denote the complement of x in a distributive lattice, if the complement exists.

Suppose that u is an expression formed using variables and operations $\wedge$, $\vee$, $'$. The *dual* of u is obtained by replacing all symbols $\wedge$ by $\vee$ and all symbols $\vee$ by $\wedge$ in u.

Theorem 2.30. (Duality Principle) ([190], 1.10) *Let $u = v$ be an equality involving variables and binary operations $\wedge, \vee$. If $u = v$ holds in all lattices, then its dual is also valid in all lattices.*

2.10 Boolean Algebras

Boolean algebras are used, for example, in design of electronic circuits, in classical logic for simplifying rules and solving systems of logical equation, in SQL language and database theory as means of analysis of queries, and in artificial intelligence. Informally, a Boolean algebra is a set with operations $\bar{}$, $\cdot$, $+$ satisfying the same laws as the complement $\bar{}$, intersection $\cap$, and union $\cup$ of sets. It is enough to notice that an algebra satisfies only 5 axioms, and then it follows that it enjoys all properties similar to the algebra of sets. The following definition gives us the axioms required.

A *Boolean algebra* is a set B, containing amongst its elements two special elements 0 and 1, with two binary operations, *addition* $(+)$ and *multiplication* $(\cdot)$, and a unary operation *complementation* $(\bar{\ })$ such that the following 5 axioms hold:

(B1)	B is closed under addition and multiplication;
(CL) (B2)	$x + y = y + x$ and $xy = yx$ for all x, y;
(DL) (B3)	$x(y + z) = (xy) + (xz)$ and $x + (yz) = (x + y)(x + z)$, for all x, y, z;
(B4)	$x + 0 = x = x \cdot 1$;
(PC) (B5)	$x \cdot \overline{x} = 0$ and $x + \overline{x} = 1$.

Many brackets can be omitted in complicated Boolean expressions if the following order or priority of operations is noted: $\bar{\ }$, then $\cdot$, then $+$. Thus, the expression $((xy) + (\overline{x}))$ can be written as $xy + \overline{x}$.

Example 2.6. Every power set $\mathcal{P}(X)$ is a Boolean algebra with $1 = X$, $0 = \emptyset$ and operations $+ = \cup$, $\cdot = \cap$, $\bar{\ } = \bar{\ }$.

Example 2.7. Let $B = \{1, 2, 3, 5, 6, 10, 15, 30\}$ be the set of positive divisors of 30. For any $x, y \in B$, we define $x+y = \text{LCM}(x, y)$, $x \cdot y = \text{GCD}(x, y)$, $\overline{x} = \frac{30}{x}$, zero is 1, and identity is 30.

It is the same as the diagram of $\mathcal{P}(\{2, 3, 5\})$ since all divisors of 30 correspond to subsets of $\{2, 3, 5\}$.

Theorem 2.31. (Stone's Representation Theorem) ([190], 3.11) *Each Boolean algebra can be embedded in the set algebra of an appropriate set. Each finite Boolean algebra is isomorphic to the set algebra of some finite set.*

To verify whether a given set with operations $+$, $\cdot$ is a Boolean algebra, we need only to check the properties $B1$ to $B5$ from the definition. $B1$ is needed, because not all subsets of $P(S)$ are Boolean algebras.

Theorem 2.32. ([190], §3) *In a Boolean algebra, the following properties hold for all $x, y, z \in B$:*

(CL) $xy = yx$ *and* $x + y = y + x$ *(commutative laws);*

(AL) $x(yx) = (xy)z$ *and* $x + (y + z) = (x + y) + z$ *(associative laws);*

(DL) $x(y + z) = xy + xz$, $x + yz = (x + y)(x + z)$ *(distributive laws);*

(IL) $xx = x$ *and* $x + x = x$ *(idempotent laws);*

(ML) $\overline{xy} = \overline{x} + \overline{y}$ *and* $\overline{x + y} = \overline{x}\,\overline{y}$ *(de Morgan's laws);*

(DCL) $\overline{\overline{x}} = x$ *(double complement law);*

(CZ) $\overline{0} = 1$ *and* $\overline{1} = 0$ *(complements of* 0 *and* 1*);*

(AbL) $x(x + y) = x$ *and* $x + (xy) = x$ *(absorption laws);*

(PC) $x\overline{x} = 0$ *and* $x + \overline{x} = 1$ *(properties of complements);*

(PZ) $x0 = 0$ *and* $x + 0 = x$ *(properties of zero);*

(PI) $x1 = x$ *and* $x + 1 = 1$ *(properties of identity).*

These are the most important properties used to prove Boolean equalities and to simplify Boolean expressions. The order of operations is: $^{-}$, then $\cdot$, then $+$; so $\overline{x} + y\overline{z}$ is the same as $(\overline{x}) + (y(\overline{z}))$.

Theorem 2.33. ([190], §3) *In every Boolean algebra the following properties hold:*

(ZU) 0 *and* 1 *are unique;*

(CU) *for each* x, $\overline{x}$ *is unique;*

(PE) $xy = x$ *if and only if* $x + y = y$ *if and only if* $x\overline{y} = 0$*;*

(WCL) *weak cancellation laws:*

(a) *if* $x + y = x + z$ *and* $\overline{x} + y = \overline{x} + z$, *then* $y = z$*;*

(b) *if* $x \cdot y = x \cdot z$ *and* $\overline{x} \cdot y = \overline{x} \cdot z$, *then* $y = z$.

A *partial order* $\leq$ can be defined on every Boolean algebra. If B is a Boolean algebra, and $a, b \in B$, then $a \leq b$ if and only if $ab = a$. This order always has properties like those of inclusion of sets.

By Theorem 2.33 the following statements are equivalent

(i) $a \le b$, (i.e. $ab = a$)
(ii) $a + b = b$
(iii) $a\overline{b} = 0$
(iv) $a' + b = 0$

Theorem 2.34. ([190], 3.5, 3.6) *Let B be a Boolean algebra. Then, for all $x, y, z \in B$,*

(1) $x \le x$ *($\le$ is reflexive);*
(2) $x \le y$, *and* $y \le z$ *imply* $x \le z$ *($\le$ is transitive);*
(3) $x \le y$, *and* $y \le x$ *imply* $x = y$ *($\le$ is antisymmetric);*
(4) $x \le y$, $x \le z$ *imply* $x \le yz$;
(5) $x \le y$ *implies that* $x \le y + z$ *for every* z;
(6) $x \le y$ *if and only if* $\overline{y} \le \overline{x}$.

This theorem tells us that the order $\le$ on every Boolean algebra has the same properties as the inclusion $\subseteq$ of sets. (1), (2), (3) mean that $\le$ is a partial order.

2.11 Rings

Rings are used, for example, in constructing error-correcting codes, solving systems of polynomial and differential equations, as well as in various symbolic rewriting algorithms with applications in computer algebra systems, in artificial intelligence etc. Formal definitions below give us small lists of axioms. As soon as we have verified these axioms for some set, we can use all theorems that are known for rings in order to answer various questions.

Throughout the word 'ring' will mean an 'associative ring' unless specified otherwise. More general concepts are of interest and have valuable applications too.

Definition 2.2. A *general ring* or a *not necessarily associative ring* is a set R with two operations $+$ (called addition) and $\cdot$ (called multiplication) which satisfy the laws for all x, y, z in R:

(R1) $0 + x = x + 0 = x$ (zero is a neutral element for addition);

(R2) $x + (y + z) = (z + y) + z$ (addition is associative);

(R3) for each x there exists a unique element $-x$ such that $x + (-x) = 0$ (negative inverse elements);

(R4) $x + y = y + x$ (addition is commutative);

(R5) $0x = x0 = 0$ (zero is an absorbing element for multiplication);

(R6) $x(y + z) = xy + xz$, $(x + y)z = xz + yz$ (two distributive laws hold, i.e., multiplication distributes over addition from both sides).

Properties (R1) through (R4) are equivalent to saying that R is an abelian group with zero 0.

Definition 2.3. A *ring* R is said to be *associative* if the following associative law is satisfied for all x, y, z in R:

(R7) $x(yz) = (xy)z$ (multiplication is associative).

Exercise 2.52. Deduce (R5) from (R1), (R2), (R3) and (R6).

Definition 2.4. A ring is said to be *commutative* if the following commutative law holds:

(CL) $xy = yx$ (multiplication is commutative).

Example 2.8. The following algebraic structures are examples of rings:

$\mathbb{Z}$, the ring of integers;

$\mathbb{R}[x]$, the ring of polynomials with real coefficients and variable x;

$\mathbb{R}_n$, the ring of $n \times n$ matrices over $\mathbb{R}$.

An element 1 of a ring R is called an *identity element* if $1x = x1 = x$, for all $x \in R$. Let R be a ring with identity element 1, and let $x \in R$. An element y of R is called an *inverse* of x if $xy = yx = 1$. We say that x is *invertible* if it has an inverse element. The inverse element of x is denoted by x^{-1}.

A *Boolean ring* is a commutative ring satisfying $2x = 0$ and $x^2 = x$ for all $x \in R$.

Theorem 2.35. ([55], Exercise 65.8) *If $(R, +, \cdot)$ is a Boolean ring with 1, and if the operations $\wedge, \vee, ^-$ are defined on R by the rules*

$$x \wedge y = xy, \quad x \vee y = x + y + xy, \quad \bar{x} = 1 + x$$

then $(R, \wedge, \vee, ^-)$ becomes a Boolean algebra. Conversely, if $(B, \wedge, \vee, ^-)$ is a Boolean algebra, and the operations $+, \cdot$ are defined by

$$xy = x \wedge y, \qquad x + y = (x \wedge \bar{y}) \vee (\bar{x} \wedge y),$$

then $(B, +, \cdot)$ turns into a Boolean ring.

Let R be a ring, and let $A, B \subseteq R$. Then we set

$$A + B = \{a + b \mid a \in A, b \in B\},$$

$$AB = \left\{ \sum_{i=1}^{n} a_i b_i \mid a_i \in A, b_i \in B \right\}.$$

Note that this meaning of AB differs from a similar notation used for groupoids and monoids.

If $T \subseteq R$, then the *subring* generated by T is the set $\langle T \rangle$ consisting of all finite sums of products $kt_1t_2 \cdots t_n$, where $k \in \mathbb{Z}$ and all $t_i \in T$. The *ideal* (*right ideal, left ideal*) generated by T in R is the set $\mathrm{id}(T)$ ($\mathrm{id}_r(T)$, $\mathrm{id}_l(T)$) consisting of all finite sums of elements of R^1TR^1 (resp., TR^1, R^1T). The ideals generated by one element are said to be *principal.* Let R^1 be the ring obtained by adjoining a unity 1 to R, i.e., a direct sum of the ideal R and subring $\langle 1 \rangle$ isomorphic to $\mathbb{Z}$. A ring R is said to be *simple* if it has no proper ideals, i.e., if the only ideals of R are 0 and the whole R.

If there exist positive integers n such that $nx = 0$ for all x in the ring R, then the smallest n with this property is called the *characteristic* of R. If there are no positive integers like this, then R is said to have characteristic zero. The characteristic of R is denoted by $\mathrm{char}(R)$.

Definition 2.5. If I is an ideal of R, then the set of all classes $x + I$, for $x \in R$, with addition $(x + I) + (y + I) = (x + y) + I$ and multiplication $(x+I)(y+I) = xy+I$ forms a ring called the *quotient ring* of R modulo I and denoted by R/I. The ring R is called *ideal extension* of I by R/I.

An *idempotent* of a ring as an element e satisfying $e^2 = e$. Two idempotents e and f are said to be *orthogonal* if $ef = 0$. An element x is *nilpotent* if $x^n = 0$ for a positive integer n. A ring is *nil* if all its elements are nilpotent. A ring R is *nilpotent* if $R^n = 0$ for some n.

Definition 2.6. A semiring is a set R with two distinct special elements 0 and 1, and with two binary operations $+$ (called addition) and $\cdot$ (called multiplication) such that the following conditions hold:

(1) $(R, +)$ is a commutative monoid with zero 0;

(2) $(R, \cdot)$ is a monoid with identity 1;

(3) $(R, +)$ is a commutative monoid with zero 0;

(4) multiplication distributes over addition from either side;

(5) 0 is a multiplicative zero for $(R, \cdot)$, i.e., $x0 = 0x = 0$ for all $x \in R$;

Thus intuitively speaking a semiring is a ring without subtraction.

Example 2.9. (i) The semiring of positive integers.

(ii) Boolean semiring $\mathbb{B} = \{0, 1\}$.

(iii) Every distributive lattice with 0 and 1 is a semiring.

(iv) The semiring of all languages with respect to $+$ and $\cdot$.

(v) The semiring of star-free languages with respect to $+$ and $\cdot$.

Let R be a semiring. A subset S of R is called a *subsemiring* of R if it contains 0 and 1 and is closed for addition and multiplication.

Example 2.10. The semiring of star-free languages with respect to $+$ and $\cdot$ is a subsemiring of the semiring of all languages.

If A and B are subsets of a semiring R, then the sets $A+B$ and AB are defined as follows:

$$A + B = \{a + b \mid a \in A, b \in B\},$$

$$AB = \left\{ \sum_{i=1}^{n} a_i b_i \mid a_i \in A, b_i \in B \right\}.$$

2.12 Fields and Vector Spaces

Informally speaking, a *field* is a set with two operations $+$ and $\cdot$ satisfying the same properties as addition and multiplication of real numbers.

Definition 2.7. A *field* is a commutative ring with identity element such that every nonzero element is invertible.

Example 2.11. The following algebraic structures are examples of fields:

$\mathbb{F}_p$, the prime field of order p;

$\mathbb{R}$, the field of real numbers;

$\mathbb{Q}$, the field of rational numbers;

$\mathbb{C}$, the field of complex numbers.

Example 2.12. The binary field, or field of order two, is denoted by $GF(2)$, or $\mathbb{Z}_2$, or $\mathbb{F}_2$. It is defined as the set $\{0, 1\}$ with two operations $+$ and $\cdot$ given by the following Cayley tables.

$\mathbb{F}_2$:

$+$	0	1
0	0	1
1	1	0

$\cdot$	0	1
0	0	0
1	0	1

Example 2.13. The ternary field, or the field of order three, is denoted by $GF(3)$,, or $\mathbb{Z}_3$, or $\mathbb{F}_3$. It is defined as the set $\{0, 1, 2\}$ with two operations $+$ and $\cdot$ given by the following Cayley tables.

$\mathbb{F}_3$:

+	0	1	2
0	0	1	2
1	1	2	0
2	2	0	1

·	0	1	2
0	0	0	0
1	0	1	2
2	0	2	1

It turns out that these operations satisfy the same laws as addition and multiplication of real numbers.

Definition 2.8. Let p be a prime number. The *field of order* p is denoted by $GF(p)$, or $\mathbb{Z}_p$, or F_p and is defined as the set $\{0, 1, \ldots, p-1\}$ with $+$ and $\cdot$ defined modulo p. This means that

- $x + y$ is the remainder of the sum of x and y on division by p;
- xy is the remainder of the product of x and y on division by p.

Properties of $\mathbb{F}_p$ are discussed in Subsection 1.5.

Definition 2.9. The minimum positive integer p such that $p \cdot 1 = 0$ in a field F is called the *characteristic* of the field F, and is denoted by $\text{char}(F)$.

Lemma 2.11. *The characteristic of every field is either zero or prime.*

Proof. Suppose to the contrary that $p = \text{char}(F)$ is not a prime. Then $p = ab$ for some integers $a, b \geq 2$. By the minimality of p, we get $a \cdot 1 \neq 0$, $b \cdot 1 \neq 0$. Hence there exists the inverse element $(b)^{-1}$ such that $b \cdot b^{-1} = 1$ in F. However, $a \cdot b = ab \cdot 1 = p \cdot 1 = 0$ by the definition of p. Multiplying this equality with b^{-1}, we get $0 = 0 \cdot b^{-1} = abb^{-1} = a \cdot 1 \neq 0$. This contradiction shows that our assumption was wrong. Thus p is a prime. □

Theorem 2.36. ([86], Theorem 3.1.1) *Every finite division ring is a field.*

A ring with identity element such that every nonzero element is invertible is called a *division ring* or a *skew field*. A commutative division ring is called a *field*. Let $\mathbb{Z}$ be the ring of integers, and let $\mathbb{Q}$, $\mathbb{R}$ and $\mathbb{C}$ denote the fields of rational, real and complex numbers, respectively.

We refer to [4] for complete explanations concerning linear spaces, and briefly review some of the most basic concepts below.

Definition 2.10. Let F be a field. A *linear space* or *vector space* over F is a set with two operations, addition of vectors, and multiplication of vectors by scalars, i.e., elements of F, where these operations satisfy the same properties as the corresponding operations on ordinary 3-dimensional vectors:

(V1) addition of vectors is commutative, i.e., $u + v = v + u$;

(V2) addition of vectors is associative;

(V3) there exists a vector 0 such that $u + 0 = u$ for all u (0 is called the zero vector);

(V4) for each $u \in V$ there exists a vector $-u$, called the negative vector of u, such that $u + (-u) = 0$;

(A5) distributive law $r(u + v) = ru + rv$ holds for all $r \in F, u, v \in V$;

(A6) distributive law $(r + s)u = ru + su$ holds for all $r, s \in F, u \in V$;

(A7) the following associative law $r(su) = (rs)u$ holds for all $r, s \in F$, $u \in V$;

(A8) $1 \cdot u = u$ for every u, where 1 is the identity element of the field F.

For example, if F is a field, then the set $M_{1,n}(F)$ of all n-sequences $(x_1, x_2, \ldots, x_n)$ of elements $x_1, \ldots, x_n \in F$ with operations

$$(x_1, x_2, \ldots, x_n) + (y_1, y_2, \ldots, y_n) = (x_1 + y_1, x_2 + y_2, \ldots, x_n + y_n),$$

$$r(x_1, x_2, \ldots, x_n) = (rx_1, rx_2, \ldots, rx_n),$$

where $r \in F$, is a linear space over F. It is denoted by F^n.

Example 2.14. The set $\mathbb{F}_2^2$ is $\{00, 01, 10, 11\}$.

Definition 2.11. The *standard basis* of F^n consists of n unit vectors

$$\begin{aligned} e_1 &= (100\ldots0) \\ e_2 &= (010\ldots0) \\ e_3 &= (001\ldots0) \\ &\vdots \\ e_n &= (000\ldots1) \end{aligned}$$

Definition 2.12. A nonempty subset V of the linear space F^n is called a *subspace* of F^n, if and only if the following two conditions are satisfied:

(1) for any vectors $u, v \in V$, the sum $u + v$ belongs to V;

(2) for any vector $u \in V$ and any number $c \in F$, the product cu belongs to V.

Conditions (1) and (2) can be replaced by one condition (3) as follows.

Definition 2.13. A nonempty subset V of the linear space F^n is called a *subspace* of F^n, if and only if the following condition is satisfied:

(3) for any vectors $u, v \in V$ and numbers $c, d \in F$ the linear combination $cu + dv$ belongs to V.

Every subspace is a linear space with respect to the same operations. Thus a nonempty subset is a subspace if and only if it is a linear space with respect to the same operations.

Definition 2.14. A nonempty subset V of the linear space F^n is called a *subspace* of F^n if and only if it is a linear space with respect to componentwise addition of sequences and multiplication of a sequence by a number.

Lemma 2.12. *Let V be a subspace of F^n. Then*

(1) 0 *belongs to* V*;*

(2) *if* $u \in V$, *then* $-u \in V$.

Example 2.15. (1) $\{(x_1, x_2) \mid x_1, x_2 \in \mathbb{F}_2,\ x_1 + x_2 = 0\}$ is a linear space, and we can draw it on the 2-dimensional coordinate system.

(2) $\{(x_1, x_2, x_3, x_4) \mid x_1, x_2, x_3, x_4 \in \mathbb{F}_2,\ x_1 + x_2 + x_3 + x_4 = 0\}$ is a linear space.

(3) $\{(x_1, x_2, x_3, x_4) \mid x_1 = x_2 = x_3 = x_4 \in \mathbb{F}_2\}$ is a linear space.

(4) $\{(x_1, x_2, x_3, x_4) \mid x_1, x_2, x_3 x_4 \in \mathbb{F}_2,\ x_1 + x_2 + x_3 + x_4 = 1\}$ is NOT a subspace of $\mathbb{F}_2^4$.

Definition 2.15. A *basis of a subspace* U is a minimal set of vectors $b_1, \ldots, b_k$ such that all vectors $v \in V$ can be expressed in the form $v = r_1 b_1 + \cdots + r_k b_k$, where $r_1, \ldots, r_k$ are elements of the field. The expression $r_1 b_1 + \cdots + r_k b_k$ is called a *linear combination* of the vectors $b_1, \ldots, b_k$.

Exercise 2.53. Find bases of all subspaces given in Example 2.15.

The set of solutions to a system of *homogeneous* linear equations

$$\begin{array}{ccccccccc} a_{11}x_1 & + & a_{12}x_2 & + & \cdots & + & a_{1n}x_n & = & 0 \\ a_{21}x_1 & + & a_{22}x_2 & + & \cdots & + & a_{2n}x_n & = & 0 \\ \vdots & & \vdots & & \cdots & & \vdots & & \vdots \\ a_{m1}x_1 & + & a_{m2}x_2 & + & \cdots & + & a_{mn}x_n & = & 0 \end{array}$$

is a linear space.

Exercise 2.54. Find a basis of the set of solutions to the following system of homogeneous linear equations with coefficients in the field $\mathbb{F}_3$

$$\begin{array}{ccccccc} x_1 & +x_2 & +2x_3 & +x_4 & = & 0 \\ -x_1 & +x_2 & +2x_3 & & = & 0 \\ x_1 & & & +2x_4 & = & 0 \end{array}$$

Solution. The matrix of the system is

$$\begin{bmatrix} 1 & 1 & 2 & 1 \\ -1 & 1 & 2 & 0 \\ 1 & 0 & 0 & 2 \end{bmatrix}$$

Let us reduce it to the row-echelon form using elementary row operations.

$$\begin{bmatrix} 1 & 1 & 2 & 1 \\ 0 & 2 & 1 & 1 \\ 0 & -1 & -2 & 1 \end{bmatrix} \begin{matrix} \\ R_2 + R_1 \\ R_3 - R_1 \end{matrix}$$

$$\begin{bmatrix} 1 & 1 & 2 & 1 \\ 0 & 2 & 1 & 1 \\ 0 & 0 & 0 & 0 \end{bmatrix} \begin{matrix} \\ \\ R_3 - R_2 \end{matrix} \qquad \text{row-echelon form}$$

$$\begin{bmatrix} 1 & 0 & 0 & 2 \\ 0 & 1 & 2 & 2 \end{bmatrix} \begin{matrix} R_1 + R_2 \\ 2R_2 \end{matrix} \quad \text{reduced row-echelon form.}$$

The reduced row-echelon form has unit columns like an identity matrix. This gives us the following system of equations:

$$\begin{aligned} x_1 \qquad\qquad\quad +2x_4 &= 0 \\ x_2 \quad +2x_3 \quad +2x_4 &= 0 \end{aligned}$$

Hence we get

$$\begin{bmatrix} x_1 \\ x_2 \\ x_3 \\ x_4 \end{bmatrix} = \begin{bmatrix} 0 \\ 1 \\ 1 \\ 0 \end{bmatrix} x_3 + \begin{bmatrix} 1 \\ 1 \\ 0 \\ 1 \end{bmatrix} x_4.$$

Therefore $(0110), (1101)$ is a basis of the space of solutions. □

A ring R is called an *algebra* over a field F, if it is a vector space over F with respect to the same addition and with scalar multiplication by the elements of F related to the ring multiplication by the laws $f(xy) = (fx)y = x(fy)$ and $f(gx) = (fg)x$, for all $f, g \in F$, $x, y \in R$.

If R is an algebra over a field F with a subset T, then the *subalgebra* $\langle T \rangle$ generated by T is the linear space spanned by all products $t_1 t_2 \cdots t_n$, where $t_i \in T$. The algebra R^1 is a direct sum of the ideal R and subalgebra $\langle 1 \rangle \cong F$. The *ideal* (*right ideal, left ideal*) generated by T in R is the linear space $\mathrm{id}(T)$ ($\mathrm{id}_r(T)$, $\mathrm{id}_l(T)$) spanned by the set $R^1 T R^1$ (resp., TR^1, $R^1 T$). If R has an identity, then every one-sided ideal of R considered as a ring is also a vector space.

2.13 Polynomial Rings and Finite Fields

Theorem 2.37. ([190], 13.1) *Let F be a finite field of characteristic p. Then F has p^n elements, where n is the dimension of F over $\mathbb{F}_p$.*

Proof. The field F is a vector space over the prime field

$$\mathbb{F}_p = \{0, 1, \ldots, p-1\}.$$

It has a basis of n elements. Each element of F can be expressed as a unique linear combination of vectors in the basis with coefficients in $\mathbb{F}_p$. Since there are precisely p^n linear combinations like this, it follows that F has p^n elements. □

Theorem 2.38. ([190], 13.6) *If p is a prime, and r is a positive integer, then there exists a finite field with p^r elements. This field is unique up to notation for its elements.*

The field consisting of p^r elements is denoted by $GF(p^r)$, or $\mathbb{F}_{p^r}$, or $\mathbb{F}_{p^r}$. We can say that every two finite fields with equal cardinalities are *isomorphic*, that is, are the same up to notation for their elements. More formal definition introduces an *isomorphism* of fields as a one-to-one and onto mapping that preserves all operations of the field.

Theorem 2.39. (Identites of Finite Fields) ([190], 13.1, 13.2) *The following equalities hold for all elements x, y of the finite field $\mathbb{F}_{p^r}$:*

(i) $px = 0$;

(ii) $x^{p^r} = x$;

(iii) $(x+y)^p = x^p + y^p$.

Equality (ii) in the case of $r = 1$, that is for the prime field $\mathbb{F}_p$, is called "Fermat's Little Theorem".

An element a of a finite field F is said to be *primitive* if every nonzero element is a power of a.

Theorem 2.40. ([190], 13.2(i)) *Every finite field has a primitive element.*

Proof. Let F be a finite field. Denote by p the characteristic of F. By Theorem 2.37, F has p^n elements. Hence the multiplicative group G of nonzero elements of F is a group of order $q = p^n - 1$. Theorem 2.11 tells us that G is the direct product of cyclic subgroups $U_1, \ldots, U_m$, where $|U_i|$ divides $|U_{i+1}|$ for $i = 1, \ldots, m-1$. It follows that the order of each element of G divides $k = |U_m|$. Hence $g^k - 1 = 0$ for all $g \in G$. The polynomial $x^k - 1$ can have at most k roots in F; whence $|G| = q - 1 \leq k$. Since $|U_m|$ divides $|G|$ by Lagrange's theorem, we have $k \leq q - 1$, and so $G = U_m$. Thus G is a cyclic group, as required. □

Exercise 2.55. ([190], 10.20, 13.3) *Let p be a prime and let x be a primitive element of $\mathbb{F}_{p^r}$. Prove that the set of primitive elements of $\mathbb{F}_{p^r}$ is given by*

$$\{x^k \mid k \textit{ coprime with } p^r - 1\}.$$

Thus the number of primitive elements is equal to the number of positive integers coprime with $p^r - 1$, that is $\phi(p^r - 1)$.

Example 2.16. The Galois field of order 4 is the set $\mathbb{F}_4 = \{0, 1, \alpha, \beta\}$ with addition and multiplication defined by the following Cayley tables.

$+$	0	1	α	β
0	0	1	α	β
1	1	0	β	α
α	α	β	0	1
β	β	α	1	0

$\cdot$	0	1	α	β
0	0	0	0	0
1	0	1	α	β
α	0	α	β	1
β	0	β	1	α

By Theorem 2.39, $\mathbb{F}_4$ satisfies the identities $2x = 0$ and $x^4 = x$.

The Hamming weight $w_H(u)$ of a word u is the number of nonzero digits in the word. The *Hamming distance* $d_H(u, v)$ between two words u and v of the same length is the number of places in which these words differ. Thus the distance from u to v is the minimum number of symbols we have to change in u if we want to make it equal to v.

Let x be a letter, and F a field. Every expression of the form $a_0 + a_1x + a_2x^2 + \cdots + a_nx^n$, where $n \in \mathbb{N}$, $a_0, a_1, a_2, \ldots, a_n \in F$, and $a_n \neq 0$, is called a *polynomial* with coefficients in F or a polynomial over F. We say that f has *degree* n and write $\deg(f) = n$. Zero is also a polynomial, and its degree is defined as $\deg(0) = -\infty$. The set of all polynomials over F with variable x is denoted by $F[x]$. Addition of polynomials and multiplication by the elements of F are defined by the laws of vector spaces. Multiplication of polynomials is defined by the laws of a ring and equality $x \cdot x = x^2$. The set $F[x]$ is a ring with respect to these operations.

A polynomial is said to be *irreducible* if it has no divisors except scalar multiples of itself and scalars. A polynomial is *monic* if its leading coefficient is 1. Monic irreducible polynomials play the same role for polynomials as primes for integers.

Theorem 2.41. (Polynomial Factorization Theorem) ([190], 11.24, [201], 2.217) *Let F be a field, and let $f(x)$ be a nonzero polynomial in $F[x]$. Then there exist pairwise distinct monic irreducible polynomials $p_1(x)$, ..., $p_k(x)$, positive integers a_1, ..., a_k and a constant $a \in F$ such that*

$$f(x) = f_1^{a_1}(x) \cdots f_k^{a_k}(x).$$

This factorization is unique up to the order of monic irreducible polynomials in the product.

It is easy to find the greatest common divisor and least common multiple of polynomials given their irreducible factorizations.

Theorem 2.42. (Greatest Common Divisor of Polynomials) *Let $p_1(x)$, ..., $p_k(x)$ be monic irreducible polynomials over a field F, and let*

$$f(x) = f_1^{a_1}(x) \cdots f_k^{a_k}(x), g(x) = f_1^{b_1}(x) \cdots f_k^{b_k}(x),$$

where $a_1, \ldots, a_k, b_1, \ldots, b_k$ are nonnegative integers. Then

$$gcd(f(x), g(x)) = f_1^{\min\{a_1,b_1\}}(x) \cdots f_k^{\min\{a_k,b_k\}}(x),$$

$$lcm(f(x), g(x)) = f_1^{\max\{a_1,b_1\}}(x) \cdots f_k^{\max\{a_k,b_k\}}(x).$$

Theorem 2.43. (Chinese Remainder Theorem for polynomials) ([190], Theorem 15.1) *Let F be a field, $m_1(x), \ldots, m_k(x) \in F[x]$ pairwise coprime polynomials, and let $g_1(x), \ldots, g_k(x) \in F[x]$ be arbitrary polynomials. Then the system of congruences*

$$f(x) \equiv g_i(x) \mod m_i(x), \text{where } i = 1, \ldots, k, \tag{2.8}$$

has a solution that is unique modulo $m_1(x) \cdots m_k(x)$.

The *ring of polynomials modulo $f(x)$* is denoted by $F[x]/(f(x))$. It consists of all polynomials of degree $< \deg(f)$ with ordinary addition and multiplication defined modulo $f(x)$. This means that we multiply two polynomials and then take the remainder on division by $f(x)$.

Theorem 2.44. (Finite Fields) ([190], 12.11) *If $f(x)$ is an irreducible polynomial of degree m over $\mathbb{F}_p$, then $\mathbb{F}_p[x]/(f(x))$ is a finite field of order p^m. All finite fields can be obtained in this way.*

A *primitive polynomial* over $\mathbb{F}_{p^r}$ is a monic irreducible polynomial $f(x)$ such that x is a primitive element of the finite field

$$\mathbb{F}_{p^r}[x]/(f(x)),$$

that is a generator of the multiplicative group of nonzero elements of $\mathbb{F}_{p^r}$. If $f(x)$ is a nonzero polynomial over $\mathbb{F}_{p^r}$ with nonzero constant term, then there exists a positive integer e such that $f(x)$ divides $x^e - 1$. The smallest positive integer e with this property is called the *order* or *exponent* or *period* of f.

Theorem 2.45. ([190], 14.22) *A monic polynomial $f(x)$ of degree m over $\mathbb{F}_{p^r}$ is primitive if and only if $f(0) \neq 0$ and the order of $f(x)$ is $p^{mr} - 1$.*

Theorem 2.46. ([190], 14.2) *If z is a root of an irreducible polynomial $f(x)$ of degree m over $\mathbb{F}_{p^r}$, then all m roots of $f(x)$ are*

$$z, z^{p^r}, z^{p^{2r}}, \ldots, z^{(m-1)p^r}.$$

Theorem 2.47. ([190], 14.3) *The product of all monic irreducible polynomials over $\mathbb{F}_{p^r}$ with degrees dividing m is equal to $x^{p^{mr}} - 1$. In particular, an irreducible polynomial divides $x^{p^{mr}} - 1$ if and only if its degree divides m.*

The field $\mathbb{F}_4$ is the quotient ring of the polynomial ring $\mathbb{F}_2[x]$ modulo $x^2 + x + 1$. Since $f(1) = 1$, $f(0) = 1$, $f(x)$ has no roots, and so it has no linear divisors. Therefore it is irreducible over $\mathbb{F}_2$. Hence $\mathbb{F}_2[x]/f(x)$ is a field of order 2^2. We get $x^2 + x + 1 = 0$ in this quotient ring. Using this let us fill in the *index table* of the field $\mathbb{F}_4$.

powers of x	vectors
0	0
1	1
x	x
x^2	$1 \quad +x$

$x^3 = 1, x^4 = x$. Indeed, $x^3 = xx^2 = x(1 + x) = x^2 + x = (1 + x) + x = 1$. Next, we find the Cayley tables of the field.

$+$	0	1	x	x^2
0	0	1	x	x^2
1	1	0	x^2	x
x	x	x^2	0	1
x^2	x^2	x	1	0

$\cdot$	0	1	x	x^2
0	0	0	0	0
1	0	1	x	x^2
x	0	x	x^2	1
x^2	0	x^2	1	x

Thus we see that this is the same field as the one defined in Example 2.16.

Example 2.17. We can define the field $\mathbb{F}_8$ using the primitive polynomial $f(x) = x^3 + x^2 + 1$ over $\mathbb{F}_2$. All 8 elements can be expressed as linear combinations of vectors of the basis x^2, x, 1, or as powers of the primitive element x, as the index table in Figure 2.16 shows.

Exercise 2.56. Compute $(x^2 + x + 1)(x^2 + x) + x^2$ in $\mathbb{F}_8$.

Solution. We get $x^4x^6 + x^2 = x^{10} + x^2 = x^3 + x^2 = (x^2 + 1) + x^2 = 1$. □

Exercise 2.57. Compute $(x^2 + 1)/(x^2 + x)$ in $\mathbb{F}_8$.

Solution. $(x^2 + 1)/(x^2 + x) = x^3/x^6 = x^{10}/x^6 = x^4 = x^2 + x + 1$. □

powers of x	vectors			
0			0	
1			1	
x		x		
x^2	x^2			
x^3	x^2		$+1$	$x^3 = -x^2 - 1$
x^4	x^2	$+x$	$+1$	$x(x^2+1) = x^3 + x = x^2 + 1 + x$
x^5		x	$+1$	$x(x^2+x+1) = x+1$
x^6	x^2	$+x$		
				$x^7 = x(x^2+x) = x^3 + x^2 = 1.$

Figure 2.16: Index table of $\mathbb{F}_8$

Exercise 2.58. Compute $(z+x^2)(z+x^2+x+1)$ in $\mathbb{F}_8$.

Solution. $(z+x^2)(z+x^2+x+1) = z^2 + (x+1)z + x^2(x^2+x+1) = z^2 + (x+1)z + x^2x^4 = z^2 + (x+1)z + (x^2+x)$. □

Exercise 2.59. Solve equation $z^2 + (x+1)z + (x^2+x) = 0$ in $\mathbb{F}_8$.

Exercise 2.60. Solve the following system of linear equations over $\mathbb{F}_8$.

$$\begin{aligned} xy + x^3z &= x^2 \\ x^2y + xz &= x^6 \end{aligned}$$

where x is the primitive element of $\mathbb{F}_8$ modulo $x^3 + x^2 + 1$.

Solution. Referring to the index table for $\mathbb{F}_8$ we use elementary row operations to reduce the augmented matrix of the system to reduced row-echelon form.

$$\begin{bmatrix} x & x^3 & x^2 \\ x^2 & x & x^6 \end{bmatrix}$$

$$\begin{bmatrix} 1 & x^2 & x \\ 1 & x^6 & x^4 \end{bmatrix} \quad \begin{matrix} x^6R_1 \\ x^5R_2 \end{matrix}$$

$$\begin{bmatrix} 1 & x^2 & x \\ 0 & x^6 - x^2 & x^4 - x \end{bmatrix} \quad R_2 - R_1$$

Since $x^6 - x^2 = \left(x^2 + x\right) - x^2 = x$ and
$x^4 - x = \left(x^2 + x + 1\right) - x = x^2 + 1 = x^3$, we get

$$\begin{bmatrix} 1 & x^2 & x \\ 0 & x & x^3 \end{bmatrix} \quad \text{row-echelon form}$$

$$\begin{bmatrix} 1 & x^2 & x \\ 0 & 1 & x^2 \end{bmatrix} \quad x^6 R_2$$

$$\begin{bmatrix} 1 & 0 & x - x^4 \\ 0 & 1 & x^2 \end{bmatrix} \quad R_1 - x^2 R_2$$

Further, $x - x^4 = x - \left(x^2 + x + 1\right) = x^2 + 1 = x^3$. Therefore

$$\begin{bmatrix} 1 & 0 & x^3 \\ 0 & 1 & x^2 \end{bmatrix} \quad \text{reduced row-echelon form}$$

Thus $y = x^3$, $z = x^2$.

To verify the answer, let us substitute solution in the system:
$x \cdot x^3 + x^3 \cdot x^2 = (x^2 + x + 1) + (x + 1) = x^2$
$x^2 \cdot x^3 + x \cdot x^2 = (x + 1) + (x^2 + 1) = x^2 + x = x^6$. □

The field $\mathbb{F}_9$ is the quotient ring of $\mathbb{F}_3[x]$ modulo the ideal generated by $f(x) = x^2 + x + 2$,

$$\mathbb{F}_9 = \mathbb{F}_3[x]/(x^2 + x + 2).$$

The fields $\mathbb{F}_9$ has nine elements. Each element can be expressed as a linear combination of the basis vectors x and 1, or as powers of the primitive element x, as the index table in Figure 2.17 shows.

Exercise 2.61. Compute $(1 + 2x)(2 + x) + 2 + 2x$ in $\mathbb{F}_9$.

Solution. We get $2 + 2x + x^2x^6 = 2 + 2x + x^8 = 3 + 2x = 2x$. □

Exercise 2.62. Compute $(1 + x)/(1 + 2x)$ in $\mathbb{F}_9$.

Solution. $(1 + x)/(1 + 2x) = x^7/x^2 = x^5 = 2x$. □

Exercise 2.63. Compute $1/2$ in $\mathbb{F}_9$.

Solution. We get $x^8/x^4 = x^4 = 2$. Verification: $2 \times 2 = 4 = 1$, since $3 = 0$ in $\mathbb{F}_3$. □

	vectors		
0	0		
1	1		
x		x	
x^2	1	$+2x$	$-(2+x)$
x^3	2	$+2x$	$x(1+2x) = 2(1+2x) + x = 5x+2$
x^4	2		$x(2+2x) = 2(1+2x) + 2x = 6x+2$
x^5		$2x$	$x \cdot 2$
x^6	2	$+x$	$x \cdot 2x = 2(1+2x) = 4x+2$
x^7	1	$+x$	$x(2+x) = (1+2x) + 2x = 4x+1$
			$x(1+x) = (1+2x) + x = 1 = x^8.$

Figure 2.17: Index table of $\mathbb{F}_9$

Exercise 2.64. Compute $(z-2x-1)(z-x-2)$ over $\mathbb{F}_9$.

Solution. We get $z^2-(2x+1+x+2)z+(2x+1)(x+2) = z^2-0z+x^2x^6 = 1+z^2$. □

Exercise 2.65. Solve equation $z^2+1=0$ in $\mathbb{F}_9$.

Solution. By substituting all 9 elements of $\mathbb{F}_9$ for z we can verify that there are two roots: $2x+1$ and $x+2$. □

The field $\mathbb{F}_{16}$ of order 16 is the quotient ring of $\mathbb{F}_2[x]$ modulo $f(x) = x^4+x^3+1$,

$$\mathbb{F}_{16} = \mathbb{F}_2[x]/(f(x) = x^4+x^3+1).$$

The filed has 16 elements. They can be expressed as linear combinations of the basis vectors $x^3, x^2, x, 1$, or as powers of the primitive element x. The index table of $\mathbb{F}_{16}$ is displayed in Figure 2.18.

Exercise 2.66. Compute $1/(x^3+1)$ in $\mathbb{F}_{16}$.

Solution. $1/(1+x^3) = x^{15}/x^4 = x^{11} = 1+x^2+x^3$. □

Exercise 2.67. Compute $(1+x)/(1+x+x^2)$ in $\mathbb{F}_{16}$.

	vectors				
0	0				
1	1				
x			x		
x^2			x^2		
x^3				x^3	
x^4	1			$+x^3$	$= 1 + x^3$
x^5	1	$+x$		$+x^3$	$x(1 + x^3) = (1 + x^3) + x = 1 + x + x^3$
x^6	1	$+x$	$+x^2$	$+x^3$	$x(1 + x + x^3) = \cdots = 1 + x + x^2 + x^3$
x^7	1	$+x$	$+x^2$		$x(1 + x + x^2 + x^3) = \cdots = 1 + x + x^2$
x^8		x	$+x^2$	$+x^3$	$x(1 + x + x^2) = x + x^2 + x^3$
x^9	1		$+x^2$		$x(x + x^2 + x^3) = \cdots = 1 + x^2$
x^{10}		x		$+x^3$	$x(1 + x^2) = x + x^3$
x^{11}	1		$+x^2$	$+x^3$	$x(x + x^3) = \cdots = 1 + x^2 + x^3$
x^{12}	1	$+x$			$x(1 + x^2 + x^3) = \cdot = 1 + x$
x^{13}		x	$+x^2$		$x(1 + x) = x + x^2$
x^{14}			x^2	$+x^3$	$x(x + x^2) = x^2 + x^3$
					$x(x^2 + x^3) = (1 + x^3) + x^3 = 1 = x^{15}.$

Figure 2.18: Index table of $\mathbb{F}_{16}$

Solution. $(1 + x)/(1 + x + x^2) = x^{12}/x^7 = x^5 = 1 + x + x^3.$ □

Lemma 2.13. *If $g(x)$ is a polynomial over a field of characteristic p, then $(g(x))^p = g(x^p)$.*

Proof easily follows from Theorem 2.39. □

Exercise 2.68. *Solve the following system of linear equations over $\mathbb{F}_{16}$.*

$$\begin{aligned} x^2y + x^{10}z &= x^3 \\ x^6y + x^4z &= x \end{aligned}$$

where x is the primitive element of $\mathbb{F}_{16}$ with primitive polynomial $x^4 + x^3 + 1$.

Solution. Referring to the index table for $\mathbb{F}_{16}$ given above we use elementary row operations and reduce the augmented matrix of the system to reduced row-echelon form.

$$\begin{bmatrix} x^2 & x^{10} & x^3 \\ x^6 & x^4 & x \end{bmatrix}$$

$$\begin{bmatrix} 1 & x^8 & x \\ 1 & x^{13} & x^{10} \end{bmatrix} \quad \begin{matrix} x^{13}R_1 \\ x^9 R_2 \end{matrix}$$

$$\begin{bmatrix} 1 & x^8 & x \\ 0 & x^{13}-x^8 & x^{10}-x \end{bmatrix} \quad R_2 - R_1$$

Since $x^{13} - x^8 = \left(x^2 + x\right) - \left(x^3 + x^2 + x\right) = x^3$ and $x^{10} - x = \left(x^3 + x\right) - x = x^3$, we get

$$\begin{bmatrix} 1 & x^8 & x \\ 0 & x^3 & x^3 \end{bmatrix} \quad \text{row-echelon form}$$

$$\begin{bmatrix} 1 & x^8 & x \\ 0 & 1 & 1 \end{bmatrix} \quad x^{12}R_2$$

$$\begin{bmatrix} 1 & 0 & x-x^8 \\ 0 & 1 & 1 \end{bmatrix} \quad R_1 - x^8 R_2$$

Since $x - x^8 = x - (x^3 + x^2 + x) = x^3 + x^2 = x^{14}$, we get

$$\begin{bmatrix} 1 & 0 & x^{14} \\ 0 & 1 & 1 \end{bmatrix} \quad \text{reduced row-echelon form}$$

$y = x^{14}$, $z = 1$.

Let us verify the answer.
$x^2 \cdot x^{14} + x^{10} \cdot 1 = x + x^{10} = x^3$
$x^6 \cdot x^{14} + x^4 \cdot 1 = x^5 + x^4 = (x^3 + x + 1) + (x^3 + 1) = x.$ □

Elements in a field can also be represented as vectors. For $\mathbb{F}_{16}$ these binary vectors are collected in the following table, where the components are taken in the basis $1, x, x^2, x^3$.

powers of x	vectors
0	0000
1	1000
x	0100
x^2	0010
x^3	0001
$x^4 = 1 + x^3$	1001
$x^5 = 1 + x + x^3$	1101
$x^6 = 1 + x + x^2 + x^3$	1111
$x^7 = 1 + x + x^2$	1110
$x^8 = x + x^2 + x^3$	0111
$x^9 = 1 + x^2$	1010
$x^{10} = x + x^3$	0101
$x^{11} = 1 + x^2 + x^3$	1011
$x^{12} = 1 + x$	1100
$x^{13} = x + x^2$	0110
$x^{14} = x^2 + x^3$	0011

Theorem 2.48. (Lattice of Subfields) ([190], 13.10) *The lattice of subfields of $\mathbb{F}_{p^m}$ is isomorphic to the lattice of divisors of m. In particular, $\mathbb{F}_{p^m}$ contains a subfield isomorphic to $\mathbb{F}_{p^r}$ iff r divides m.*

The mapping $\theta : \mathbb{F}_{p^r} \to \mathbb{F}_{p^r}$ defined by the rule

$$\theta(x) = x^p$$

is an automorphism of R called the *Frobenius automorphism.*

Theorem 2.49. (Frobenius Automorphism Theorem) ([190], 13.13) *The automorphism group of every finite field $\mathbb{F}_{p^r}$ is a cyclic group of order r generated by the Frobenius automorphism θ.*

2.14 Matrix Rings

Let R be a set, and let m, n be positive integers. The set of all $m \times n$ matrices with entries in R is denoted by $M_{m,n}(R)$. The set of all square $n \times n$ matrices over R is denoted by $M_n(R)$.

Let R be a ring, $r \in R$, and let $A \in M_{k,\ell}(R)$, $B \in M_{m,n}(R)$ be the matrices

$$A = \begin{bmatrix} a_{11} & a_{12} & \dots & a_{1\ell} \\ a_{21} & a_{22} & \dots & a_{2\ell} \\ \vdots & \vdots & \dots & \vdots \\ a_{k1} & a_{k2} & \dots & a_{k\ell} \end{bmatrix} \quad \text{and} \quad B = \begin{bmatrix} b_{11} & b_{12} & \dots & b_{1n} \\ b_{21} & b_{22} & \dots & b_{2n} \\ \vdots & \vdots & \dots & \vdots \\ b_{m1} & b_{m2} & \dots & b_{mn} \end{bmatrix}$$

The *product* rA is always defined as the matrix

$$A = \begin{bmatrix} ra_{11} & ra_{12} & \dots & ra_{1\ell} \\ ra_{21} & ra_{22} & \dots & ra_{2\ell} \\ \vdots & \vdots & \dots & \vdots \\ ra_{k1} & ra_{k2} & \dots & ra_{k\ell} \end{bmatrix}$$

and is also called the *scalar product.* The *sum* $A + B$ is defined if and only if $k = m$ and $\ell = n$, i.e., if the matrices have the same dimensions. Then the sum is the matrix

$$A + B = \begin{bmatrix} a_{11} + b_{11} & a_{12} + b_{12} & \dots & a_{1n} + b_{1n} \\ a_{21} + b_{21} & a_{22} + b_{22} & \dots & a_{2n} + b_{2n} \\ \vdots & \vdots & \dots & \vdots \\ a_{k1} + b_{k1} & a_{k2} + b_{k2} & \dots & a_{kn} + b_{kn} \end{bmatrix}$$

The *product* AB of the matrices A and B is defined if and only if $\ell = m$, i.e., if the length of rows of A matches the height of columns of B. Then the product is equal to the matrix

$$C = \begin{bmatrix} c_{11} & c_{12} & \dots & c_{1n} \\ c_{21} & c_{22} & \dots & c_{2n} \\ \vdots & \vdots & \dots & \vdots \\ c_{k1} & c_{k2} & \dots & c_{kn} \end{bmatrix}$$

such that each entry c_{ij} is the sum of products of entries in i-th row of A with j-th column of B:

$$c_{ij} = a_{i1}b_{1j} + a_{i2}b_{2j} + \dots + a_{im}b_{1m}.$$

Let R be a ring, and let $A \in M_{k,\ell}(R)$ be the matrix

$$A = \begin{bmatrix} a_{11} & a_{12} & \dots & a_{1\ell} \\ a_{21} & a_{22} & \dots & a_{2\ell} \\ \vdots & \vdots & \dots & \vdots \\ a_{k1} & a_{k2} & \dots & a_{k\ell} \end{bmatrix}$$

The *transpose* A^t of A is the matrix

$$A^t = \begin{bmatrix} a_{11} & a_{21} & \dots & a_{\ell 1} \\ a_{12} & a_{22} & \dots & a_{\ell 2} \\ \vdots & \vdots & \dots & \vdots \\ a_{1k} & a_{2k} & \dots & a_{\ell k} \end{bmatrix}$$

It is obtained by writing all rows of A down the columns of A^t.

Lemma 2.14. For all matrices $A, B \in M_{k,\ell}$ the following equality holds

$$(AB)^t = B^t A^t. \tag{2.9}$$

Matrix notation enables us to write concise easy formulas. For example, the system of linear equations

$$\begin{array}{ccccccccc} a_{11}x_1 & + & a_{12}x_2 & + & \cdots & + & a_{1n}x_n & = & b_1 \\ a_{21}x_1 & + & a_{22}x_2 & + & \cdots & + & a_{2n}x_n & = & b_2 \\ \vdots & & \vdots & & \dots & & \vdots & & \vdots \\ a_{m1}x_1 & + & a_{m2}x_2 & + & \cdots & + & a_{mn}x_n & = & b_m \end{array} \tag{2.10}$$

can we recorded in the more compact form

$$AX^t = B^t, \tag{2.11}$$

where

$$A = \begin{bmatrix} a_{11} & a_{12} & \cdots & a_{1n} \\ a_{21} & a_{22} & \cdots & a_{2n} \\ \vdots & \vdots & \dots & \vdots \\ a_{m1} & a_{m2} & \cdots & a_{mn} \end{bmatrix}$$

is the matrix of the system,

$$X^t = [x_1, \dots, x_n]^t$$

is the the column vector of variables, and

$$B^t = [b_1, \ldots, b_n]^t$$

is the column vector of constant terms. The following *augmented matrix* of the system is used in finding solutions:

$$[A|B] = \begin{bmatrix} a_{11} & a_{12} & \cdots & a_{1n} & b_1 \\ a_{21} & a_{22} & \cdots & a_{2n} & b_2 \\ \vdots & \vdots & \cdots & \vdots & \vdots \\ a_{m1} & a_{m2} & \cdots & a_{mn} & b_m \end{bmatrix}$$

The system is said to be *homogeneous* if all constant terms are zero. If we transpose the matrix equation (2.11) according to (2.9), then we get another version, which may be more convenient in the cases where we prefer to write down row vectors:[1]

$$XA^t = B. \tag{2.12}$$

A matrix M is called a *permutation matrix* if each row and every column of M has exactly one entry equal to 1, and all entries of M belong to the set $\{0, 1\}$. A permutation

$$f = \begin{pmatrix} 1 & 2 & \ldots & n \\ f(1) & f(2) & \ldots & f(n) \end{pmatrix}$$

can be defined by the matrix product rule

$$(f(1), f(2), \ldots, f(n)) = (1, 2, \ldots, n)M, \tag{2.13}$$

where M is the permutation matrix with $M_{1,f(1)} = \cdots = M_{n,f(n)} = 1$ and all other entries equal to 0.

A *linear semigroup* is a semigroup of matrices with entries in a field. Denote by e_{ij} or $e_{i,j}$ the *standard elementary matrix* with 1 in the (i, j) entry and all other entries equal to 0. The set

$$\{e_{ij} \mid 1 \leq i, j \leq n\} \cup \{0\}$$

[1]It is important to be aware of both forms (2.11) and (2.12), for example, because different computer algebra systems use different notations for matrices and vectors, and in some cases we may have to be careful when reading computer output and may have to transpose matrices before inputing their entries in a computer algebra system in order to receive correct results.

forms a semigroup with respect to ordinary matrix multiplication. It is denoted by B_n and is called a *Brandt semigroup*.

Theorem 2.50. ([121]) *Let F be a field, and let $M_n(F)$ be the set of all $n \times n$ matrices over F. For $k = 0, 1, \ldots, n$, denote by I_k the set of all matrices of rank $\leq k$. Then*

$$0 = I_0 \subset I_1 \subset \cdots \subset I_n = M_n(F)$$

are the only ideals of the multiplicative semigroup $M_n(F)$. For every $k = 1, \ldots, n$, the set $I_k \backslash I_{k-1}$ is a disjoint union of subsets $G_{\alpha\beta}$, indexed by the elements α, β of a certain set Λ_k, and such that for all $\alpha, \beta, \gamma, \delta \in \Lambda_k$ the following conditions hold:

(i) *either $G_{\alpha\beta}$ is a linear group, or $G_{\alpha\beta}^2 \subseteq I_{k-1}$;*

(ii) $G_{\alpha\beta} M_n(F) G_{\gamma\delta} \subseteq G_{\alpha\delta} \cup I_{k-1}$;

(iii) *$G_{\alpha *} \cup I_{k-1}$ is a right ideal of $M_n(F)$, where $G_{\alpha *} = \cup_{\lambda \in \Lambda_k} G_{\alpha\lambda}$;*

(iv) *$G_{*\beta} \cup I_{k-1}$ is a left ideal of $M_n(F)$, where $G_{*\beta} = \cup_{\alpha \in \Lambda_k} G_{\alpha\lambda}$;*

(v) *$G_{\alpha\beta} \cup I_{k-1}$ is a left ideal of $G_{\alpha *} \cup I_{k-1}$ and a right ideal of $G_{*\beta} \cup I_{k-1}$.*

2.15 Linear Codes

Let X be an alphabet. An (n, m) *encoding function* of a block code is a mapping $f : X^m \to X^n$. The elements of X^m are called *messages*, and their images in X^n are called *codewords*. The set of all codewords is called a *code* or *block code*. Additional structure is usually defined on a code in order to develop efficient encoding and decoding algorithms. The first step is to introduce linear codes.

Linear codes give an example of application of vector spaces. Let $\mathbb{F}_2$ be the field of order 2. Denote by $\mathbb{F}_2^n$ the set of all binary n-sequences. It contains 2^n sequences. It is a vector space of dimension n over $\mathbb{F}_2$.

Definition 2.16. An (n, m) *linear code* over $\mathbb{F}_2$ is a linear subspace C of $\mathbb{F}_2^n$, where C has dimension m over $\mathbb{F}_2$.

Example 2.18. The set

$$C = \left\{ \begin{array}{l} (100110) = b_1, \\ (010011) = b_2, \\ (001111) = b_3, \\ (110101) = b_1 + b_2, \\ (011100) = b_2 + b_3, \\ (111001) = b_1 + b_3, \\ (111010) = b_1 + b_2 + b_3, \\ (000000) = 0 \end{array} \right\}$$

is a linear subspace of $\mathbb{F}_2^6$. The dimension of C is 3, because b_1, b_2, b_3 is a basis of C. A basis is a linearly independent spanning set, or equivalently, a maximal linearly independent set, or a minimal spanning set.

The dimension helps to find the number of digits in the messages before encoding.

Definition 2.17. The *generator* or *encoding matrix* of the code is a matrix with rows forming a basis of the code.

For example, the following matrix is a generator matrix of the code in Example 2.18:

$$G = \begin{bmatrix} 100110 \\ 010011 \\ 001111 \end{bmatrix}$$

Definition 2.18. A matrix G is in *standard coding form* if

$$G = [I_m | B_{m \times (n-m)}],$$

where I_m is the $m \times m$ identity matrix.

Exercise 2.69. Given the set of generators $\{1001, 1100, 0110, 0011\}$, find a basis and generator matrix for the code.

Solution. Here is a basis of this code

$$\{1100, 0110, 0011\}.$$

The generator matrix with this basis is

$$\begin{bmatrix} 1100 \\ 0110 \\ 0011 \end{bmatrix}$$

□

Definition 2.19. If $x = x_1x_2 \dots x_n$ and $y = y_1y_2 \dots y_n$ are codewords, then their *scalar product* or *dot product* is

$$xy = x_1y_1 + x_2y_2 + \cdots + x_ny_n.$$

The vectors x and y are said to be *orthogonal* if $xy = 0$. Then we write $x \perp y$. The *orthogonal complement* $C^\perp$ of C in $\mathbb{F}_2^n$ is the set

$$C^\perp = \{y \mid y \perp x \text{ for all } x \in C\}.$$

Definition 2.20. If C is a code, then $C^\perp$ is called the *dual code.*

It is easy to verify that the orthogonal complement of a linear code is a linear code too.

Definition 2.21. The *parity-check matrix* of a code C is a generator matrix of the dual code $C^\perp$.

Lemma 2.15. *If $G = [I_m | B_{m\times(n-m)}]$ is a generator matrix of the code C in standard coding form, then the parity-check matrix H of C can be found with the following formula*

$$H = [-(B^t)_{(n-m)\times m} | I_{n-m}] \tag{2.14}$$

Proof. Indeed, $GH^t = [-I_mB + BI_{n-m}] = 0$. Therefore the rows of G are orthogonal to the rows of H. It follows that the rows of H generate $C^\perp$. □

Exercise 2.70. Find a parity-check matrix of the code with generator matrix

$$G = \begin{bmatrix} 100110 \\ 010011 \\ 001111 \end{bmatrix}$$

Solution.

$$H = [-B^t|I_3] = [B^t|I_3] = \begin{bmatrix} 101100 \\ 111010 \\ 011001 \end{bmatrix}$$

It remains to check that the rows of G and H are orthogonal. □

In order to define a code we may specify

(1) a generator matrix G;

(2) a parity-check matrix H;

(3) a system of homogeneous linear equations $xH^t = 0$;

(4) an encoding function: message $x_1x_2 \mapsto$ codeword $y_1y_2y_3y_4$.

Definition 2.22. Let C be a code with the parity-check matrix H, and let x be a vector. Then the *syndrome* of x is the product xH^t.

It follows from definitions that a vector belongs to a code if and only if it has zero syndrome.

Exercise 2.71. Find a system of linear equations whose set of solutions is the code with generator matrix

$$G = \begin{bmatrix} 100110 \\ 010011 \\ 001111 \end{bmatrix}$$

Solution. As above we find the parity-check matrix and write down equations

$$\begin{aligned} xH &= (x_1x_2x_3x_4x_5x_6) \begin{bmatrix} 101100 \\ 111010 \\ 011001 \end{bmatrix}^t \\ &= [x_1 + x_3 + x_4, x_1 + x_2 + x_3 + x_5, x_2 + x_3 + x_6] \\ &= 0. \end{aligned}$$

Thus we get the system

$$\begin{cases} x_1 + x_3 + x_4 = 0 \\ x_1 + x_2 + x_3 + x_5 = 0 \\ x_2 + x_3 + x_6 = 0 \end{cases}$$

□

Exercise 2.72. Let C be the code with generator matrix

$$G = \begin{bmatrix} 10001 \\ 11101 \\ 00010 \end{bmatrix}$$

Find a generator matrix of the dual code $C^{\perp}$.

Solution. Let $v = (x_1x_2x_3x_4x_5) \in C^{\perp}$. Then $g_1 \perp v$, $g_2 \perp v$, $g_3 \perp v$, where g_1, g_2, g_3 are the rows of G. In other words

$$\begin{cases} x_1 & & & & & & & + & x_5 & = & 0 \\ x_1 & + & x_2 & + & x_3 & & & + & x_5 & = & 0 \\ & & & & & + & x_4 & & & = & 0 \end{cases}$$

$$\begin{cases} x_1 & & & & & & & + & x_5 & = & 0 \\ & + & x_2 & + & x_3 & & & & & = & 0 \\ & & & & & + & x_4 & & & = & 0 \end{cases} \quad \text{row-echelon form}$$

Free variables are x_3, x_5. Formulas defining all vectors orthogonal to C are

$$\begin{aligned} x_1 &= -x_5 \\ x_2 &= -x_3 \\ x_4 &= 0 \end{aligned}$$

If we substitute a basis, say $(1,0), (0,1)$, for free variables (x_3, x_5), then we get a basis for $C^{\perp}$.

x_1	x_2	x_4	x_3	x_5
0	1	0	1	0
1	0	0	0	1

The basis for $C^{\perp}$ is $(x_1x_2x_3x_4x_5) : \{(01100), (10001)\}$. A generator matrix for $C^{\perp}$ is

$$\begin{bmatrix} 01100 \\ 10001 \end{bmatrix}$$

□

Exercise 2.73. Find a generator matrix for the ternary code with the spanning set

$$\begin{array}{c}(10111)\\(12011)\\(01200)\\(10211)\end{array}$$

Solution. Form a matrix with these vectors as rows. Reduce to the row-echelon form using elementary row operations, and remove zero rows.

$$\begin{bmatrix}(10111)\\(02200)\\(01200)\\(00100)\end{bmatrix}\begin{array}{l}\\R_2-R_1\\\\R_4-R_1\end{array}$$

$$\begin{bmatrix}(10111)\\(02200)\\(00100)\\(00100)\end{bmatrix}\begin{array}{l}\\\\R_3+R_2\\\\\end{array}$$

$$\begin{bmatrix}(10111)\\(02200)\\(00100)\\(00000)\end{bmatrix}\begin{array}{l}\\\\\\R_4-R_3\end{array}$$

All nonzero rows of this matrix form a basis for the code. Thus a generator matrix of the code is

$$\begin{bmatrix}(10111)\\(02200)\\(00100)\end{bmatrix}$$

□

Exercise 2.74. Given the parity-check matrix $H = \begin{bmatrix}1122\\1012\end{bmatrix}$ of a code C over $\mathbb{F}_3$, find a generator matrix of the code.

Solution. The code is the set of all vectors $x = (x_1, x_2, x_3, x_4)$ such that $xH^t = 0$. Therefore the rows of the generator matrix G form a basis of the set of solutions to the following system

$$\begin{array}{lllllll}x_1 & +x_2 & +2x_3 & +2x_4 & = & 0\\x_1 & & +x_3 & +2x_4 & = & 0\end{array}$$

$$\begin{bmatrix}1 & 1 & 2 & 2\\1 & 0 & 1 & 2\end{bmatrix}$$

$$\begin{bmatrix} 1 & 1 & 2 & 2 \\ 0 & 2 & 2 & 0 \end{bmatrix} \begin{matrix} \\ R_2 - R_1 \end{matrix} \qquad \text{row-echelon form}$$

$$\begin{bmatrix} 1 & 0 & 1 & 2 \\ 0 & 1 & 1 & 0 \end{bmatrix} \begin{matrix} R_1 - \frac{1}{2}R_2 \\ \frac{1}{2}R_2 \end{matrix} \qquad \text{reduced row-echelon form}$$

$$\begin{bmatrix} x_1 \\ x_2 \\ x_3 \\ x_4 \end{bmatrix} = \begin{bmatrix} 2 \\ 2 \\ 1 \\ 0 \end{bmatrix} x_3 + \begin{bmatrix} 1 \\ 0 \\ 0 \\ 1 \end{bmatrix} x_4.$$

Therefore (2210), (1001) is a basis of the set of solutions.

ANSWER: $\begin{bmatrix} 2210 \\ 1001 \end{bmatrix}$. □

Encoding with generator matrix. If G is a generator matrix, v a message, then to encode v we multiply it by G and get vG.

Example 2.19. Let $G = \begin{bmatrix} 100110 \\ 010001 \\ 001000 \end{bmatrix}$ be a generator matrix of a binary code. Encode the message 111 000 101 010.

Solution. We divide the message into blocks and multiply them with the generator matrix

$$(111)G = 111111,$$

$$(000)G = 000000,$$

$$(101)G = 101110,$$

$$(010)G = 010001.$$

The encoded message is 111111000000101110010001. □

Decoding. If G is in standard coding form, then the encoded block always begins with the original message, and so decoding without errors is easy.

Encoding with parity-check matrix is used only when the parity-check matrix has unit rows which form an identity matrix. If the parity-check matrix is in standard coding form, then we put the digits of the message on the first places of the codeword, and find the remaining digits using the parity-check equations. More generally, we put the digits of the message on the places complementing the unit rows, and compute the digits corresponding to the unit rows.

Example 2.20. Encode 01 and 11 using a linear code with parity-check matrix $\begin{bmatrix} 1010 \\ 1101 \end{bmatrix}$.

Solution. Parity-check equations are

$$[x_1 x_2 x_3 x_4] \begin{bmatrix} 1 & 1 \\ 0 & 1 \\ 1 & 0 \\ 0 & 1 \end{bmatrix} = 0.$$

Hence

$$\begin{cases} x_1 + x_3 = 0 \\ x_1 + x_2 + x_4 = 0 \end{cases}$$

Use x_1, x_2 as message digits, and compute x_3, x_4 from

$$\begin{cases} x_3 = -x_1 \\ x_4 = -x_1 - x_2 \end{cases}$$

Therefore $01 \to 0101, 11 \to 1110$. □

Next we introduce standard decoding arrays with the help of the following example.

Example 2.21. Let C be the code with generator matrix

$$G = \begin{bmatrix} 1010 \\ 0101 \end{bmatrix}$$

To find C take all possible linear combinations of rows of G, and write them down in the first row of the standard decoding array

$$C = \{0000, 1010, 0101, 1111\}.$$

The vector of minimum weight not in the code is $s_1 = 1000$. We add it to all vectors of the code and get the second row of the standard decoding array. Next vector of minimum weight not included in the array is $s_2 = 0100$. The sums of s_2 with all vectors in C forms the third row of the array. Finally, the vector of minimum weight not in the array is $s_3 = 1100$. The sums of s_3 with all codewords form the fourth row of the array. After that all vectors have been included in the standard decoding array.

$$\begin{array}{cccc} 0000 & 1010 & 0101 & 1111 \\ 1000 & 0010 & 1101 & 0111 \\ 0100 & 1110 & 0001 & 1011 \\ 1100 & 0110 & 1001 & 0011 \end{array}$$

The vectors $0, s_1, s_2, s_3$ are called *coset leaders.* All elements in the same row of the standard array have the same syndrome. When we receive a message $w = u + c$, we can compute the syndrome $wH^t = uH^t + cH^t = cH^t$, find a coset leader with the same syndrome, and subtract it from w to get the codeword at the top of the column with w.

Matrix encoding with decoding by the standard array leads to the decoding of each received word into the nearest codeword (the codeword with the smallest Hamming distance to the received word). For a linear code, maximum likelihood decoding is completely described by the list of coset leaders and their syndromes vH^t.

Thus matrix encoding with decoding by the standard array corrects precisely those error-patterns c which are coset leaders. Therefore the standard array for e-error-correcting code must have all sequences of weight $\leq e$ as coset leaders, and maybe some other sequences.

It's easier to find the distance of a linear code, than the distance of an arbitrary code, as the following theorem shows.

Theorem 2.51. ([218], §2) *The distance of a linear code equals the minimum weight of all its nonzero codewords.*

Proof. Let m be the minimum weight of a nonzero codeword u. Then

$m = d_H(u, 0)$. Let d be the distance of the code C, i.e., the minimum $d_H(v, w)$ for $v, w \in C$. Clearly, $m \geq d$.

Since the code is linear, $v - w$ is a codeword. It has the weight $w_H(v - w) = d$. Therefore $d \geq m$. Thus $m = d$. □

Definition 2.23. A code C is *equivalent* to a code W if there exists a function f from C to W such that

(i) f is one-to-one and onto;

(ii) f preserves distance, i.e., $d_H(u, v) = d_H(f(u), f(v))$ for all $u, v \in C$.

Let G be a generator matrix of the linear code C. Then any of the following operations on G will transform it into a matrix G^*, which generates an equivalent code C^*:

(row scaling) we may multiply all elements of a row by a nonzero element of the field;

(row interchange) we may interchange the rows of G;

(row addition) we may add any row of G multiplied by a scalar k to any other row of G;

(column interchange) we may rearrange the columns of G.

The first three operations are called *elementary row operations.* They don't change the code at all. On the other hand, column addition, that is addition of a column to a column, can change the distance and weight of a code. For example, let us take a look at the code

$$C = \{(000), (111)\}.$$

It has minimum distance 3, and it's generator matrix is

$$G = \left[\begin{array}{c} 111 \end{array}\right].$$

If we add the first column to the second one, then we get new matrix

$$G' = \left[\begin{array}{c} 101 \end{array}\right]$$

It generates another code with minimum distance 2.

Let G be a generator matrix of a linear code. The following steps transform G into a matrix G^*, in standard coding form, which generates an equivalent code C^*.

Step 1. Find row-echelon form using elementary row operations.

Step 2. Find reduced row-echelon form using elementary row operations.

Step 3. Find the standard coding form using column interchange.

Example 2.22. Transform

$$G = \begin{bmatrix} 101011 \\ 110101 \\ 101100 \end{bmatrix}$$

into standard coding form.

Solution.

$$\begin{bmatrix} 101011 \\ 010110 \\ 000111 \end{bmatrix} \begin{matrix} R_1 \\ R_2 - R_1 \\ R_3 - R_1 \end{matrix} \quad \text{row-echelon form}$$

$$\begin{bmatrix} 101011 \\ 010001 \\ 000111 \end{bmatrix} \begin{matrix} R_1 \\ R_2 - R_3 \\ R_3 \end{matrix} \quad \text{reduced row-echelon form}$$

$$\begin{bmatrix} 1 & 0 & 0 & 1 & 1 & 1 \\ 0 & 1 & 0 & 0 & 0 & 1 \\ 0 & 0 & 1 & 0 & 1 & 1 \\ C_1 & C_2 & C_4 & C_3 & C_5 & C_6 \end{bmatrix} \quad \text{standard coding form}$$

□

Encoding and decoding. If the generator matrix $G = [I|B]$ is in standard coding form, then the parity-check matrix is given by our standard formula (2.14), we reproduce it here:

$$H = \begin{bmatrix} -B^t \\ I \end{bmatrix}. \tag{2.15}$$

In this case the encoding with G is compatible with encoding using H. For example, let $G = \begin{bmatrix} 10ab \\ 01cd \end{bmatrix}$. Encoding with G we get:

$$(x_1x_2)G = (x_1, x_2, ax_1 + cx_2, bx_1 + dx_2).$$

Encoding with H uses the parity-check equations

$$xH^t = (x_1x_2x_3x_4)\begin{bmatrix} -a & -b \\ -c & -d \\ 1 & 0 \\ 0 & 1 \end{bmatrix} = 0,$$

$$\begin{cases} -ax_1 & - & cx_2 & - & x_3 & = & 0 \\ -bx_1 & - & dx_2 & - & x_4 & = & 0, \end{cases}$$

$$\begin{cases} x_3 & = & ax_1 + cx_2, \\ x_4 & = & bx_1 + dx_2. \end{cases}$$

For matrices in reduced row-echelon form, encodings with G and H are compatible if and only if the unit columns of G complement the unit rows of H. For example, we can compare two encoding methods using the following generator and parity-check matrices. Let's look at a generator matrix of a simple code C for the purposes of this example.

$$G = \begin{bmatrix} 1a0d0 \\ 0b1e0 \\ 0c0f1 \end{bmatrix}$$

By rearranging columns, we get a standard coding generator matrix

$$G^* = \begin{bmatrix} 100ad \\ 010be \\ 001cf \end{bmatrix}$$

of another equivalent code C^*. The standard formula (2.14) for parity-check matrices gives

$$H^* = \begin{bmatrix} -a & -d \\ -b & -e \\ -c & -f \\ 1 & 0 \\ 0 & 1 \end{bmatrix}$$

Returning to the original order of columns produces

$$H = \begin{bmatrix} -a & 1 & -b & 0 & -c \\ -d & 0 & -e & 1 & -f \end{bmatrix}$$

Notice that this is the required parity-check matrix of the original code.

Exercise 2.75. Let C be the binary code with generator matrix

$$G = \begin{bmatrix} 10001 \\ 11101 \\ 00010 \end{bmatrix}$$

Find an equivalent code C^* with generator matrix G^* in standard coding form, and use it to find a generator matrix $G^{\perp}$ of the dual code $C^{\perp}$.

Solution. First, find a generator matrix in reduced row-echelon form.

$$\begin{bmatrix} 10001 \\ 01100 \\ 00010 \end{bmatrix} \begin{matrix} R_1 \\ R_2 - R_1 \\ R_3 \end{matrix} \quad \begin{matrix} \\ \text{row-echelon form,} \\ \text{reduced} \end{matrix}$$

Next, find a generator matrix in standard coding form.

$$\begin{array}{c} \begin{bmatrix} 1 & 0 & 0 & 0 & 1 \\ 0 & 1 & 0 & 1 & 0 \\ 0 & 0 & 1 & 0 & 0 \end{bmatrix} \\ \begin{matrix} C_1 & C_2 & C_4 & C_3 & C_5 \\ & I_3 & & & B \end{matrix} \end{array} \quad \text{standard coding form}$$

The formula (2.14) gives the parity- check matrix

$$\begin{aligned} H^* &= [-B^t | I_2] \\ &= \begin{bmatrix} 0 & 1 & 0 & 1 & 0 \\ 1 & 0 & 0 & 0 & 1 \end{bmatrix} \\ & \quad \begin{matrix} C_1 & C_2 & C_4 & C_3 & C_5 \\ & -B^t & & & I_2 \end{matrix} \end{aligned}$$

Here $G^*(H^*)^t = 0$. Reorder the rows of H^* and get

$$H = \begin{bmatrix} 0 & 1 & 1 & 0 & 0 \\ 1 & 0 & 0 & 0 & 1 \end{bmatrix}$$
$$\begin{matrix} C_1 & C_2 & C_3 & C_4 & C_5 \end{matrix}$$

Now $GH^t = 0$ with the original matrix G. Therefore H is a parity-check matrix of C and a generator matrix of the dual code. □

2.16 Cyclic Codes

Given a sequence of n elements $b_0, b_1, b_2, \ldots, b_{n-2}, b_{n-1}$ of the field F, we associate the following polynomial with it:

$$g(x) = b_0 + b_1 x + b_2 x^2 + \cdots + b_{n-2} x^{n-2} + b_{n-1} x^{n-1}.$$

If $b_0, b_1, \ldots, b_{n-1}$ is a message, then the associated polynomial $g(x)$ is called a *message polynomial.* If b_0, b_1, ..., b_{n-1} is a codeword, then the associated polynomial $g(x)$ is called a *code polynomial.*

The algebra of polynomials modulo $1 - x^n$ is the quotient algebra $\mathbb{F}_q[x]/(1 - x^n)$, i.e., the set of all polynomials of degree less than n with ordinary addition and multiplication defined modulo $1 - x^n$. For example, in $\mathbb{F}_2[x]/(1 - x^7)$ we get

$$\begin{aligned}
&(1 + x + x^2)(1 + x + x^5 + x^6) = \\
&= 1 + 2x + 2x^2 + x^3 + x^5 + 2x^6 + 2x^7 + x^8 \\
&= 1 + x^3 + x^5 + x^8 \\
&= 1 + x + x^3 + x^5.
\end{aligned}$$

Definition 2.24. A *cyclic code* is a linear subspace C of $\mathbb{F}_{p^r}^n$ such that, for every n-sequence

$$(b_0, b_1, \ldots, b_{n-2}, b_{n-1}) \in C,$$

the *cyclic shift* $(b_{n-1}, b_0, b_1, \ldots, b_{n-2})$ belongs to C, too.

Cyclic subspaces of $GF(q)^n$ correspond to ideals of the ring

$$\mathbb{F}_q[x]/(1 - x^n).$$

Proposition 2.2. (Cyclic Codes Ideal Theorem) ([217], §5) *Let f be the mapping from $\mathbb{F}_{p^r}^n$ to $\mathbb{F}_{p^r}[x]/(1 - x^n)$ defined by the rule*

$$f(a_0, a_1, a_2, \ldots, a_{n-1}) = a_0 + a_1 x + a_2 x^2 + \cdots + a_{n-1} x^{n-1}$$

for $a_0, a_1, a_2, \ldots, a_{n-1} \in \mathbb{F}_{p^r}$. A code $C \subseteq \mathbb{F}_{p^r}^n$ is cyclic if and only if its image

$$f(C) = \{f(x) \mid x \in C\}$$

is an ideal of the ring $\mathbb{F}_q[x]/(1-x^n)$.

Proof. It suffices to note that multiplying a polynomial by x results in a cyclic shift of the corresponding sequence. □

Example 2.23. The subspace spanned by (11111) is cyclic and corresponds to the ideal generated by $1+x+x^2+x^3+x^4$ in $\mathbb{F}_2[x]/(1-x^5)$.

Example 2.24. The subspace $\{(x_0, x_1, x_2) \mid x_0 + x_1 + x_2 = 0\}$ is cyclic and corresponds to the ideal generated by $1+x$ in $\mathbb{F}_2[x]/(1-x^3)$.

Exercise 2.76. Find a generator polynomial of the ideal corresponding to the linear space

$$\{(x_0, x_1, x_2, x_3) \mid x_0 = x_1,\ x_2 = x_3\}.$$

Definition 2.25. An (n, m) *polynomial code* is an ideal in

$$\mathbb{F}_q[x]/(1-x^n)$$

which has dimension m as a vector space.

Polynomial codes encode messages by polynomial multiplication.

Definition 2.26. Let I be an ideal of $\mathbb{F}_q[x]/(1-x^n)$. The *standard generator polynomial* of I is a monic polynomial of minimal degree in I.

Theorem 2.52. (Generator of Cyclic Code) ([190], 18.6) *Let I be an ideal of $\mathbb{F}_q[x]/(1-x^n)$, and let $g(x)$ be the standard generator polynomial of I.*

(i) *$g(x)$ is unique;*

(ii) *$g(x)$ divides all polynomials of I in $\mathbb{F}_q[x]/(1-x^n)$, and so generates I as ideal;*

(iii) *if a generator of I divides $1-x^n$ in $\mathbb{F}_q[x]$, then it is equal to $g(x)$;*

(iv) *$g(x)$ divides $1-x^n$ in $\mathbb{F}_q[x]$;*

(v) *every monic divisor of* $1 - x^n$ *generates a distinct ideal in*

$$\mathbb{F}_q[x]/(1 - x^n).$$

If $g(x)h(x) = 1 - x^n$ in $F[x]$ and $g(x)$ is a generator polynomial of ideal I, then $h(x)$ is called a *parity-check polynomial* of I. The parity-check polynomial of an ideal I generates the set

$$\mathrm{Ann}(I) = \{r \in R_n \mid Ir = 0\}$$

called the *annihilator* of I in R_n. In general, $I^\perp \neq \mathrm{Ann}(I)$. The *reciprocal polynomial* of

$$h(x) = h_0 + h_1 x + \cdots + h_{k-1}x^{k-1} + h_k x^k$$

is the polynomial

$$h^\perp(x) = x^k h(x^{-1}) = h_k + h_{k-1}x + \cdots + h_1 x^{k-1} + h_0 x^k.$$

It has the same coefficients as $h(x)$ but in reversed order. Note that only the coefficients from the constant term to the leading terms are rewritten in reversed order to find $h^\perp(x)$.

Orthogonality of polynomials. Let

$$f(x) = f_0 + f_1 x + \cdots + f_{n-2}x^{n-2} + f_{n-1}x^{n-1},$$

$$g(x) = g_0 + g_1 x + \cdots + g_{n-2}x^{n-2} + g_{n-1}x^{n-1}.$$

Consider the vectors of coefficients of these polynomials:

$$\vec{a} = (f_0, f_1, \ldots, f_{n-1}),$$

$$\vec{b} = (g_0, g_1, \ldots, g_{n-1}).$$

Then $f(x)g(x) = 0$ in $R_n = F[x]/(1 - x^n)$ if and only if $\vec{a}$ is orthogonal to the reversed vector of $\vec{b}$ and its every cyclic shift. For example, if $n = 4$, then we get

$$\begin{aligned} f(x)g(x) &= (f_0g_3 + f_1g_2 + f_2g_1 + f_3g_0)x^3 \\ &+ (f_0g_0 + f_1g_3 + f_2g_2 + f_3g_1) \\ &+ (f_0g_1 + f_1g_0 + f_2g_3 + f_3g_2)x \\ &+ (f_0g_2 + f_1g_1 + f_2g_0 + f_3g_3)x^2 \end{aligned}$$

Obviously, $f(x)g(x) = 0$ if and only if all the coefficients of this polynomial are zero. If the constant term of fg is zero, this has the following meaning:

$$f_0g_0 + f_1g_3 + f_2g_2 + f_3g_1 = 0 \text{ means } a \perp (g_0, g_3, g_2, g_1).$$

Similarly, looking at other coefficients we get

$$f_0g_1 + f_1g_0 + f_2g_3 + f_3g_2 = 0 \text{ means } a \perp (g_1, g_0, g_3, g_2),$$

$$f_0g_2 + f_1g_1 + f_2g_0 + f_3g_3 = 0 \text{ means } a \perp (g_2, g_1, g_0, g_3),$$

$$f_0g_3 + f_1g_2 + f_2g_1 + f_3g_0 = 0 \text{ means } a \perp (g_3, g_2, g_1, g_0).$$

Generator of the dual code. If $g(x)h(x) = 1 - x^n$ lies in $F[x]$ and $g(x)$ is a generator polynomial of a code C, then the reciprocal polynomial $h^\perp$ is a generator polynomial of the dual code $C^\perp$.

It follows that the dual code is also cyclic. Indeed, since $g(x)h(x) = 0$, we see as above that the coefficient vector of $g(x)$ is orthogonal to the reversed vector of $h(x)$, i.e. to $h^\perp(x) = x^k h(x^{-1})$. Thus $h^\perp$ generates $C^\perp$.

Example 2.25. Given a generator polynomial $g(x) = 1 + x^2 + x^3$ of a code C in $R_7 = \mathbb{F}_2[x]/(1 - x^7)$, find a generator polynomial of the dual code $C^\perp$.

The check polynomial $h(x) = \frac{1-x^7}{g(x)} = 1 + x^2 + x^3 + x^4$. The reciprocal polynomial $h^\perp = x^4 h(x^{-1}) = 1 + x + x^2 + x^4$ generates $C^\perp$.

Basis and dimension. Let $g(x)h(x) = 1 - x^n$ in $R_n = F[x]/(1 - x^n)$, where $g(x) = g_0 + g_1 x + \cdots + g_m x^m$ is a generator polynomial of a code C, and $h(x)$ has degree $n - m$. Then the set

$$g(x), xg(x), \ldots, x^{n-m-1}g(x)$$

is a basis of C. It follows that C has dimension $n - m$ and the following $(n - m) \times n$ generator matrix

$$\begin{bmatrix} g(x) \\ xg(x) \\ x^2 g(x) \\ \vdots \\ x^{n-m-1}g(x) \end{bmatrix} =$$

$$\begin{bmatrix} g_0 & g_1 & g_2 & \cdots & g_{m-1} & g_m & 0 & 0 & 0 & \cdots & 0 \\ 0 & g_0 & g_1 & \cdots & g_{m-2} & g_{m-1} & g_m & 0 & 0 & \cdots & 0 \\ 0 & 0 & g_0 & \cdots & g_{m-3} & g_{m-2} & g_{m-1} & g_m & 0 & \cdots & 0 \\ \vdots & \vdots & \vdots & \cdots & \vdots & \vdots & \vdots & \cdots & & & \\ 0 & 0 & 0 & \cdots & g_0 & g_1 & \cdots & g_{m-1} & g_m & 0 & 0 \\ 0 & 0 & 0 & \cdots & 0 & g_0 & g_1 & \cdots & g_{m-1} & g_m & 0 \\ 0 & 0 & 0 & \cdots & 0 & 0 & g_0 & g_1 & \cdots & g_{m-1} & g_m \end{bmatrix}$$

Polynomial encoding. Let C be a linear cyclic code of length n and dimension k with generator polynomial $g(x)$. The k information digits $a_0, a_1, \ldots, a_{k-2}, a_{k-1}$ to be encoded can be thought of as a polynomial

$$a(x) = a_0 + a_1 x + \cdots + a_{k-2}x^{k-2} + a_{k-1}x^{k-1}.$$

Then $a(x)$ is encoded as $a(x)g(x)$.

Example 2.26. Let $g(x) = 1 + x + x^3$ and $n = 7$ for a binary cyclic code. Then $k = 7 - 3 = 4$. In order to encode the message 0101 we take its message polynomial

$$a(x) = 0 + x + 0x^2 + x^3$$

and multiply it by $g(x)$ modulo $1 - x^7$:

$$\begin{aligned} c(x) &= a(x)g(x) \\ &= (x + x^3)(1 + x + x^3) \\ &= x + x^2 + x^3 + 2x^4 + x^6 \\ &= x + x^2 + x^3 + x^6. \end{aligned}$$

Therefore $c = 0111001$.

Let us compare the answer with the standard generator matrix encoding. A generator matrix is found as follows:

$$G = \begin{bmatrix} g(x) \\ xg(x) \\ x^2g(x) \\ x^3g(x) \end{bmatrix} = \begin{bmatrix} 1 & 1 & 0 & 1 & 0 & 0 & 0 \\ 0 & 1 & 1 & 0 & 1 & 0 & 0 \\ 0 & 0 & 1 & 1 & 0 & 1 & 0 \\ 0 & 0 & 0 & 1 & 1 & 0 & 1 \end{bmatrix}$$

To encode we multiply by the generator matrix and get the same codeword

$$(0101)G = 0111001$$

Polynomial codes are better since it easier to store a generator polynomial than a generator matrix.

Decoding. To decode a received message, we divide it by the generator polynomial. If we know the codeword $c(x)$ without errors, then $c(x) = a(x)g(x)$ and to find the message polynomial $a(x)$ we can divide $c(x)$ by generator polynomial $g(x)$.

Example 2.27. Decode 0111001 using the binary code with generator polynomial $g(x) = 1 + x + x^3$.

$$c(x) = \begin{array}{ccccccc} 0 & 1 & 1 & 1 & 0 & 0 & 1 \\ & x & +x^2 & +x^3 & & & +x^6 \end{array}$$

Dividing $c(x)$ by $g(x)$, we get

$$\begin{array}{r|rrrrr} & & & x^3 & & +x \\ \hline x^3 \;\; +x \;\; +1 & x^6 & & +x^3 & +x^2 & +x \\ & x^6 & +x^4 & +x^3 & & \\ \hline & & x^4 & & +x^2 & +x \\ & & x^4 & & +x^2 & +x \\ \hline & & & & & 0 \end{array}$$

The message polynomial is $x + x^3$, which means that the message is 0101.

Example 2.28. Decode the codeword 1110010 using the binary code with generator polynomial $g(x) = 1 + x + x^3$.

Solution. The code polynomial is

$$c(x) = \begin{array}{ccccccc} 1 & 1 & 1 & 0 & 0 & 1 & 0 \\ 1 & x & +x^2 & & & +x^5 & \end{array}$$

Dividing $c(x)$ by $g(x)$, we get

$$\begin{array}{r|llll} & & & x^2 & & +1 \\ \hline x^3 + x + 1 & x^5 & & +x^2 & +x & +1 \\ & x^5 & +x^3 & +x^2 & & \\ \hline & & x^3 & & +x & +1 \\ & & x^3 & & +x & +1 \\ \hline & & & & & 0 \end{array}$$

The message polynomial is $x^2 + 1$, and the message is 0101. □

Exercise 2.77. Find a parity-check matrix H for the $(7, 4)$ binary cyclic code with generator polynomial $g(x) = 1 + x + x^3$.

Solution. We have already found the generator matrix

$$G = \begin{bmatrix} 1 & 1 & 0 & 1 & 0 & 0 & 0 \\ 0 & 1 & 1 & 0 & 1 & 0 & 0 \\ 0 & 0 & 1 & 1 & 0 & 1 & 0 \\ 0 & 0 & 0 & 1 & 1 & 0 & 1 \end{bmatrix}$$

Using elementary row operations, reduce it to a matrix in standard coding form as follows

$$G = \begin{bmatrix} 1 & 1 & 0 & 0 & 1 & 0 & 1 \\ 0 & 1 & 1 & 0 & 1 & 0 & 0 \\ 0 & 0 & 1 & 0 & 1 & 1 & 1 \\ 0 & 0 & 0 & 1 & 1 & 0 & 1 \end{bmatrix} \begin{array}{l} R_1 + R_4 \\ R_2 \\ R_3 + R_4 \\ R_4 \end{array}$$

$$G = \begin{bmatrix} 1 & 1 & 0 & 0 & 1 & 0 & 1 \\ 0 & 1 & 0 & 0 & 0 & 1 & 1 \\ 0 & 0 & 1 & 0 & 1 & 1 & 1 \\ 0 & 0 & 0 & 1 & 1 & 0 & 1 \end{bmatrix} \begin{array}{l} R_1 \\ R_2 + R_3 \\ R_3 \\ R_4 \end{array}$$

$$G = \begin{bmatrix} 1 & 0 & 0 & 0 & 1 & 1 & 0 \\ 0 & 1 & 0 & 0 & 0 & 1 & 1 \\ 0 & 0 & 1 & 0 & 1 & 1 & 1 \\ 0 & 0 & 0 & 1 & 1 & 0 & 1 \end{bmatrix} \begin{matrix} R_1 + R_2 \\ R_2 \\ R_3 \\ R_4 \end{matrix}$$

This matrix is in standard coding form, and we may use the formula

$$G = [I|B] \mapsto H = \begin{bmatrix} -B \\ I \end{bmatrix}.$$

Hence

$$H = \begin{bmatrix} 1 & 0 & 1 & 1 & 1 & 0 & 0 \\ 1 & 1 & 1 & 0 & 0 & 1 & 0 \\ 0 & 1 & 1 & 1 & 0 & 0 & 1 \end{bmatrix}$$

is a parity-check matrix. □

2.17 Universal Algebras

Let n be a non-negative integers. An *n-ary operation* on a set A is a mapping from A^n to A. The integer n is called the *arity* of the operation. Note that every *nullary* operation $f : A^n \to A$ can be identified with the element $f(\emptyset)$ in A. This means that each nullary operation chooses a certain fixed element in A.

A *universal algebra* is a set A with a family $\Omega = (\omega_i)_{i \in I}$ of n_i-are operations ω_i. The set Ω is called the *signature* of the algebra A, and the sequence $\tau = (n_i)_{i \in I}$ is called the *type* of Ω. For a positive integer n, denote by Ω_n the set of all n-ary operations in Ω. Hence $\Omega = \cup_{n=0}^{\infty} \Omega_n$.

Let (A, Ω) be an algebra. A subset $B \subseteq A$ is called a *subalgebra* if it is closed under all operations in Ω, i.e., if $\omega(b_1, \ldots, b_n) \in B$ for all $\omega \in \Omega$ and all $b_1, \ldots, b_n \in B$, where n is the arity of ω. If B is a subalgebra of A, then we write $B \leq A$.

An equivalence relation ϱ on A is said to be *compatible* with operations in Ω and is called a *congruence* on (A, Ω), if

$$a_1 \varrho b_1, \ldots, a_n \varrho b_n \Longrightarrow \omega(a_1, \ldots, a_n) \varrho \omega(b_1, \ldots, b_n) \qquad (2.16)$$

for all $\omega \in \Omega$ and all $a_1, \ldots, a_n, b_1, \ldots, b_n \in A$, where n is the arity of ω. If ϱ is a congruence on (A, Ω), then the set A/ϱ of equivalence classes of ϱ forms an algebra of the same signature Ω with operations defined by

$$\omega(a_1^{\varrho}, \ldots, a_n^{\varrho}) = \omega(a_1, \ldots, a_n)^{\varrho} \tag{2.17}$$

for all $\omega \in \Omega$ and all $a_1, \ldots, a_n \in A$, where n is the arity of ω. This algebra is called the *factor algebra* or *quotient algebra* of A with respect to ϱ or modulo ϱ.

A mapping h from (A, Σ) to (B, Σ) is called a *morphism* or a *homomorphism* if

$$h(\omega(a_1, \ldots, a_n)) = \omega(h(a_1), \ldots, h(a_n)) \tag{2.18}$$

for all $\omega \in \Omega$ and all $a_1, \ldots, a_n \in A$, where n is the arity of ω.

Theorem 2.53. (Homomorphism Theorem) ([190], 40.3) *Let h be a homomorphism from (A, Ω) to (B, Ω). Then the relation Ker(h) on defined by $a\,\mathrm{Ker}(h) b \Leftrightarrow h(a) = h(b)$ is a congruence on A and the homomorphic image $h(A)$ is isomorphic to $A/\sim$:*

$$h(A) \cong A/\,\mathrm{Ker}(h).$$

The following two theorems have easy proofs using Theorem 2.53.

Theorem 2.54. (First Isomorphism Theorem) *Let (A, Ω) be a universal algebra with a subalgebra (B, Ω), and let ϱ be congruences on A. Then the restriction $\varrho \cap 1_B$ of ϱ on B is a congruence on B, and the quotient algebra $B/(\varrho \cap 1_B)$ is isomorphic to the image of B is A/ϱ:*

$$B/(\varrho \cap 1_B) \cong \{x^{\varrho} \mid x \in B\} \subseteq A/\varrho.$$

Theorem 2.55. ([90], Theorem 1.5.8, [190], 28.13) (Second Isomorphism Theorem) *Let $\varrho \subseteq \delta$ be congruences on a universal algebra (A, Ω). Then the relation*

$$\delta/\varrho = \{(x^{\varrho}, y^{\varrho}) \mid (x, y) \in \delta\}$$

is a congruence on the quotient algebra A/ϱ and $(A/\varrho)/(\delta/\varrho)$ is isomorphic to A/δ:

$$(A/\varrho)/(\delta/\varrho) \cong A/\delta.$$

The homomorphism $f : A \to B$ is called an *endomorphism* if $A = B$. A homomorphism is called an *isomorphism* (*monomorphism*, *epimorphism*) if it is a bijection (resp., injection, surjection). An *automorphism* is an isomorphism $f : A \to A$ onto the same set A.

Let Ω be a family of operations, and let X be an alphabet. A *term* over Ω and X is defined recursively by the rules:

(i) all nullary operations and all variables in X are terms;

(ii) if ω is an n-ary operation in Ω and $t_1, \ldots, t_n$ are terms, then $\omega(t_1, \ldots, t_n)$ is a term too;

(iii) Every term can be constructed in a finite number of steps using (i) and (ii).

The set of all terms over Ω and X is denoted by $T_X(\Omega)$ and is called the *term algebra* or over Ω and X of type τ, or *free algebra generated by* X over Ω.

Example 2.29. Let $X = \{x, y, z\}$ and $\Omega = \Omega_0 \cup \Omega_1 \cup \Omega_2 \cup \Omega_3$, where $\Omega_0 = \{\alpha\}$, $\Omega_1 = \{\beta\}$, $\Omega_2 = \{\gamma\}$, $\Omega_3 = \{\delta, \mu\}$. Then the following expressions are terms of $F_X(\Omega)$:

$$x,\ y,\ z,\ \alpha,\ \beta(x),\ \gamma(y, \alpha),\ \delta(x, \gamma(y, x), \alpha),\ \beta(\gamma(y, z), \mu(x, \alpha, y)).$$

Let X be a subset of A. The intersection of all subalgebras of A containing X is the smallest subalgebra containing X. It is called the *subalgebra generated by* X. It can be expressed in terms of term algebra: the subalgebra generated by a set X is equal to the set of all terms where the elements of X are the variables. The set X is called the generating set of the subalgebra.

A mapping $f : A \to A$ is called an *elementary translation* of the Ω-algebra A, if there exist a positive integer n, an operation $\omega \in \Omega_n$, an index j with $1 \leq j \leq n$, and elements $c_1, \ldots, c_{j-1}, c_{j+1}, \ldots, c_n$ such that

$$f(a) = \omega(c_1, \ldots, c_{j-1}, a, c_{j+1}, \ldots, c_n)$$

holds for all $a \in A$. The set of all elementary translations of A is denoted by $\mathrm{ET}(A)$.

Lemma 2.16. (Elementary Translations Lemma) ([67], Lemma 3.16) *An equivalence relation ϱ on A is a congruence if and only iff ϱ is invariant with respect to all elementary translations, i.e.,*

$$a \varrho b \text{ implies } f(a) \varrho f(b) \text{ for all } a, b \in A, f \in \mathrm{ET}(A). \tag{2.19}$$

Proof. The 'only if' part is obvious. In order to prove the 'if' part, suppose that (2.19) holds. Take an arbitrary positive integer n, $\omega \in \Omega_n$, and elements $a_1, \ldots, a_n, b_1, \ldots, b_n \in A$ such that $a_1 \varrho b_1, \ldots, a_n \varrho b_n$. For $1 \leq j \leq n$, consider the following elementary transitions

$$f_j(x) = \omega(b_1, \ldots, b_{j-1}, x, a_{j+1}, \ldots, a_n).$$

We get

$$\begin{aligned} \omega(a_1, a_2, \ldots, a_n) &= f_1(a_1) \, \varrho \, f_1(b_1) \\ &= f_2(a_2) \, \varrho \, f_2(b_2) \\ &\vdots \\ &= f_n(a_n) \, \varrho \, f_n(b_n) \\ &= \omega(b_1, b_2, \ldots, b_n). \end{aligned}$$

Hence $\omega(a_1, a_2, \ldots, a_n) \, \varrho \, \omega(b_1, b_2, \ldots, b_n)$, which means that ϱ is a congruence. □

An *equation* over Ω is a pair (t_1, t_2) of terms over Ω. The equation (t_1, t_2) can be also written as $t_1 = t_2$. We say that the equation $t_1 = t_2$ *holds* in an algebra (A, Ω) if the values of t_1 and t_2 coincide after each replacement of variables by elements of A.

A class $\mathcal{K}$ of algebras of signature Ω is said to be *equationally definable* and is called a *variety* if there exists a set E of equations over Ω such that $\mathcal{K}$ consists of all algebras over Ω satisfying all equations in E.

Theorem 2.56. (Birkhoff's Theorem) ([190], 40.7) *A class $\mathcal{K}$ of algebras of signature Ω is a variety if and only if $\mathcal{K}$ is closed with respect to subalgebras, homomorphic images and direct products.*

A subset B of an algebra (A, Ω) in variety $\mathcal{V}$ is called a *free basis* of B with respect to Ω if, for every algebra (B, Ω) in $\mathcal{V}$, each mapping from A to B can be uniquely extended to a homomorphism from A to B. An algebra (A, Ω) is said to be *free* of $\mathcal{V}$ if it has a free basis.

Theorem 2.57. (Free Universal Algebra Theorem) ([190], 40.9) *Let $\mathcal{V}$ be a variety over Ω, and let $\sim_{\mathcal{V}}$ be the relation defined on the term algebra $T_X(\Omega)$ by the rule*

$$t_1 \sim_{\mathcal{V}} t_2 \Leftrightarrow t_1 = t_2 \text{ holds in all algebras of } \mathcal{V}.$$

Then $\sim_{\mathcal{V}}$ is a congruence on $T_X(\Omega)$, the quotient algebra

$$F_X(\mathcal{V}) = T_X(\Omega)/\sim_{\mathcal{V}}$$

belongs to the variety $\mathcal{V}$, and it is the free algebra in $\mathcal{V}$ with free basis

$$\{x^{\sim_{\mathcal{V}}} \mid x \in X\}.$$

Each free algebra in $\mathcal{V}$ is isomorphic to some $F_X(\mathcal{V})$, and every algebra A in $\mathcal{V}$ is the homomorphic image of a free algebra $F_X(\mathcal{V})$ if and only if A is generated by a subset of cardinality not exceeding that of X.

The congruence $\sim_{\mathcal{V}}$ is called *equational congruence.* If E is a set of equations over Ω, then

$$\mathcal{K}_E = \{A \in \mathcal{K}(\Omega) \mid \text{ all equations of } E \text{ hold in } A\}$$

is a variety defined by the set E of equations.

If we take a class $\mathcal{K}$ of algebras of signature Ω, then the set of all equations over Ω that hold in all algebras of $\mathcal{K}$ is called the *equational theory* of $\mathcal{K}$ and is denoted by $\mathrm{Th}(\mathcal{K})$. The variety $\mathcal{V}_{\mathrm{Th}(\mathcal{K})}$ defined by the equational theory of $\mathcal{K}$ is called the *variety generated by* $\mathcal{K}$.

2.18 Exercises

1 Construct a binary Huffman encoding for the alphabet

$$A = \{x, y, z, t, u\}$$

with probabilities $0.30, 0.23, 0.20, 0.15, 0.12$, respectively. Find the average length of encoding of one letter chosen randomly in a text.

2 Construct a ternary Huffman encoding for the alphabet

$$A = \{a, b, c, d, e, f\}$$

with probabilities $0.25, 0.23, 0.21, 0.14, 0.11, 0.06$, respectively. Find the average length of encoding of one letter chosen randomly in a text. Ternary encodings use the alphabet $\{0, 1, 2\}$.

3 Construct a binary Huffman encoding for the alphabet

$$A = \{a, b, c, d, e\}$$

with probabilities $0.33, 0.25, 0.23, 0.17, 0.02$, respectively. Find the average length of encoding of one letter chosen randomly in a text.

4 Construct a quaternary Huffman encoding for the alphabet

$$A = \{a, b, c, d, e, f, g, h\}$$

with probabilities $0.22, 0.20, 0.19, 0.13, 0.12, 0.08, 0.05, 0.01$, respectively. Find the average length of encoding of one letter chosen randomly in a text. Quaternary encodings use the alphabet $\{0, 1, 2, 3\}$.

5 A monoid S is said to be *cancellative* if, for all $x, y, z \in S$,

$$xy = xz \quad \text{implies} \quad y = z \text{ and}$$
$$yx = zx \quad \text{implies} \quad y = z.$$

Prove that every free monoid A^* is cancellative.

6 A monoid S is called *equidivisible* if, for all $s, t, u, v \in S$, the equality $st = uv$ implies that either

(i) there exists x in S^1 such that $s = ux$ and $v = xt$; or

(ii) there exists y in S^1 such that $u = sy$ and $t = yv$.

Prove that the free monoid is equidivisible.

7 Let A^* be a free monoid and let S be a submonoid of A^*. Then S is said to be *free* if there exists a code X in A such that $S = X^*$. Prove that S is free if and only if, for each $w \in S$,

$$wS \cap S \neq \emptyset \text{ and } Sw \cap S \neq \emptyset \text{ imply } w \in S. \tag{2.20}$$

8 Let A^+ be a free semigroup and let S be a subsemigroup of A^+. Then S is said to be *free* if there exists a code X in A such that $S = X^+$. Prove that S is free if and only if, for each $w \in S$, implication (2.20) holds.

9 **(a)** Construct a binary Huffman code and use it for optimal compression of the text "free monoid". (Ignore the space signs in text.)

(b) Construct a quaternary Huffman code and use it for optimal compression of the text "monoids are semigroups".

(c) Write down a complete proof explaining that the algorithm for constructing a binary Huffman code gives a prefix code.

10 **(a)** Construct a ternary Huffman code and use it for optimal compression of the text "all groups are monoids!". (Ignore the space signs in text.)

(b) Write down a complete proof explaining that the algorithm for constructing a Huffman code gives a prefix code.

(c) Write down a proof showing that the algorithm for constructing a Huffman code gives us a code that achieves optimal compression.

11 **(a)** Write down the definitions of a group and an ω-factorization of an infinite sequence.

(b) Write down the Ramsey Theorem on ω-factorizations of words in finite semigroups.

(c) Use the pigeonhole principle to prove Ramsey Theorem for all finite groups.

12 Find all subgroups of the symmetric group S_3.

13 Explain how Fermat's Little Theorem follows from Lagrange's Theorem.

14 **(a)** Write down the Principal Theorem on Finite Abelian Groups.

(b) Find the primary components of the cyclic group

$$\mathbb{Z}_6 = \{1, g, g^2, g^3, g^4, g^5\}.$$

(c) Represent the following groups as direct products of cyclic groups with the order of each group dividing the order of the next one

$$\mathbb{Z}_2 \times \mathbb{Z}_4 \times \mathbb{Z}_3 \times \mathbb{Z}_6,$$

$$\mathbb{Z}_6 \times \mathbb{Z}_{15} \times \mathbb{Z}_{18}.$$

15 **(a)** Write down the Principal Theorem on Finite Abelian Groups.

(b) Find the primary components of the cyclic group

$$\mathbb{Z}_{10} = \{1, g, g^2, g^3, \ldots, g^9\}.$$

(c) Represent the following groups as direct products of cyclic groups with the order of each group dividing the order of the next one

$$G = \mathbb{Z}_6 \times \mathbb{Z}_9 \times \mathbb{Z}_{10} \times \mathbb{Z}_{15},$$

$$H = \mathbb{Z}_2 \times \mathbb{Z}_4 \times \mathbb{Z}_6 \times \mathbb{Z}_6 \times \mathbb{Z}_{18}.$$

16 **(a)** Prove that $\mathbb{Z}_{10}$ is isomorphic to the direct product of its primary components.

(b) Prove that if $p_1, \ldots, p_m$ are pairwise distinct primes, then the direct product

$$\mathbb{Z}_{p_1} \times \cdots \times \mathbb{Z}_{p_m}$$

is isomorphic to the cyclic group $\mathbb{Z}_{p_1 \cdots p_m}$.

17 **(a)** Find the primary components of the abelian group

$$G = \langle a, b, c \mid a^2 = b^2 = c^3 = 1, ab = ba, ac = ca, bc = cb \rangle$$

(b) Represent the abelian group

$$G = \langle a, b, c \mid a^3 = b^3 = c^5 = 1, ab = ba, ac = ca, bc = cb \rangle$$

as the direct product of cyclic groups with the order of each group dividing the order of the next one.

(c) Represent the abelian group

$$G = \langle a, b, c \mid a^2 = b^4 = c^5 = 1, ab = ba, ac = ca, bc = cb \rangle$$

as the direct product of cyclic groups with the order of each group dividing the order of the next one.

18 **(a)** Find the number of Archimedean components of the commutative semigroup by the following Cayley table.

	e_0	g_0	e_1	g_1	e_2	g_2
e_0	e_0	g_0	e_0	g_0	e_0	g_0
g_0	g_0	e_0	g_0	e_0	g_0	e_0
e_1	e_0	g_0	e_1	g_1	e_1	g_1
g_1	g_0	e_0	g_1	e_1	g_1	e_1
e_2	e_0	g_0	e_1	g_1	e_2	g_2
g_2	g_0	e_0	g_1	e_1	g_2	e_2

(b) Find the semilattice decomposition of the semigroup in (a).

19 **(a)** Find the number of Archimedean components of the commutative semigroup with the following Cayley table.

	e_0	g_0	h_0	e_1	g_1	h_1
e_0	e_0	g_0	h_0	e_0	g_0	h_0
g_0	g_0	h_0	e_0	g_0	h_0	e_0
h_0	h_0	e_0	g_0	h_0	e_0	g_0
e_1	e_0	g_0	h_0	e_1	g_1	h_1
g_1	g_0	h_0	e_0	g_1	h_1	e_1
h_1	h_0	e_0	g_0	h_1	e_1	g_1

(b) Find the semilattice decomposition of the semigroup in (a).

20 **(a)** Let $\mathbb{Z}_6$ be the cyclic group with identity e and generator g. Draw the Cayley graph $\text{Cay}(G,T)$, where $T = \{g, g^3\}$.

(b) Prove that the Cayley graph $\text{Cay}(G,T)$ of a group G is undirected if and only if $x \in T$ implies $x^{-1} \in T$.

21 **(a)** In terms of the index m and the period n of a finite monogenic semigroup S, write down conditions necessary and sufficient for S to be a group.

(b) In terms of the index m and the period n, write down conditions necessary and sufficient for a finite monogenic semigroup to be nilpotent.

22 Find the semilattice decomposition of the multiplicative semigroup consisting of all integers of the form $2^a 3^b$, where a, b are positive integers.

23 Find the semilattice decomposition of the band with the following Cayley table.

	e_0	g_0	h_0	e_1	g_1	h_1
e_0	e_0	e_0	e_0	e_0	e_0	e_0
g_0	g_0	g_0	g_0	g_0	g_0	g_0
h_0	h_0	h_0	h_0	h_0	h_0	h_0
e_1	e_0	e_0	e_0	e_1	e_1	e_1
g_1	g_0	g_0	g_0	g_1	g_1	g_1
h_1	h_0	h_0	h_0	h_1	h_1	h_1

24 Find the semilattice decomposition of the following two bands.

	e_0	g_0	e_1	g_1	e_2	g_2
e_0	e_0	e_0	e_0	e_0	e_0	g_0
g_0	g_0	g_0	g_0	g_0	g_0	g_0
e_1	e_0	e_0	e_1	e_1	e_0	e_0
g_1	g_0	g_0	g_1	g_1	g_0	g_0
e_2	e_0	e_0	e_0	e_0	e_2	e_2
g_2	g_0	g_0	g_0	g_0	g_2	g_2

	e_0	g_0	e_1	g_1	e_2	g_2
e_0	e_0	g_0	e_0	g_0	e_0	g_0
g_0	e_0	g_0	e_0	g_0	e_0	g_0
e_1	e_0	g_0	e_1	g_1	e_0	g_0
g_1	e_0	g_0	e_1	g_1	e_0	g_0
e_2	e_0	g_0	e_0	g_0	e_2	g_2
g_2	e_0	g_0	e_0	g_0	e_2	g_2

25 Find the semilattice decomposition of the following semigroups which are unions of groups.

	e_0	g_0	e_1	g_1	e_2	g_2
e_0	e_0	g_0	e_0	g_0	e_0	g_0
g_0	g_0	e_0	g_0	e_0	g_0	e_0
e_1	e_0	g_0	e_1	g_1	e_1	g_1
g_1	g_0	e_0	g_1	e_1	g_1	e_1
e_2	e_0	g_0	e_1	g_1	e_2	g_2
g_2	g_0	e_0	g_1	e_1	g_2	e_2

	e_0	g_0	e_1	g_1	e_2	g_2
e_0	e_0	g_0	e_0	g_0	e_0	g_0
g_0	g_0	e_0	g_0	e_0	g_0	e_0
e_1	e_0	g_0	e_1	g_1	e_0	g_0
g_1	g_0	e_0	g_1	e_1	g_0	e_0
e_2	e_0	g_0	e_0	g_0	e_2	g_2
g_2	g_0	e_0	g_0	e_0	g_2	e_2

Chapter 3

Automata and Languages

3.1 Finite State Automata

Finite automata are well-known tools of language theory. Every finite state automaton accepts or recognizes a certain language. Before a formal definition, let's take a look at the following example of Australian public pay-telephone illustrated as a finite state automaton.

The telephone accepts 5c, 10c and 20c coins, and allows a call to be made for 40c. It has 9 states, q_0, q_5, q_{10}, q_{15}, q_{20}, q_{25}, q_{30}, q_{35}, q_{40}, where the subscript indicates the amount of money received. The state q_0, when the customer begins to insert coins, is called the *initial* state (or *start* state); the state q_{40}, when the phone allows a call, is the *terminal* state. The telephone *accepts* or *recognizes* any sequence of coins that adds up to precisely 40c. It can be described pictorially with a transition diagram displayed in Figure 3.1.

The *language* accepted by the machine is the following set of input sequences:

$$\{(20,20),(20,10,10),(20,10,5,5),(20,5,10,5),(20,5,5,10),(20,5,5,5,5),$$

$$(10,20,10),(10,20,5,5),(10,10,20),(10,10,10,10),$$

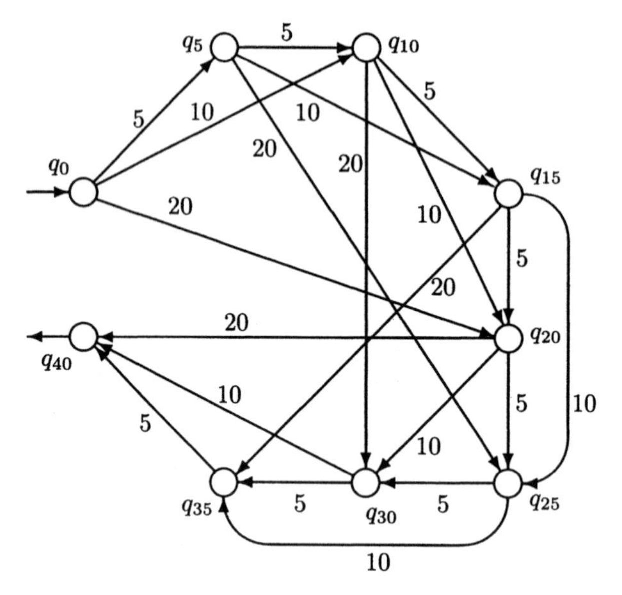

Figure 3.1: Australian pay-phone

$$(10,10,10,5,5),(10,10,5,10,5),(10,10,5,5,10),(10,10,5,5,5,5),$$

$$(10,5,10,10,5),(10,5,10,5,10),(10,5,5,10,10),$$

$$\ldots,$$

$$(5,5,5,5,20),(5,5,5,5,10,10),(5,5,5,5,10,5,5),$$

$$(5,5,5,5,5,10,5),(5,5,5,5,5,5,10),(5,5,5,5,5,5,5,5)\}$$

which correspond to walks from the initial to the terminal state.

A similar coin telephone accepting dimes and quarters and making a call for \$1 is shown in Figure 3.2.

A *finite state automaton* or *acceptor* is a 5-tuple (Q, X, δ, q_0, T), where

- Q is a set of states;
- X is the input alphabet;

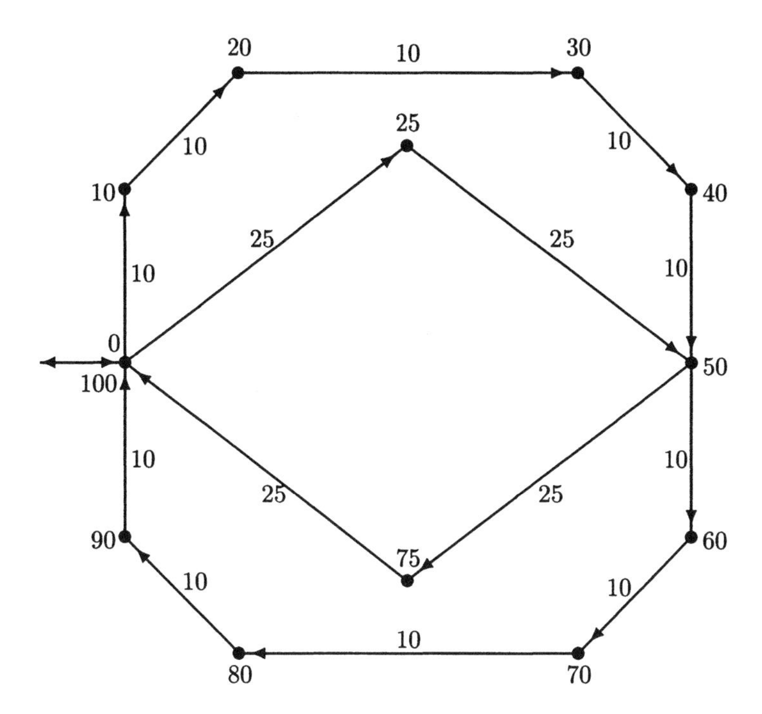

Figure 3.2: Simplified coin telephone

- δ is the next-state function or transition function;
- q_0 is the start state or initial state;
- T is the set of terminal states or final states.

Transition table stores all values of the transition function.

δ	x_1	$\dots$	x_n
q_1	$\delta(q_1, x_1)$	$\dots$	$\delta(q_1, x_n)$
$\vdots$	$\vdots$		$\vdots$
q_n	$\delta(q_n, x_1)$	$\dots$	$\delta(q_n, x_n)$

For example, the transition table of the automaton described in Figure 3.1 is

	5	10	20
q_0	q_5	q_{10}	q_{20}
q_5	q_{10}	q_{15}	q_{25}
q_{10}	q_{15}	q_{20}	q_{30}
q_{15}	q_{20}	q_{25}	q_{30}
q_{20}	q_{25}	q_{30}	q_{35}
q_{25}	q_{30}	q_{35}	q_{40}
q_{30}	q_{35}	q_{40}	
q_{35}	q_{40}		
q_{40}			

The missing values in the table show where the transition function is undefined.

Transition diagram is a directed graph associated with a finite state automaton. The vertices of the graph correspond to the states of the automaton. If there is a transition from state s to state t on input x, then we draw an arc labeled x from state s to state t in the transition diagram.

The acronym FSA is often used for the term 'finite state automaton'. If the transition function is defined for all states and all input symbols, that is, if $\delta(q, x)$ is a state for all $q \in Q$, $x \in X$, then the automaton is a *complete deterministic automaton.* The following two more general concepts can be converted to complete deterministic finite state automata and simplify the task of constructing FSA. If $\delta(q, x)$ is undefined for some q and some x, then the automaton is said to be *incomplete.* If $\delta(q, x)$, has several values, that is if δ is not a function, but a relation, then we say that the automaton is *nondeterministic.*

Extended transition function. Inductively, we can extend δ to a transition function δ^* from $Q \times X^*$ to Q. The extended function acts on words. Formally, we define

(1) $\delta^*(q, 1) = q$ for all $q \in Q$;

(2) $\delta^*(q, wa) = \delta(\delta^*(q, w), a)$ for all $q \in Q$, $w \in X^*$, $x \in X$.

Thus (1) says that without reading an input symbol the finite automaton cannot change state, and (2) tells us how to find the state after reading a nonempty input word wx. That is, first we find the state $p = \delta^*(q, w)$ after reading w, and then compute the state $\delta(p, x)$.

We can rewrite (2) as $\delta^*(q, x_1 x_2 \ldots x_k) = \delta[(\ldots\delta(\delta(q, x_1), x_2)\ldots), x_k]$ for all $x_1, x_2, \ldots, x_k \in X$.

We use the same symbol δ also for δ^*, since they coincide where they are both defined. Thus $\delta(q, w)$ is the state achieved by the automaton if, being in state q, it reads the word w. Notation $q \cdot w$ or qw is also used for $\delta(q, w)$.

Example 3.1. Suppose that the transition function δ of the automaton $\mathcal{A} = (Q, X, \delta, q_0)$, where $Q = \{q_0, q_1, q_3\}$ and $X = \{x, y\}$, is given by the table

δ	x	y
q_0	q_0	q_1
q_1	q_2	q_1
q_2	q_1	q_0

Let us find $\delta(q_0, w)$ for $w = xyyx$. If in the start state q_0 the automaton reads the word $xyyx$, then we get

$$\begin{aligned}
\delta(q_0, xyyx) &= \delta(\delta(q_0, x), yyx) \\
&= \delta(q_0, yyx) \\
&= \delta(\delta(q_0, y), yx) \\
&= \delta(q_1, yx) \\
&= \delta(\delta(q_1, y), x) \\
&= \delta(q_1, x) \\
&= q_2.
\end{aligned}$$

3.2 Languages Accepted by Automata

The *language accepted by an automaton* is the set of all words accepted or recognized by the automaton. A word w is said to be *accepted* by an automaton $U = (Q, X, \delta, q_0, T)$ if $\delta(q_0, w) \in T$. The language accepted by the automaton U is therefore the set

$$L(U) = \{w \mid w \in X^*, \delta(q_0, w) \in T\}.$$

The automaton may be represented as a finite control device reading letters of a finite word recorder on a tape and moving from left to right only in one direction.

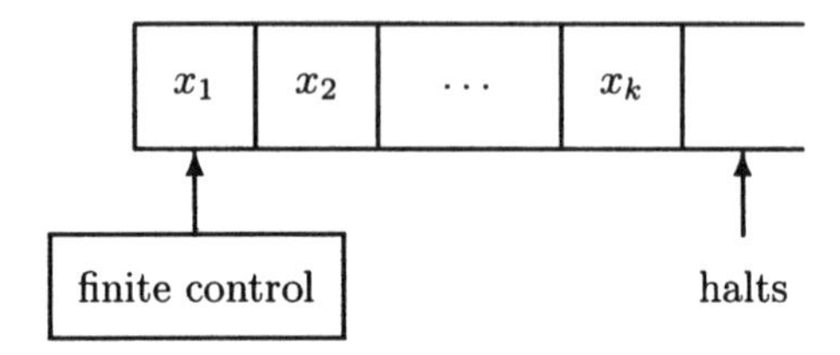

Then $L(U)$ is the set of words such that U halts in a terminal state of the control. If the automaton U is represented by its state diagram, then $L(U)$ is the set of words labeling all walks from q_0 to terminal states.

Definition 3.1. A language is said to be *recognizable* if it is accepted by a finite state automaton.

Example 3.2. Consider the following automaton F that recognizes unsigned fixed point binary numbers. We define these as nonempty words of digits $0, 1$ optionally followed by a decimal point and another nonempty word of digits. An integer cannot end with a decimal point and a number between 0 and 1 must have exactly one 0 before its decimal point. The states of F are as follows:

q_0 — initial state;

q_1 — digits beginning with a nonzero digit have been read;

q_2 — the first digit is zero;

q_3 — the decimal point has been read;

q_4 — terminal state.

The state diagram of F is

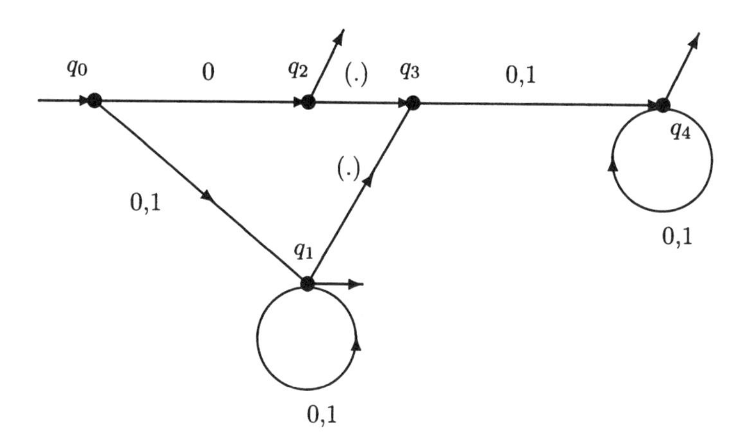

Clearly, q_1 and q_2 are terminal states, as well as q_4. This is an incomplete automaton. We could complete it by adding a state for all errors; in this way all incomplete automata can be made complete.

Lemma 3.1. (Pumping Lemma) ([195], §3.2) *For each recognizable language L, there exists a positive integer $n = n(L)$ such that if z is an arbitrary word in L and $|z| \geq n$, then we can always write z as $z = uvw$ so that $|uv| \leq n$, $|v| \geq 1$, and $uv^k w \in L$ for all $k \in \mathbb{N}$. Besides, $n(L)$ does not exceed the number of states of the smallest finite state automaton accepting L.*

Example 3.3. Lemma 3.1 shows that the language

$$L = \{x^{k^2} \mid k \in \mathbb{N}\}$$

is not recognizable.

3.3 Operations on Languages, Regular Expressions

The languages accepted by finite state automata are easily described by simple expressions called regular expressions.

Operations on languages. Let X be a finite set of symbols, and let L_1 and L_2 be sets of words from X^*. The *concatenation* of L_1 and L_2 is the set

$$L_1L_2 = \{xy \mid x \in L_1, y \in L_2\}.$$

This means that the words in L_1L_2 are formed by choosing a word in L_1 and following it by a word in L_2, in all possible combinations.

Let L be a set of words over X. Denote $L^0 = \{1\}$ and $L^i = LL^{i-1}$ for $i \geq 1$. The *Kleene's closure* of L is the set

$$L^* = \bigcup_{i=0}^{\infty} L^i,$$

and the *positive closure* of L is the set

$$L^+ = \bigcup_{i=1}^{\infty} L^i.$$

Thus L^* and L^+ are the submonoid and subsemigroup generated by L in X^*.

Regular expressions tell us how a language is obtained from singleton sets $\{x\}$ and the empty set $\emptyset$ using regular operations $+$, $\cdot$ and $*$. In regular expressions, letters denote singleton sets, $+$ stands for union and $*$ for Kleene's $*$-operation. The following examples illustrate the conventions used to denote finite sets in regular expressions: $\{x\} = x$, $\{x, y\} = x \cup y$. Examples are also given in Table 3.1.

Example 3.4. Let $X = \{x, y\}$. Then $(xy)^*$, $(x^* \cup y^*) \cdot \{x, y\}^*$, $\{xy, yx\}^*$ can be described by regular expressions. Similarly, $\{x, y\}^+$ is defined by a regular expression, because $\{x, y\}^+ = x\{x, y\}^* \cup y\{x, y\}^*$.

Regular Expression	**Language**
$x + y + z + t$	$\{x, y, z, t\}$
$x(y + z)$	$\{xy, xz\}$
x^*	$\{1, x, xx, xxx, \ldots\}$
$(x + y)^* z$	$\{z, xz, yz, xxz, xyz, yxz, yyz, \ldots\}$

Table 3.1: Examples of regular expressions.

Definition 3.2. Let X be an alphabet. Regular expressions and their sets are defined recursively by the rules:

(1) $\emptyset$ is a regular expression and denotes the empty set;

(2) 1 is a regular expression for $\{1\}$;

(3) For each x in X, x is a regular expression for $\{x\}$;

(4) if r and s are regular expressions denoting the languages R and S, respectively, then $(r + s)$, rs, and (r^*) are regular expressions that denote the sets $R \cup S$, RS and R^*, respectively.

Definition 3.3. A language is said to be *regular* if it can be defined with a regular expression.

As a practical example of the use of regular expressions, we mention that they have been implemented in various modern programming languages and computer algebra systems (often with additional adjustments and modifications). Besides, several operating systems, screen editors and word processing programs have can not only find a particular word, but also find a word that matches a specified regular expression. Best known of these search tools is the Unix *grep* command which stands for "get regular expression".

Example 3.5. The expression $(x+y)^*$ stands for the set of all words in letters x and y. Similarly, $x(x+y)^*y$ denotes the set of all words in x and y that begin with x and end with y.

Order of operations. In writing regular expressions we can omit many parentheses if we assume that $*$ has higher precedence than concatenation or $+$, and that concatenation has higher precedence than $+$. For example, $(x(y^*))+x$ may be written as xy^*+x, and the expression rr^* is abbreviated as r^+.

When necessary to distinguish between a regular expression r and the language denoted by r, we use $L(r)$ for the latter. When there is no confusion, we use r for both the regular expression and its language.

Example 3.6. (1) The order of operations in $x+yz^*t$ can be illustrated with brackets $x+((y(z^*))t)$.

(2) all C++ identifiers are defined by the regular expression

$$(A+B+\ldots+Z+a+b+\ldots+z)\times$$
$$(A+B+\ldots+Z+a+b+\ldots+z+0+1+\ldots+9+\_)^*$$

Exercise 3.1. Write down the regular expressions defining the following languages.

(a) The set of all words in x and y.

(b) The set of all words over the alphabet $\{x, y\}$ all of which contain the subword xx.

(c) The set of all words over $\{x, y\}$ with at least one letter y in every word.

(d) The set of all words over $\{x, y\}$ containing at most one letter y.

(e) The set of all words over $\{x, y\}$ ending in xyy.

(f) The set of all words consisting of arbitrarily many x's followed by an arbitrary number of y's followed by arbitrarily many z's.

(g) The set of all words consisting of arbitrarily many (but at least one) x's followed by arbitrarily many (but at least one) y's followed by arbitrarily many (but at least one) z's.

Exercise 3.2. Describe the languages defined by the following regular expressions:

(a) $x + y + z$;

(b) $xyzx$;

(c) $(x + y + z)^*$;

(d) $(x^*y^*)^*$;

(e) $(x^*yx^*y)^*x^*$;

(f) $1^* = \epsilon^* = \lambda^*$;

(g) $\emptyset^*$.

Theorem 3.1. (Kleene's Theorem) ([195], §3.2) *Let X be a finite alphabet, and let $L \subseteq X^*$. Then L is recognized by FSA iff L can be defined by a regular expression.*

Thus, a language in X^* is recognized by FSA if and only if it can be obtained from finite subsets of X^* by a finite number of applications of $\cup$ (union), $\cdot$ (multiplication) and $*$ (Kleene's closure, Kleene's $*$-operation).

The intersection of several regular languages is also regular, and the complement of a regular language is regular. The easiest way to prove these facts is to use finite state automata recognizing regular languages.

3.4 Algorithm 1: Language Accepted by FSA

This section contains first algorithm which, given a finite state automaton $\mathcal{A} = (Q, X, \delta, q_0, T)$, finds a regular expression for the language $L(\mathcal{A})$ accepted by $\mathcal{A}$. Let us begin by introducing notation required for the algorithm. The set of labels of sequences of edges in all walks from p to q is denoted by

$$L(p, q) = \{w \in X^* \mid pw = q\}. \tag{3.1}$$

For a subset R of Q, the set of words labeling walks from p to q with all vertices visited by the walk belonging to R is denoted by

$$L(p, R, q) = \{w \in X^* \mid pw = q,\ pz \in R \text{ for all prefixes } z \text{ of } w\}. \tag{3.2}$$

A walk from p to q that does not visit p, q on the way will be called a *proper walk* from p to q. For $V \subseteq Q$, $p, q \notin V$, the set of words in X^* labeling edges of proper walks from p to q with all vertices visited by the walk belonging to $V \cup \{p, q\}$ is denoted by

$$Z(p, V, q) = \{w \in X^* \mid pw = q,\ pz \in V \text{ for all proper prefixes } z \text{ of } w\}. \tag{3.3}$$

The set of letters labeling edges from p to q is designated by

$$X(p, q) = \{a \in X \mid pa = q\}. \tag{3.4}$$

Algorithm 1. To find a regular expression for the language accepted by the automaton $\mathcal{A} = (Q, X, \delta, q_0, T)$, reduce $L(\mathcal{A})$ to simpler languages using the following reduction formulas, where $p, q \in Q$, $p \neq q$:

$$L(\mathcal{A}) = \bigcup_{t \in T} L(q_0, t), \tag{3.5}$$

$$L(p, R, q) = [Z(p, R \backslash \{p\}, p)]^* \cdot Z(p, R \backslash \{p, q\}, q) \cdot L(q, R \backslash \{p\}, q), \tag{3.6}$$

$$L(p, R, p) = [Z(p, R \backslash \{p\}, p)]^*, \tag{3.7}$$

$$Z(p, V, q) = X(p, q) \cup \bigcup_{r, s \in V} [X(p, r) \cdot L(r, V, s) \cdot X(s, q)]. \tag{3.8}$$

Exercise 3.3. Use reduction formulas to find regular expressions for the languages recognized by finite state automata with the following transition diagrams.

(i) $\mathcal{A}_1$:

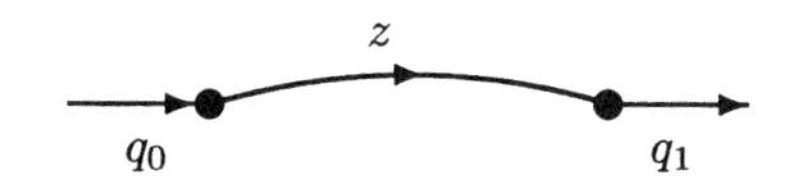

(ii) $\mathcal{A}_2$:

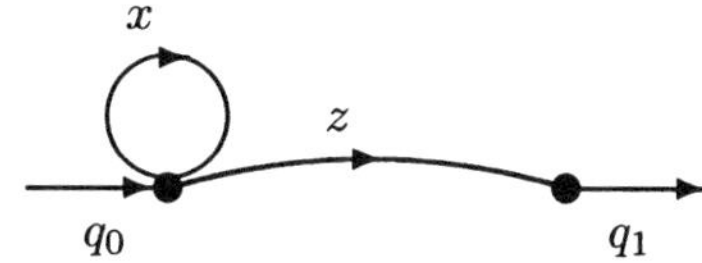

(iii) $\mathcal{A}_3$:

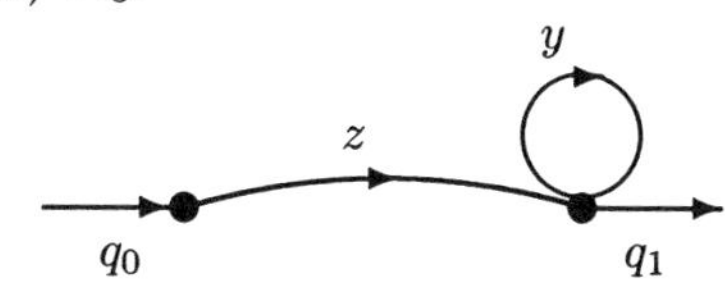

(iv) $\mathcal{A}_4$:

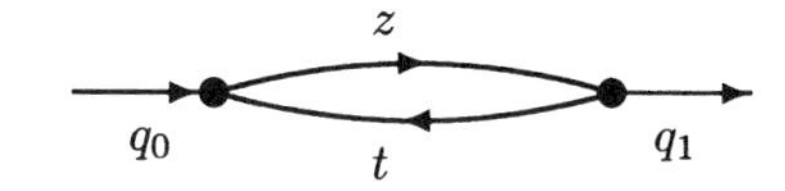

(v) $\mathcal{A}_5$:

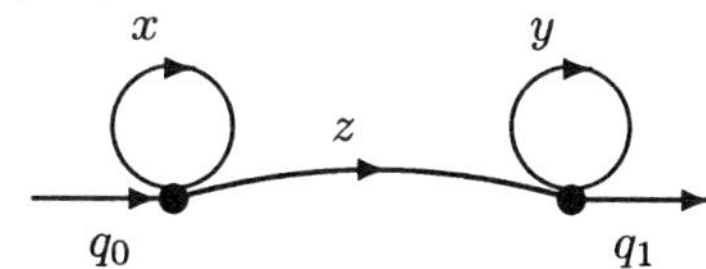

(vi) $\mathcal{A}_6$:

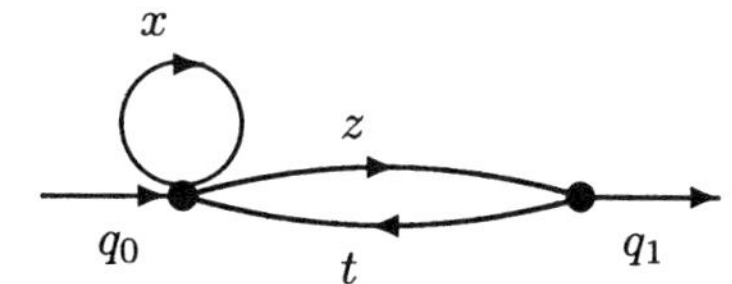

(vii) $\mathcal{A}_7$:

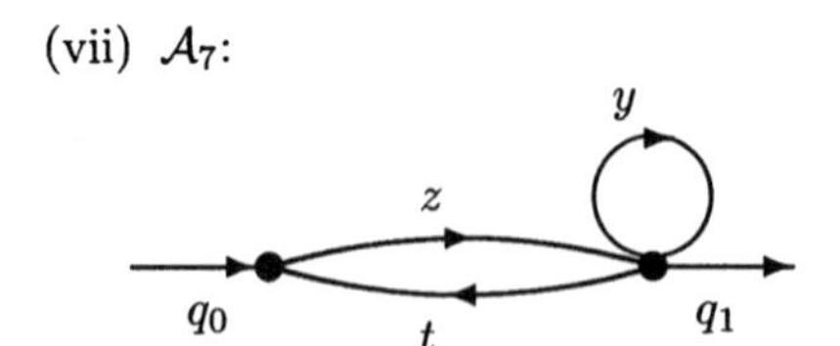

(viii) $\mathcal{A}_8$:

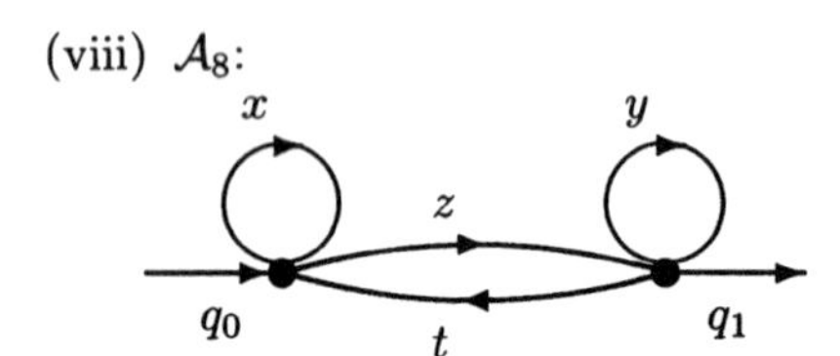

Example 3.7. Find the language recognized by the following automaton.

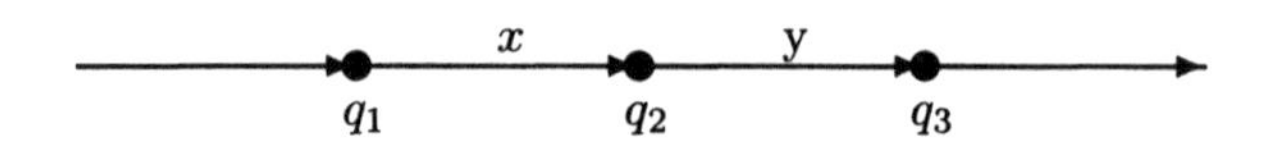

Solution. $L(\mathcal{A}) = L(q_1, q_3)$, and we get

$$
\begin{aligned}
L(q_1, q_3) &= L(q_1, \{q_1, q_2, q_3\}, q_3) \\
L(q_1, \{q_1, q_2, q_3\}, q_3) &= [Z(q_1, \{q_2, q_3\}, q_1)]^* \cdot Z(q_1, \{q_2\}, q_3) \cdot \\
&\quad \cdot L(q_3, \{q_2, q_3\}, q_3),
\end{aligned}
$$

$$
\begin{aligned}
Z(q_1, \{q_2, q_3\}, q_1) &= X(q_1, q_2) \cdot L(q_2, \{q_2, q_3\}, q_2) \cdot X(q_2, q_1) \\
&\quad \cup X(q_1, q_2) \cdot L(q_2, \{q_2, q_3\}, q_3) \cdot X(q_3, q_1) \\
&\quad \cup X(q_1, q_3) \cdot L(q_3, \{q_2, q_3\}, q_2) \cdot X(q_2, q_1) \\
&\quad \cup X(q_1, q_3) \cdot L(q_3, \{q_2, q_3\}, q_3) \cdot X(q_3, q_1) \\
&\quad \cup X(q_1, q_1) \\
&= 1,
\end{aligned}
$$

$$
\begin{aligned}
Z(q_1, \{q_2\}, q_3) &= X(q_1, q_2) \cdot L(q_2, \{q_2\}, q_2) \cdot X(q_2, q_3) \\
&\quad \cup X(q_1, q_3) \\
&= x1y \cup \emptyset \\
&= xy,
\end{aligned}
$$

$$L(q_3, \{q_2, q_3\}, q_3) = [Z(q_3, \{q_2\}, q_3)]^*,$$

$$\begin{aligned} Z(q_3, \{q_2\}, q_3) &= X(q_3, q_2) \cdot L(q_2, \{q_2\}, q_2) \cdot X(q_2, q_3) \\ &\quad \cup X(q_3, q_3) \\ &= \emptyset. \end{aligned}$$

Therefore $L(\mathcal{A}) = \emptyset^* \cdot xy \cdot \emptyset^* = xy$. □

Example 3.8. Describe the language recognized by the following automaton.

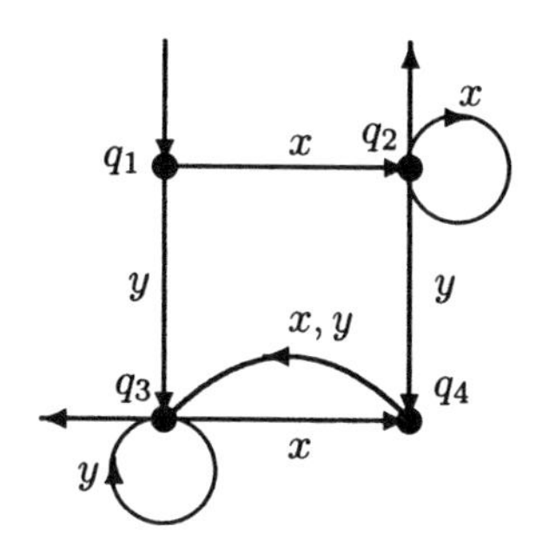

Solution. $L(\mathcal{A}) = L(q_1, q_2) \cup L(q_1, q_3)$,

$$\begin{aligned} L(q_1, q_2) &= L(q_1, \{q_1, q_2, q_3, q_4\}, q_2) \\ &= [Z(q_1, \{q_2, q_3, q_4\}, q_1)]^* \cdot Z(q_1, \{q_3, q_4\}, q_2) \cdot L(q_2, \{q_2, q_3, q_4\}, q_2), \end{aligned}$$

$$\begin{aligned} [Z(q_1, \{q_2, q_3, q_4\}, q_1)]^* &= 1^* = 1, \\ Z(q_1, \{q_3, q_4\}, q_2) &= \{x\}, \\ L(q_2, \{q_2, q_3, q_4\}, q_2) &= [Z(q_2, \{q_3, q_4\}, q_2)]^* = x^*. \end{aligned}$$

Therefore $L(q_1, q_2) = 1 \cdot x \cdot x^* = x^+$.

$$\begin{aligned} L(q_1, q_3) &= L(q_1, \{q_1, q_2, q_3, q_4\}, q_3) \\ &= [Z(q_1, \{q_2, q_3, q_4\}, q_1)]^* \cdot Z(q_1, \{q_2, q_4\}, q_3) \cdot L(q_3, \{q_2, q_3, q_4\}, q_3) \\ &= 1^* \cdot y \cdot [y + x(x + y)]^* \end{aligned}$$

Thus $L(\mathcal{A}) = x^+ \cup y[y + xx + xy]^*$. □

Example 3.9. Consider the automaton M with initial state q_0, terminal states $T = \{q_2\}$ and transition function δ given by

δ	x	y
q_0	q_1	q_0
q_1	q_3	q_2
q_2	q_1	q_1
q_3	q_3	q_2

The state diagram is

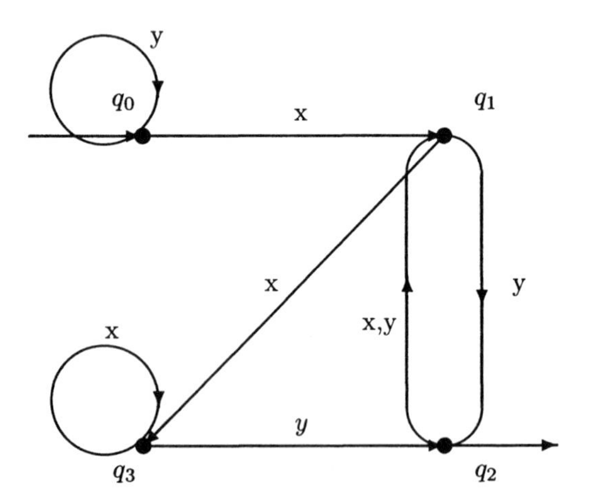

Every walk from q_0 to q_2 can be divided into three parts: a walk from q_0 to the last occurrence of q_0, then a walk to the first occurrence of q_2, and then a few walks from q_2 to q_2. Therefore the language accepted by the automaton is

$$L(M) = y^*x(y + xx^*y)((x + y)(y + xx^*y))^*.$$

3.5 Constructing Automata for Languages

We have already learnt an algorithm which proves that every language accepted by a finite automaton can be defined with a regular expression. In the next few sections we are going to complete the proof of Kleene's Theorem. All what remains to verify is that every language defined by a regular expression is accepted by FSA. References concerning three main methods of transforming regular expressions into finite automata can be found in [255], §3.2.

However, first we need to introduce a few concepts that make this verification easier. Namely, the more general concepts of nondeterministic automata, incomplete automata, and automata with ε-transitions help to construct FSA.

Nondeterministic automata are more convenient when we need to construct an automaton recognizing a given language, because it is easier to start by defining a more flexible nondeterministic automaton and then use the standard step of replacing is with an equivalent deterministic one.

Definition 3.4. If the transition function δ is from $Q \times X$ to $\mathcal{P}(Q)$, the set of all subsets of Q, then we say that the automaton

$$(Q, X, \delta, q_0, T)$$

is *nondeterministic*. This means that the next state qx is ambiguous; it can be chosen from the set $\delta(q, a)$. The language recognized by a nondeterministic automaton consists of all words such that there exists at least one walk labeled by the word ending in a terminal state. A nondeterministic automaton (Q, X, δ, q_0, T) *accepts* or *recognizes* a word w if $T \cap q_0 w \neq \emptyset$.

Example 3.10. Consider the automaton $U(Q, X, \delta, i, T)$, where $Q = \{q_0, q_1\}$, $X = \{x, y\}$, q_1, $T = \{q_2\}$ and δ is given by

δ	x	y
q_0	$\{q_0, q_1\}$	q_1
q_1	$\emptyset$	$\{q_0, q_1\}$

The state diagram is in Figure 3.3.

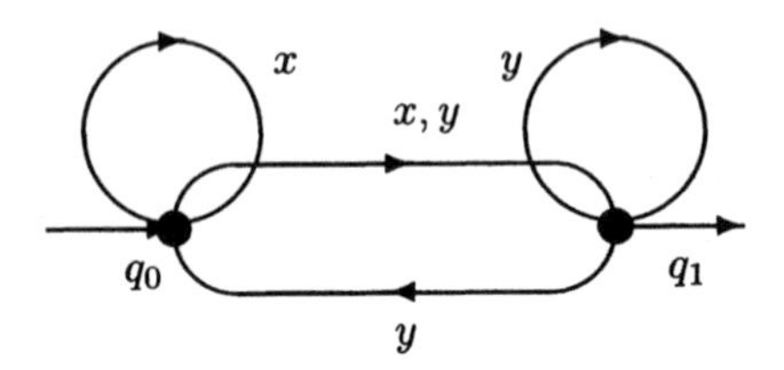

Figure 3.3: Automaton of Example 3.10.

We can form a deterministic automaton U' accepting the same language as the nondeterministic automaton U above. The states of U' are all subsets of Q. The transition function is defined by $\psi(P, a) = \bigcup_{q \in P} \delta(q, a)$ where $P \in \mathcal{P}(Q)$. It is shown in Figure 3.4.

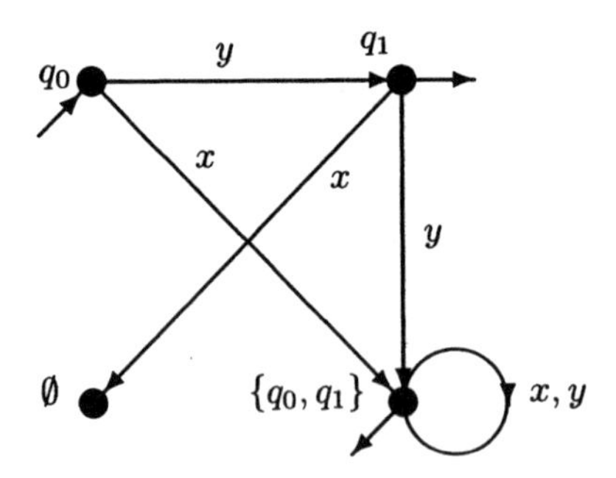

Figure 3.4: Automaton of Example 3.10.

The terminal states are the sets containing at least one terminal element, i.e.,

$$T' = \{\{q_1\}, \{q_0, q_1\}\}.$$

The initial state $i' = q_0$. We can complete the automaton by adding one more state that collects all missing edges, or use the nonterminal state of $\emptyset$ for collecting these edges.

Theorem 3.2. ([255], Lemma 2.2) *A language is accepted by an incomplete nondeterministic automaton if and only if it is accepted by a complete* *deterministic automaton.*

It is often easier to construct an automaton for required task with the use of ε*-transitions*. They are depicted in transition diagram as edges labeled by ε. The assumption is that the automaton can change state as indicated by the ε-transitions without reading any input, that is reading only the empty word ε as an input.

Every automaton with ε-transitions can be replaced by an ordinary nondeterministic automaton. To find it, we eliminate all ε-transitions by repeating the following steps.

(1) Choose an ε-transition (q_1, q_2) and delete it.

(2) If $q_2 \in T$, then make the state q_1 terminal too by adding it to Q if it is not there yet.

(3) For each edge (q_2, q_i) with a label $x_i \in X$ and $q_i \in Q$, introduce a new edge (q_1, q_i) with the same label x_i that replaces the combined effect of the edges (q_1, q_2) and (q_2, q_i).

Let X be a finite alphabet. We are going to verify that, for every word w in X^+, the language $\{w\}$ is recognizable.

Indeed, suppose that $w = x_1x_2 \ldots x_n$, where $x_1, x_2, \ldots, x_n \in X$. Then the incomplete deterministic automaton given in Figure 3.5 recognizes the language $\{w\}$.

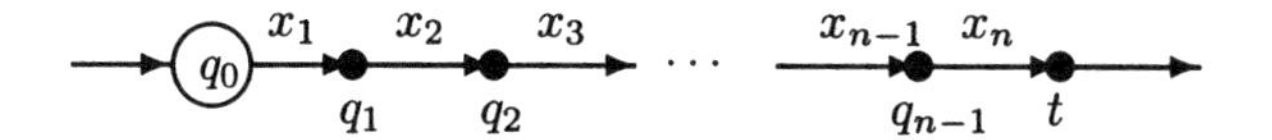

Figure 3.5: FSA recognizing $\{w\}$.

If $w = 1$, then $\{1\}$ is accepted by the automaton in Figure 3.6. $T = \{i\}$; $ia = 0a = 0$ for all $x \in X$.

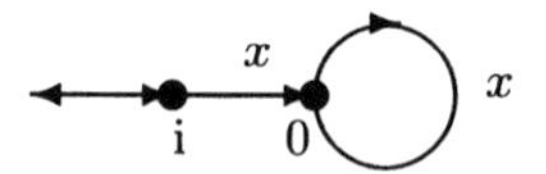

Figure 3.6: FSA recognizing $\{1\}$.

The empty set is accepted by the following imcomplete finite state automaton displayed in Figure 3.7. Strictly speaking there is no need even to include the terminal state in this FSA. We can add one state and complete it as shown in Figure 3.8.

Figure 3.7: Incomplete FSA accepting $\emptyset$.

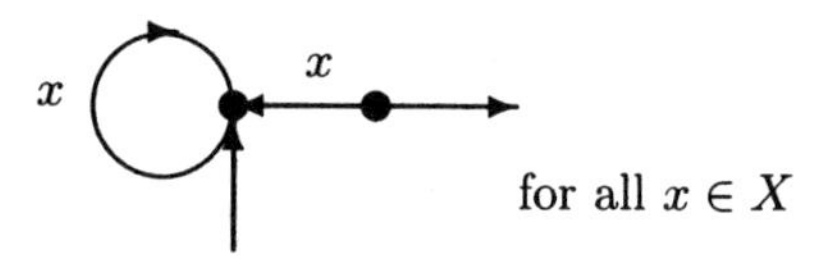

Figure 3.8: FSA accepting $\emptyset$.

Let X be a finite alphabet, and let L_1,L_2 be languages in X^* accepted by FSA. We claim that then there exists FSA accepting $L_1 \cup L_2$.

Indeed, suppose that $L_1 = L(\mathcal{A}_1)$, $L_2 = L(\mathcal{A}_2)$, for the automata $\mathcal{A}_1 = (Q_1, X, \delta_1, i_1, T_1)$ and $\mathcal{A}_2 = (Q_2, X, \delta_2, i_2, T_2)$. Define a new automaton $\mathcal{A} = (Q, X, \delta, i, T)$, where

$$Q = Q_1 \times Q_2,$$
$$\delta((q_1, q_2), a) = (\delta_1(q_1, a), \delta_2(q_2, a)),$$
$$i = (i_1, i_2) \text{ and}$$
$$T = (T_1 \times Q_2 \cup Q_1 \times T_2).$$

If $w \in L_1$, then $i_1 w \in T_1$, and so $(i_1, i_2)w = (i_1 w, i_2 w) \in T_1 \times Q_2 \subseteq T$. Similarly, if $w \in L_2$, then $(i_1, i_2)w \in Q_1 \times T_2 \subseteq T$. Therefore $L_1 \cup L_2 \subseteq L(\mathcal{A})$.

Conversely, if $w \in L(\mathcal{A})$, then $i_1 w \in T_1$ or $i_2 w \in T_2$, and so $w \in L_1 \cup L_2$. Thus $L(\mathcal{A}) = L_1 \cup L_2$.

Next notice that that all finite set are accepted by FSA. This easily follows from the three results above, because every nonempty finite set is the union of one-element sets.

Example 3.11. The automaton with transition diagram shown in Figure 3.9 recognizes the language

$$L = \{x^2y, xy^2, y^2, yxy\}.$$

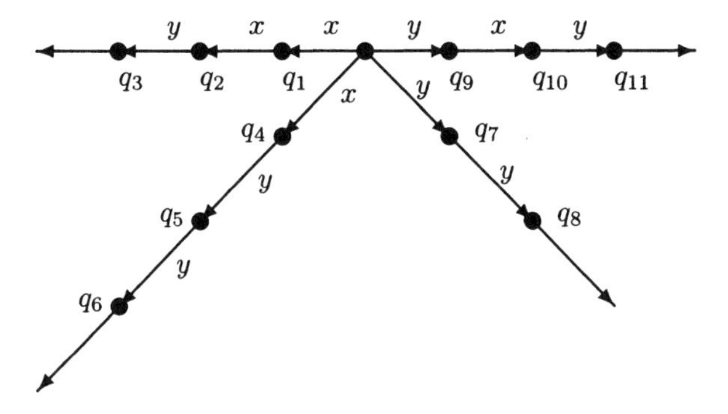

Figure 3.9: FSA recognizing $\{x^2y, xy^2, y^2, yxy\}$.

The set of terminal states is $T = \{q_3, q_6, q_8, q_{11}\}$.

Next, we are going to verify the following three facts:

- if L_1 and L_2 are accepted by FSA, then there exists FSA accepting $L_1 \cup L_2$;
- if L_1, L_2 are accepted by FSA, then there exists FSA accepting $L_1 L_2$;

- if L is accepted by FSA, then there exists FSA accepting L^*.

First, we discuss how, given two automata $\mathcal{A}_1$ and $\mathcal{A}_2$, we can construct a new automaton that recognizes $L(\mathcal{A}_1) \cup L(\mathcal{A}_2)$.

One method for solution of this problem has already been explained above. Here we briefly describe another method. We can identify the start states of $\mathcal{A}_1$ and $\mathcal{A}_2$ and form a nondeterministic automaton as shown in Figures 3.10 and 3.11. If we have two automata $\mathcal{A}_1$ and $\mathcal{A}_2$ as in Figure 3.10, then the new nondeterministic automaton recognizing the union $L(\mathcal{A}_1) \cup L(\mathcal{A}_2)$ is shown in Figure 3.11.

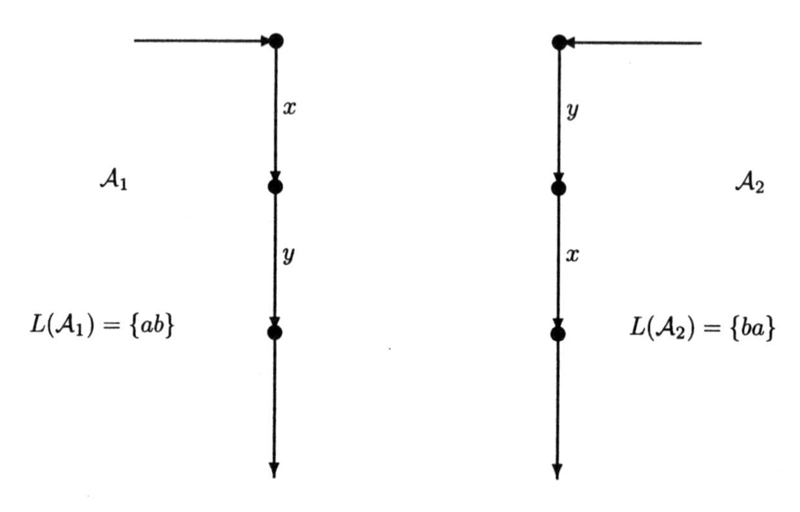

Figure 3.10: Automata $\mathcal{A}_1$ and $\mathcal{A}_2$.

Another example is given in Figures 3.12 and Figures 3.13. If we have two automata $\mathcal{A}_1$ and $\mathcal{A}_2$ as in Figure 3.12, then the new nondeterministic automaton recognizing the union $L(\mathcal{A}_1) \cup L(\mathcal{A}_2)$ is given in Figure 3.13.

Second, we show how, given two automata $\mathcal{A}_1$ and $\mathcal{A}_2$, we can construct a new automaton that recognizes concatenation $L(\mathcal{A}_1)L(\mathcal{A}_2)$.

Let us solve this problem for automata in Figure 3.14.

METHOD 1. We can construct a nondeterministic automaton by

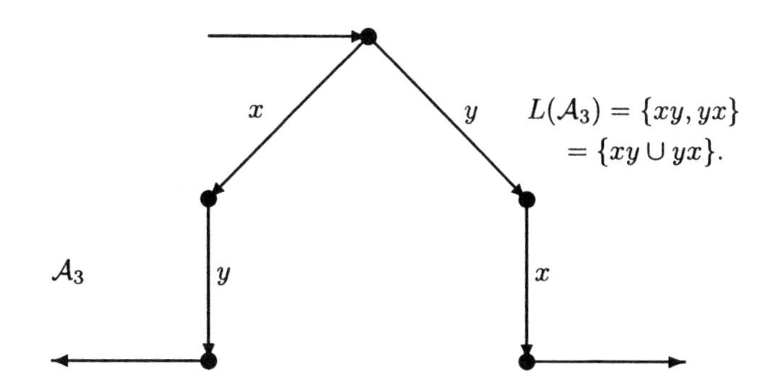

Figure 3.11: The automaton $\mathcal{A}_3$ recognizing $L(\mathcal{A}_1) \cup L(\mathcal{A}_2)$.

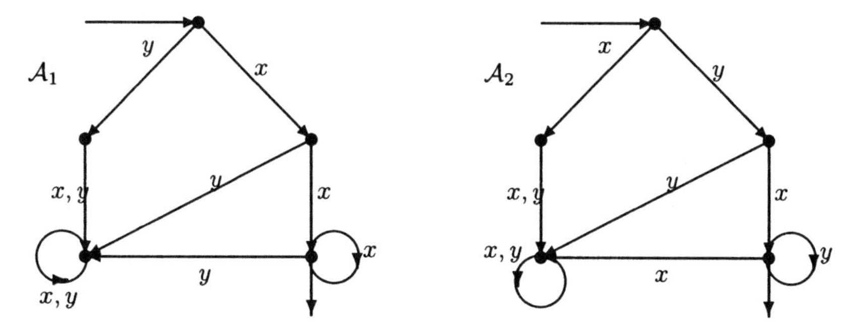

Figure 3.12: $L(\mathcal{A}_1) = xxx^*$ and $L(\mathcal{A}_2) = yxy^*$.

connecting $\mathcal{A}_1$ and $\mathcal{A}_2$ in series. If an arrow of $\mathcal{A}_1$ ends in a terminal state, then we add an arrow with the same label and start vertex, leading to the initial state of $\mathcal{A}_2$, as shown in Figure 3.15.

Example 3.12. Construct an automaton that recognizes the concatenation of the languages accepted by the automata $\mathcal{A}_1$ and $\mathcal{A}_2$ shown in Figure 3.16.

METHOD 2. As shown in Figure 3.17, we

(i) Delete the "terminal arrow" in $\mathcal{A}_1$ and the "initial arrow" in $\mathcal{A}_2$.

(ii) For each arrow $q_i q_j$ in $\mathcal{A}_1$ such that $q_j \in T$, add a new arrow $q_i i_2$, where i_2 is the initial state of $\mathcal{A}_2$.

Third, we show how, given an automaton $\mathcal{A}$, we can construct a new

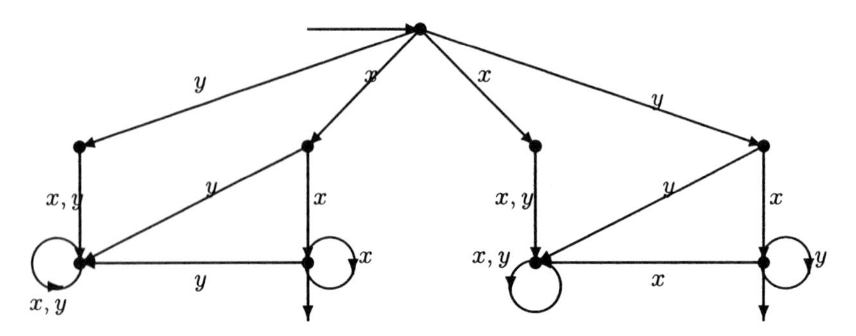

Figure 3.13: $L(\mathcal{A}_3) = xxx^* \cup yxy^*$.

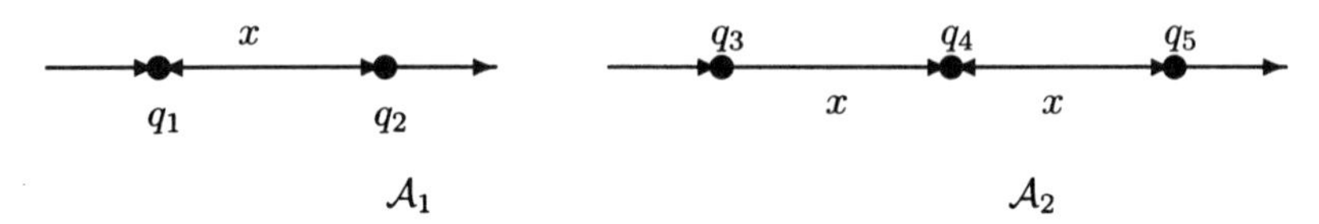

Figure 3.14: Automata $\mathcal{A}_1$ and $\mathcal{A}_2$.

FSA recognizing Kleene's closure $(L(\mathcal{A}))^+$.

This is done by connecting $\mathcal{A}$ to itself in a loop. More specifically, for each arrow leading to a terminal vertex, we add a new arrow with the same label leading to the initial vertex. This gives a nondeterministic automaton $\mathcal{A}^+$, with $L(\mathcal{A}^+) = (L(\mathcal{A}))^+$.

Example 3.13. Given the automaton $\mathcal{A}_1$ with transition diagram in Figure 3.18, it is shown in Figure 3.19 how to construct a nondeterministic automaton $\mathcal{A}_2$ recognizing the language $L(\mathcal{A})^+$.

Example 3.14. Construct deterministic automata that recognize the union and the intersection of languages recognized by the automata with transition diagrams in Figure 3.20. Solution is illustrated in Figure 3.21.

The following method can be used to solve this problem. Take the set of all pairs (q_1, r_1), (q_1, r_2), (q_1, r_3), (q_2, r_1), (q_2, r_2), (q_2, r_3) as states.

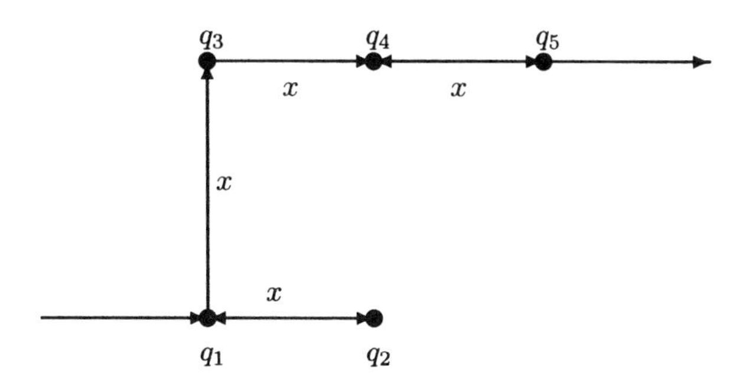

Figure 3.15: $L(\mathcal{A}_3) = L(\mathcal{A}_1)L(\mathcal{A}_2)$.

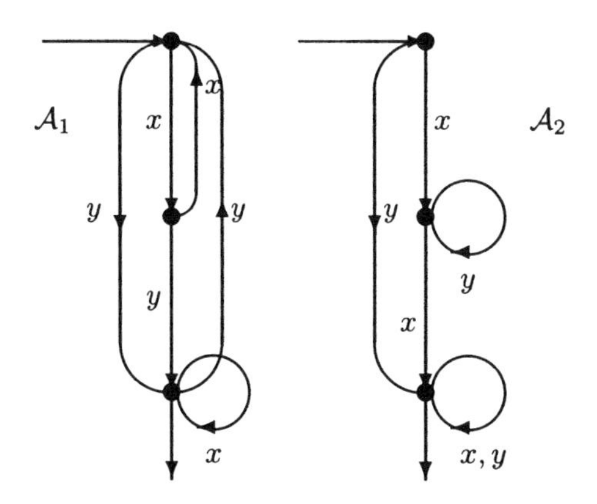

Figure 3.16: Automata of Example 3.12.

There is an arrow from $q_i r_j$ to $q_m r_n$ exactly when there is an arrow from q_i to q_m in $\mathcal{A}_1$ or one from r_j to r_j in $\mathcal{A}_2$.

(i) The automaton recognizing $L(\mathcal{A}_1) \cup L(\mathcal{A}_2)$.

$T = \{q_i r_j \mid q_i$ is a terminal state in $\mathcal{A}_1$ OR r_j is a terminal state in $\mathcal{A}_2\}$,

i.e., $T = \{q_1 r_2, q_2 r_1, q_2 r_2, q_2 r_3\}$

(ii) The automaton recognizing $L(\mathcal{A}_1) \cap L(\mathcal{A}_2)$.

$T = \{q_i r_j \mid q_i$ is a terminal state in $\mathcal{A}_1$ AND r_j is a terminal state in $\mathcal{A}_2\}$.

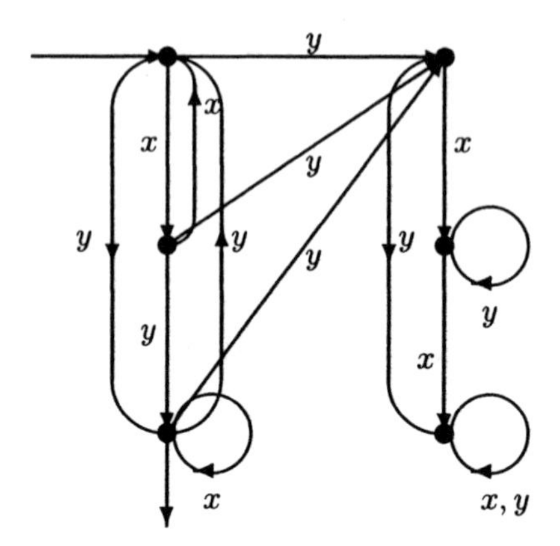

Figure 3.17: Example 3.12: $L(\mathcal{A}_3) = L(\mathcal{A}_1)L(\mathcal{A}_2)$.

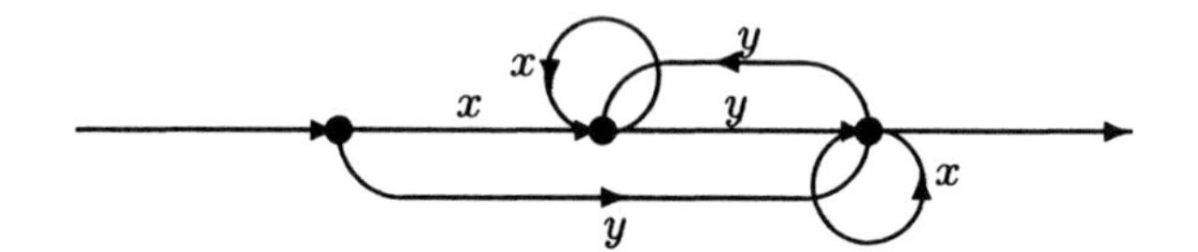

Figure 3.18: Automaton $\mathcal{A}_1$.

In our example we get $T = \{q_2 r_2\}$.

Kleene's Theorem tells us that languages accepted by finite state automata are precisely the languages defined by regular expressions.

Two regular expressions R and S are equivalent if they define the same language. In this case we write $R \equiv S$. Here is a set of rules and identities for regular expressions.

Algebraic laws for regular expressions. If R, S, T are any regular expressions over alphabet Q, then

(a) $R + R \equiv R$, $\emptyset + R \equiv R + \emptyset \equiv R$;

(b) $R + S \equiv S + R$;

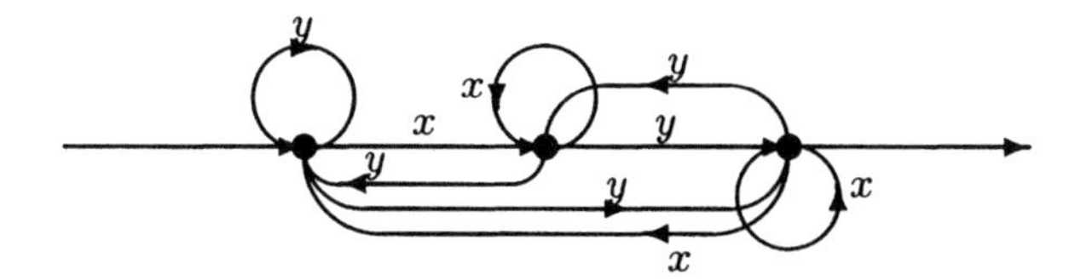

Figure 3.19: $L(\mathcal{A}_2) = L(\mathcal{A}_1)^+$.

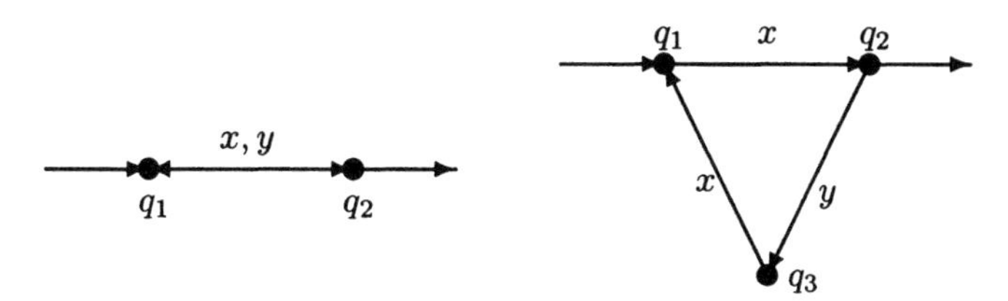

Figure 3.20: $L(\mathcal{A}_1) = x(yx)^*$, $L(\mathcal{A}_2) = x(yx^2)^*$ in Example 3.14.

(c) $(R + S) + T \equiv R + (S + T) \equiv R + S + T$;

(d) $(R \cdot S) \cdot T \equiv R \cdot (S \cdot T) \equiv R \cdot S \cdot T$;

(e) $1 \cdot R \equiv R \cdot 1 \equiv R$, $\emptyset \cdot R \equiv R \cdot \emptyset \equiv \emptyset$;

(f) $(S + T) \cdot R \equiv S \cdot R + T \cdot R$;

(g) $R \cdot (S + T) \equiv R \cdot S + R \cdot T$;

(h) $R^* \cdot R^* \equiv R^*$;

(i) $(R^*)* \equiv R^*$;

(j) $R \cdot R^* \equiv R^* \cdot R$;

(k) $R^* \equiv 1 + R + R^2 + \ldots + R^{k+1} \cdot R^*$ for any $k \geq 0$;

(l) $1^* \equiv 1$ and $\emptyset^* \equiv 1$;

(m) $(R^* + S^*)^* \equiv (R^* \cdot S^*)^* \equiv (R + S)^*$;

(n) $(R \cdot S)^* \cdot R \equiv R \cdot (S \cdot R)^*$;

(o) $(R^* \cdot S)^* \cdot R^* \equiv (R + S)^*$;

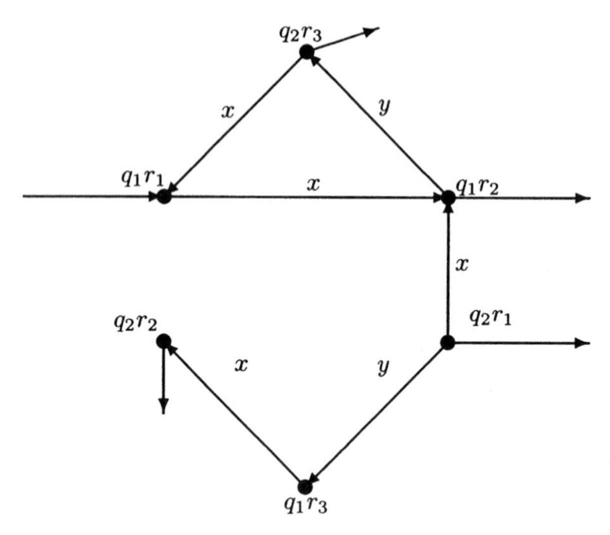

Figure 3.21: Example 3.14: $L(\mathcal{A}_3) = L(\mathcal{A}_1) \cup L(\mathcal{A}_2)$.

(p) $(R^* \cdot S)^* \equiv (R + S)^* \cdot S + 1$ and $(R \cdot S^*)^* \equiv R \cdot (R + S)^* + 1$;

(q) $R \equiv S^* \cdot T \Longrightarrow R \equiv S \cdot R + T$;

(r) If $1 \notin S$, then $R \equiv S \cdot R + T \Longrightarrow Q \equiv R^* \cdot S$.

Most of these results are straightforward. It is important to emphasize that concatenation is not commutative, in contrast with our intuition associated with symbol $\cdot$ being used for multiplication of numbers too. We can always use finite state automata and minimal automata for a complete algorithm verifying whether two regular expressions define equal languages.

3.6 Algorithm 2: Language Accepted by FSA

Algorithm 2. This algorithm allows us to take the transition diagram of any automaton and use it to build, step by step, a regular expression defining the same language. The algorithm can be summarized as follows.

Step 1. Change the transition diagram produced by the previous

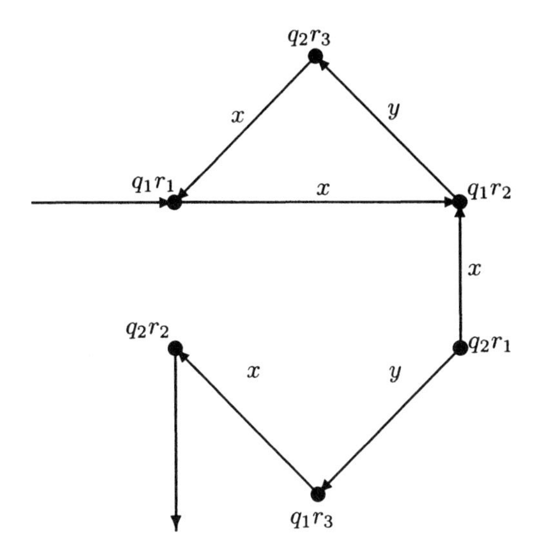

Figure 3.22: Example 3.14: $L(\mathcal{A}_4) = L(\mathcal{A}_1) \cap L(\mathcal{A}_2)$.

step to an equivalent transition diagram of a nondeterministic automaton with just one final state, which differs from the start state. To this end, if there are several terminal states, then introduce a new terminal state with duplicates of all old edges entering old terminal states, then make old terminal state nonterminal.

Step 2. If there is a state with more than one loop and the loops have labels $r_1, r_2, \ldots, r_n$, then replace all the loops on that state by a single loop bearing the label $r_1 + r_2 + \ldots + r_n$.

Step 3. If there are states connected by more than one edge in the same direction with labels $r_1, r_2, \ldots, r_n$, then replace these parallel edges by a single edge with the label $r_1 + r_2 + \ldots + r_n$.

Step 4. If the transition diagram does not consist of only two states with a single edge, then choose one state q other than the start state or final state. Remove q by applying the following operation.

For each pair $(q_1, q)(q, q_2)$ of edges with labels r_1, r_2, where $q_1, q_2 \neq q$, introduce a new edge (q_1, q_2) with the label

$$r = \begin{cases} r_1 r_0^* r_2 & \text{if there was a loop } (q,q) \text{ with label } r_0, \\ r_1 r_2 & \text{otherwise.} \end{cases}$$

Then remove the state q and all edges incident on it.

Step 5. If the resulting transition diagram has only two states, the start state q_0 and the terminal state q_1, with exactly two edges between them, (q_0, q_1) labeled by r_{01} and (q_1, q_0) labeled by r_{10}, and possibly with loops, (q_0, q_0) labeled by r_0 and (q_1, q_1) labeled by r_1, then we have to eliminate the backward edge (q_1, q_0). It suffices to replace the label on (q_1, q_1) by $r_1 + r_{10} r_0^* r_{01}$.

Step 6. If the resulting transition diagram has only a start state q_1 and a final state q_2 connected by one edge with label r, then we are done. The answer is given by

$$r' = \begin{cases} r & \text{if there are no loops } (q_1,q_1), (q_2,q_2), \\ r_1^* r & \text{if } (q_1,q_1) \text{ has label } r_1 \text{ and there is no loop } (q_2,q_2) \\ r r_2 * & \text{if } (q_2,q_2) \text{ has label } r_2 \text{ and there is no loop } (q_1,q_1) \\ r_1^* r r_2 * & \text{if } (q_1,q_1), (q_2,q_2) \text{ have labels } r_1, r_2. \end{cases} \tag{3.9}$$

Otherwise, go to Step 2.

Since each step eliminates either a state or at least one edge, it follows that the algorithm always terminates, i.e., reaches a stage where we are left with two states and one edge.

The following exercise illustrates Step 5 of Algorithm 2.

Exercise 3.4. Find the language accepted by the automaton in Figure 3.23.

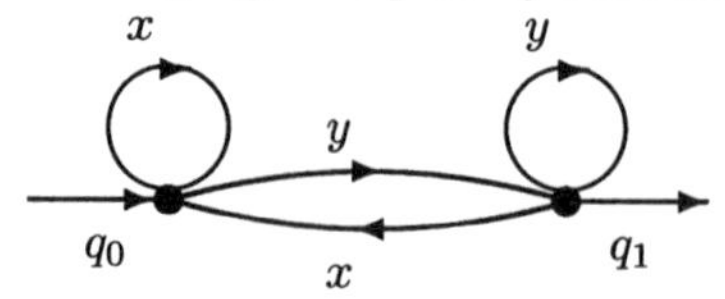

Figure 3.23: Automaton of Exercise 3.4

Answer: $L(\mathcal{A}) = x^* y (y + x x^* y)^*$. It simplifies to $(x + y)^* y$.

Exercise 3.5. Find the language accepted by the automaton in Figure 3.24.

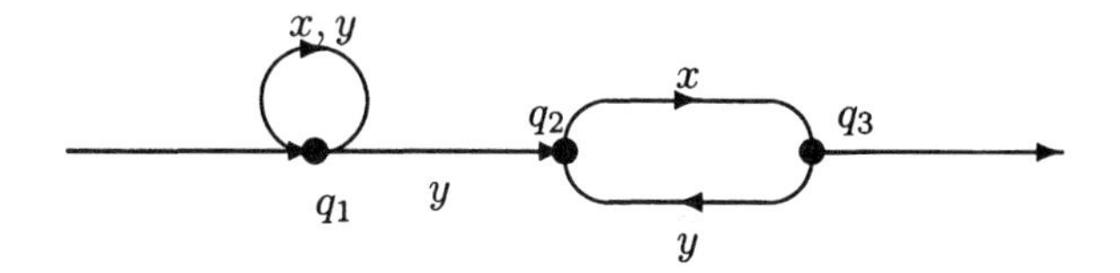

Figure 3.24: Automaton of Example 3.5.

Answer: $(x+y)^*yx(yx)^*$.

Exercise 3.6. Find the language accepted by the automaton in Figure 3.25.

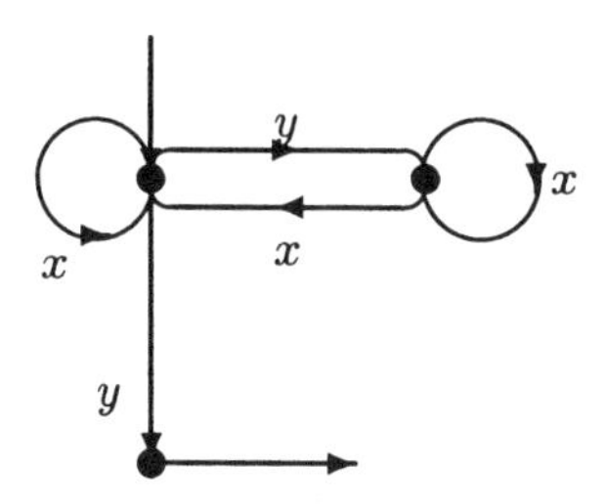

Figure 3.25: Automaton of Example 3.6.

Solution. $(x+yx^*y)^*y$. □

Exercise 3.7. Find the language accepted by the automaton in Figure 3.26.

Solution. $(x+yx^*y)^*+y = 1+(x+yx^*y)(x+yx^*y)^*+y$. □

Exercise 3.8. Find the language accepted by the automaton in Figure 3.27.

Solution. $((x+y)y^*x+y)y+x$. □

3.7 Minimization Algorithm

It is possible to make some finite state automata simpler by identifying equivalent states. The formal concept of congruences is required for finding

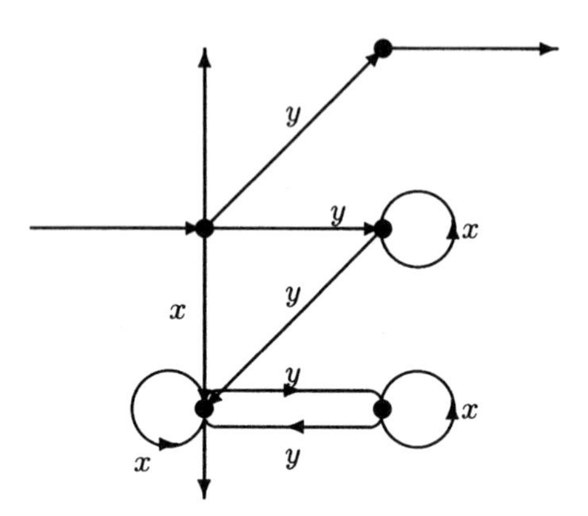

Figure 3.26: Automaton of Example 3.7.

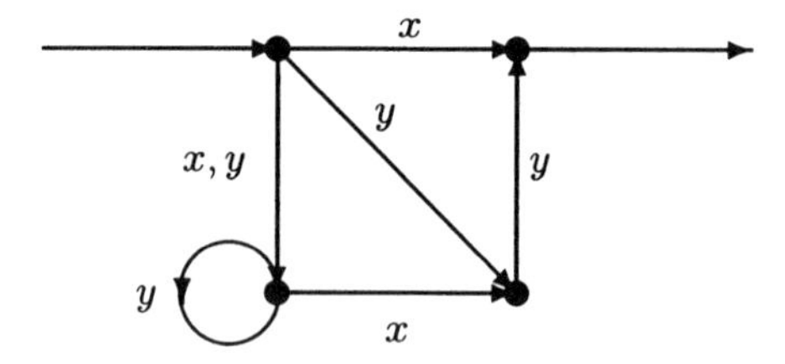

Figure 3.27: Automaton of Example 3.8.

minimal automata. The computation of a minimal automaton can be carried out in several ways. We are going to learn the *reduction algorithm* that starts with a given automaton and computes successive approximations of the Nerode equivalence. A careful implementation of this algorithm due to Hopcroft [1] takes time $O(N \log N)$ for an N-state automaton.

Congruences on automata. Let $\mathcal{A} = (Q, X, \delta, q_0, T)$ be a finite state automaton. An equivalence relation ϱ on the set Q of states is called a *congruence* if it satisfies the following conditions:

(C1) $(a, b) \in \varrho$ implies $(a \cdot x, b \cdot x) \in \varrho$, for all $x, b \in Q$, $x \in X$;

(C2) if $(a, b) \in \varrho$ and $x \in T$, then $y \in T$.

If condition (C1) holds, then we say that the relation ϱ is *compatible* with the next-state function. If condition (C2) holds, then we say that the relation ϱ *saturates* the set T of terminal states, i.e., T is the union of some classes of the relation.

Every equivalence relation compatible with the next-state function defines the *quotient automaton* in a standard fashion: the set of states consists of all equivalence classes of ϱ, and the next-state function is defined on representatives of the classes. The quotient automaton recognizes the same language if and only if the relation saturates the set of terminal states. Thus (C1) allows us to define the quotient automaton, that is to identify some states of the original automaton, and (C2) makes sure that it accepts the same language.

Denote by $\mathrm{Con}(\mathcal{A})$ the set of all congruences on $\mathcal{A}$. Given congruences ρ, δ on $\mathcal{A}$, the *meet* $\varrho \bigwedge \delta$ and *join* $\varrho \bigvee \delta$ denote their intersection and the transitive closure of their union, respectively. The set of all congruences on any automaton forms a lattice with respect to $\bigwedge$ and $\bigvee$. Therefore $\mathrm{Con}(\mathcal{A})$ is a sublattice of the lattice of all equivalence relations on Q.

The smallest congruence of this lattice is the equality relation ι and the largest congruence on the automaton $\mathcal{A}$ is the *Nerode equivalence* η_T described by

$$\eta = \eta_T \quad = \quad \{(a, b) \in Q \times Q \mid a \cdot u \in T \text{ iff } b \cdot u \in T \text{ for all } u \in X^*\}. \tag{3.10}$$

Denote the equality relation on $\mathcal{A}$ by ι. A congruence on the automaton is said to be *proper* if it is distinct from ι and η_T.

The minimization algorithm we are going to learn computes successive approximations of the Nerode equivalence. In general, simplification of a finite state automaton involves identifying 'equivalent states' that can be combined without affecting the action of the automaton on input words.

***-Equivalence of states.** Two states of a finite state automaton are said to be **-equivalent* (read 'star equivalent') if any word accepted by the automaton when it starts from one of the states is accepted by the automaton when it starts from the other state.

k-Equivalence of states. As a practical matter, you can tell whether or not two states s and t of a finite state automaton are *-equivalent, by using an iterative procedure based on a simpler kind of equivalence of states called k-equivalence. Two states are *k-equivalent* if any word *of length less than or equal to k* that is accepted by the automaton when it starts from one of the states is accepted by the automaton when it starts from the other state.

Let $\mathcal{A}$ be a finite state automaton with transition function δ. Given any states s and t in Q, we define

(a) s is 0-equivalent to t iff either s and t are both accepting states or are both nonaccepting states;

(b) for every integer $k \geq 1$, s is k-equivalent to t iff s and t are $(k-1)$-equivalent and, for any input symbol m, $\delta(s, m)$ and $\delta(t, m)$ are also $(k-1)$-equivalent.

Therefore we can use dynamic programming and find k-equivalence classes by subdividing the $(k-1)$-equivalence classes according to the action of the transition function on the members of the classes.

If $\mathcal{A}$ is a finite state automaton, then for some integer $K \geq 0$, the set of K-equivalence classes of states of A equals the set of $K+1$-equivalence classes of states of A, and for all such K these are both equal to the set of $*$-equivalence classes of states of $\mathcal{A}$.

Minimization Algorithm. Let $\mathcal{A}$ be a finite state automaton with set of states S, transition function δ, relation R_* of $*$-equivalence of states, and relations R_k of k-equivalence of states. The preceding definitions and explanations show that the following steps find a minimal automaton accepting the same language as $\mathcal{A}$.

Step 1. Find the set of 0-equivalence classes of S.

Step 2. For $k = 1, 2, 3$ and so on, $(k-1)$-equivalence classes have been found, then subdivide the $(k-1)$-equivalence classes as described above to find the k-equivalence classes of S. Stop subdividing when you notice that for some integer K, the set of $(K+1)$-equivalence classes is equal to the set of K-equivalence classes. At this point, conclude that the set of K-equivalence classes is the set of $*$-equivalence classes.

Step 3. Construct the minimal automaton $\bar{\mathcal{A}}$, whose states are the $*$-equivalence classes of A and whose transition function $\bar{\delta}$ is given by

$$\bar{\delta}([s], m) = [\bar{\delta}(s, m)]$$

for any state of $\bar{\mathcal{A}}$ and any input symbol m, where s is any state in $[s]$.

Example 3.15. Find 0-equivalence classes, 1-equivalence and 2-equivalence classes for the states of the automaton shown in Figure 3.28.

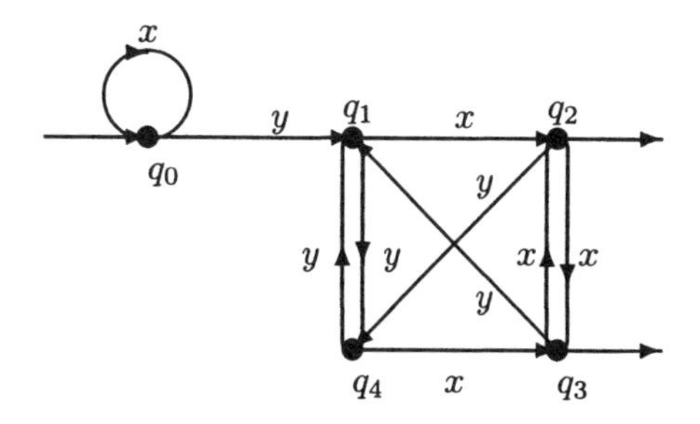

Figure 3.28: Automaton of Example 3.15.

0-equivalence classes. There are two sets of 0-equivalent states:

(N) the non-accepting states $\{q_0, q_1, q_4\}$;

(T) and the accepting states $\{q_2, q_3\}$.

1-equivalence classes. We have

name of class	class	type of element after input:	
		x	y
B	$\{q_0\}$	N	N
C	$\{q_1,q_4\}$	T	N
D	$\{q_2,q_3\}$	T	N

For example, q_1 is not equivalent to q_0 because when a x is input to the automaton in state q_1 it goes to state q_2, whereas when a x is input to the automaton in state q_0 it goes to state q_0, and q_2 and q_0 are not 0-equivalent.)

2-equivalence classes. We have

name of class	class	type of element after input:	
		x	y
E	$\{q_0\}$	B	C
F	$\{q_1,q_4\}$	D	C
G	$\{q_2,q_3\}$	D	C

Finally, we identify all states which are in the same class, and get the minimal automaton as in Figure 3.29.

Example 3.16. Given the automaton with transition diagram in Figure 3.30, we can find an equivalent minimal automaton as displayed in Figure 3.31.

Exercise 3.9. Find a minimal automaton equivalent to the automaton in Figure 3.32.

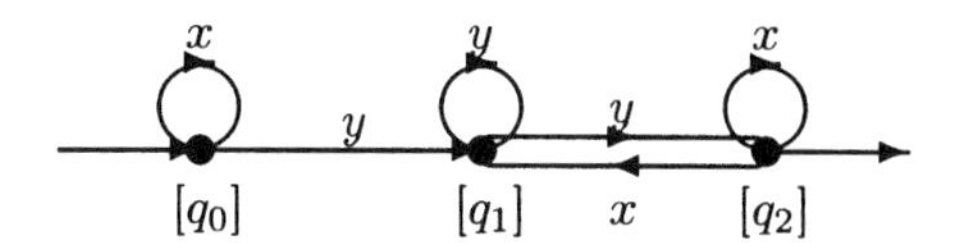

Figure 3.29: Minimal automaton of Example 3.15.

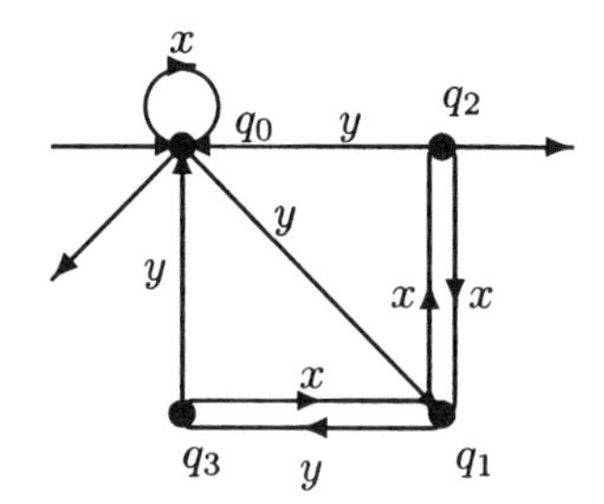

Figure 3.30: Automaton of Example 3.16.

Exercise 3.10. Find a minimal automaton equivalent to the automaton with transition diagram in Figure 3.33.

3.8 Syntactic and Transformation Monoids

Definition 3.5. *Transformation monoid* of an automaton is the monoid generated in the monoid of all transformations (or functions on the set of states, with respect to composition) by all transformations (of the set of states) defined by words of the free monoid. *Transformation semigroup* of an automaton is the semigroup generated in the monoid of all functions (on the set of states, with respect to composition) by all transformations (of the set of states) defined by words of the free semigroup.

Let S be a semigroup with a subset T, and let $x \in S$. Then the *context* of x with respect to T is denoted by $\mathrm{Cont}_T(x)$ or $\mathrm{Cont}_{T,S}(x)$ and is defined as the set

$$\mathrm{Cont}_T(x) = \{(a, b) \in S^1 \times S^1 \mid axb \in T\}.$$

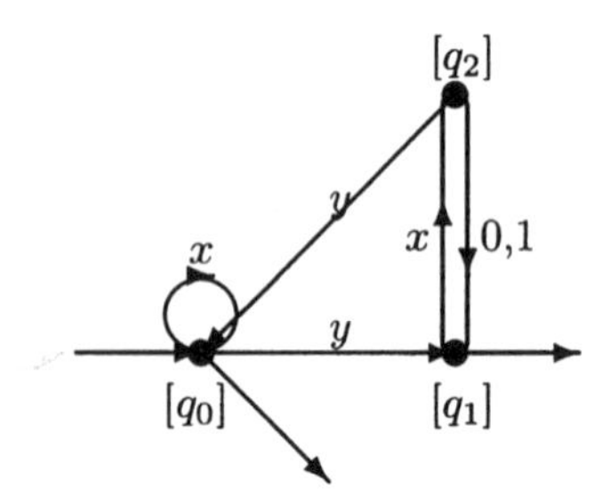

Figure 3.31: Minimal automaton for Example 3.16

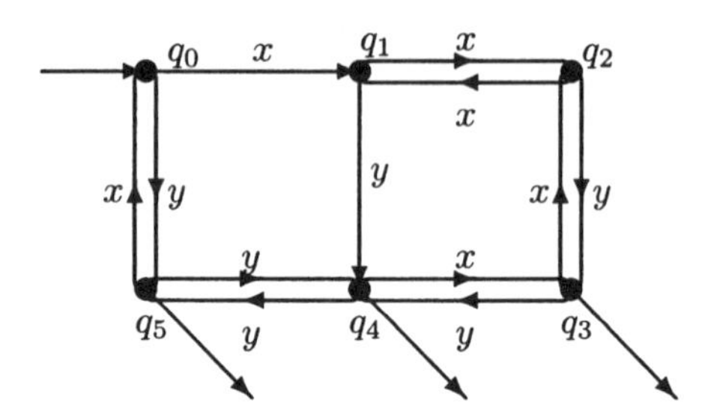

Figure 3.32: Automaton of Exercise 3.9.

Definition 3.6. The *syntactic semigroup* $\mathrm{Syn}(L)$ of a language $L \subseteq X^+$ is the quotient of X^+ by the *Myhill congruence* μ_L on X^+:

$$\mu_L = \{(w_1, w_2) \in X^+ \times X^+ \mid \mathrm{Cont}_{L,X^+}(w_1) = \mathrm{Cont}_{L,X^+}(w_2)\}.$$

The Myhill congruence is also called a *syntactic congruence.*

Definition 3.7. *Syntactic monoid* of a language is the quotient of the free monoid modulo the Myhill congruence on X^*:

$$\mu_L = \{(w_1, w_2) \in X^* \times X^* \mid \mathrm{Cont}_{L,X^*}(w_1) = \mathrm{Cont}_{L,X^+}(w_2)\}.$$

Syntactic monoids can be used to recognize languages in the following sense. A language L of X^+ is said to be *recognized* by a semigroup S, if there exists a homomorphism ϕ: $X^+ \to S$ and a subset T of S such that $L = \phi^{-1}(T)$.

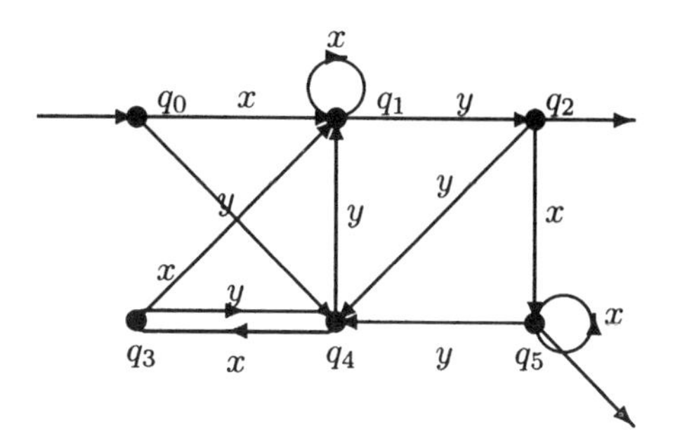

Figure 3.33: Automaton of Exercise 3.10.

Lemma 3.2. ([58], Proposition 7.2.1s) *A language L is recognized by a semigroup S if and only if there exists a subsemigroup T of S and an epimorphism mapping T onto the syntactic semigroup of L.*

Let S be a semigroup with a subset T, and let $x \in S$. Then the *right context* of x with respect to T is denoted by $\mathrm{RCont}_T(x)$ or $\mathrm{RCont}_{T,S}(x)$ and is defined as the set

$$\mathrm{RCont}_T(x) = \{a \in S^1 \mid xa \in T\}.$$

The right context is also denoted by $D_x T$ and is called the *derivative* of T with respect to x. The *minimal automaton* of a language $L \subseteq X^*$ is the quotient of X^* by the *Nerode equivalence*

$$\eta_L = \{(w_1, w_2) \in X^+ \times X^+ \mid \mathrm{RCont}_{L,X^+}(w_1) = \mathrm{RCont}_{L,X^+}(w_2)\}.$$

Obviously, for every language, the Myhill congruence is contained in the Nerode equivalence. Therefore the mininal automaton can be regarded as a quotient of the syntactic monoid.

Example 3.17. The transformation semigroup of the automaton in Figure 3.23 is the left zero semigroup with two elements $T_x = \begin{pmatrix} q_0 & q_1 \\ q_0 & q_0 \end{pmatrix}$, $T_y = \begin{pmatrix} q_0 & q_1 \\ q_1 & q_1 \end{pmatrix}$ and identity element 1 adjoined. It is the syntactic semigroup of the language X^*y.

Exercise 3.11. Find the syntactic monoids of the following languages:

(i) $L_1 = xy^*$;

(ii) $L_1 = x(x+y)^+y$;

(iii) $L_1 = xy^* + yx^*$;

(iv) $L_1 = (xy)^*$;

(v) $L_1 = (xy)^* + (yx)^*$.

Various important properties of languages can be characterized in terms of their syntactic semigroups.

Lemma 3.3. ([58], Proposition 2.3) *A semigroup S is the syntactic semigroup of some language if and only if S contains a subset T such that every two distinct elements of S have different contexts with respect to T.*

A subset T with this property is called a *disjunctive set.*

Let's take a look at the Cayley graph $\mathcal{A} = \mathrm{Cay}(\mathbb{Z}_3, \{x\})$ regarded as a FSA. We choose the start state e and terminal state x^2, and sketch its transition diagram in Figure 3.34.

The language accepted by this automaton is

$$\begin{aligned} L(\mathcal{A}) &= x^2(x^3)^* \\ &= \{x^{2+3n} \mid n \in \mathbb{N}_0\} \end{aligned}$$

Next, we look at the Cayley graph $\mathrm{Cay}(\mathbb{Z}_6, \{x, x^2\})$ as a language recognizer. Choosing the start state e and terminal states states x^3, x^4, we sketch the transition diagram of this FSA in Figure 3.35.

The language accepted by this automaton equals

$$\begin{aligned} L(\mathcal{A}) &= (x^3 + x^4)(x^6)^* \\ &= \{(x^3 + x^4)x^{6n} \mid n \in \mathbb{N}_0\} \end{aligned}$$

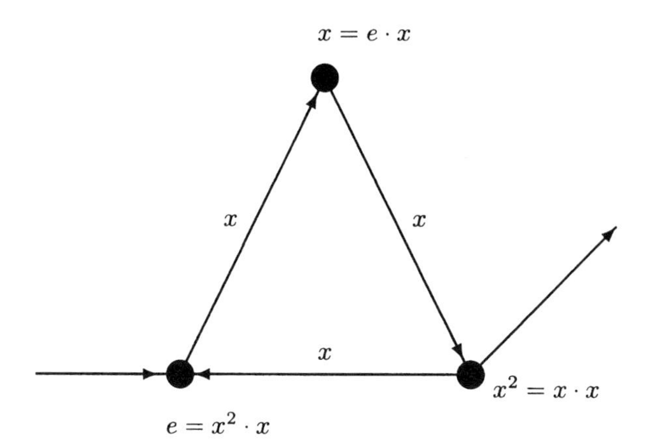

Figure 3.34: Cayley graph of $\mathbb{Z}_3$ as FSA

Cayley graphs viewed as acceptors are as powerful as FSA. Indeed, if L is recognised by an FSA, then L is also recognised by a Cayley graph of the syntactic monoid $\mathrm{Syn}(L)$. However, Cayley graphs are not always minimal recognisers.

Exercise 3.12. Find all automata with the following syntactic monoids:

(i) $\mathbb{Z}_2$;

(ii) $\mathbb{Z}_3$.

Exercise 3.13. Prove that the following monoids are syntactic:

(i) every finite group;

(ii) the power semigroup $\mathcal{P}([n])$ of all subsets of the set $[n] = \{1, 2, \ldots, n\}$ with respect to intersection.

Exercise 3.14. Show that the following semigroups are not syntactic, if they have more than 4 elements:

(i) left zero semigroup, that is a semigroup satisfying the identity $xy = x$;

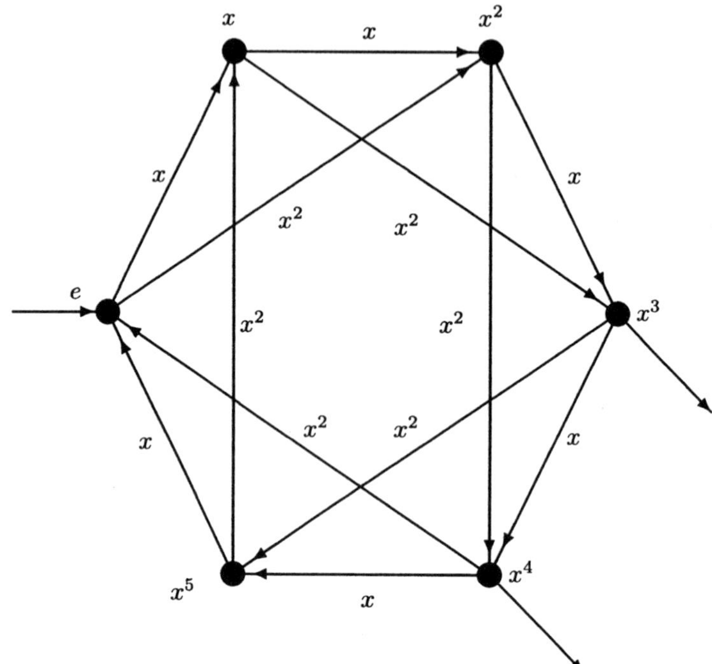

Figure 3.35: Cayley graph of $\mathbb{Z}_6$ as FSA

(ii) right zero semigroup, i.e., a semigroup satisfying the identity $xy = y$;

(ii) rectangular band, that is a semigroup satisfying the identities $x^2 = x$ and $xyx = x$.

Lemma 3.4. *Every rectangular semigroup is isomorphic to the direct product of a left zero semigroup and a right zero semigroup.*

Exercise 3.15. A rectangular band is a syntactic semigroup if and only if it is neither left nor right zero semigroup.

Theorem 3.3. ([180], Exercise 7.6) *Every finite semigroup which is a semilattice of groups is a syntactic semigroup. In particular, every finite commutative semigroup which is a union of groups is a syntactic semigroup.*

Theorem 3.4. ([180], Proposition 8.1.2) *The set of invertible elements in the syntactic monoid of every finite prefix code is a cyclic group.*

Theorem 3.5. ([180], Corollary 8.1.3) *If the syntactic monoid of a finite prefix code $C \subseteq X^n$ is a group G, then G is a cyclic group of order n and $C = X^n$.*

Let S and T be semigroups. A mapping $f : S \to T$ is called an *anti-isomorphism* if it is a bijection and $f(xy) = f(y)f(x)$ for all $x, y \in S$.

Theorem 3.6. (Syntactic and Transformation Monoids Theorem) ([190], 29.8) *Syntactic monoid of a language is anti-isomorphic to the transformation monoid of its minimal automaton.*

Theorem 3.7. (Myhill-Nerode Theorem) ([195], §3.2, [255], Theorem 4.7) *For each language $L \subseteq X^*$, the following conditions are equivalent:*

(i) *L is regular;*

(ii) *L is recognizable;*

(iii) *the Myhill congruence η_L has finite index, i.e., the syntactic monoid of L is finite;*

(iv) *the Nerode equivalence μ_L has finite index.*

A relation ρ is called a *right congruence* if

$$(x, y) \in \rho, z \in X \Leftrightarrow (xz, yz) \in \rho. \tag{3.11}$$

In this case we also say that the relation is *right invariant.*

Every right congruence on a free monoid X^* defines an automaton $(X^*/\rho, X, \delta)$, where the set of states consists of all equivalence classes of X^*/ρ and the transition function is given by the concatenation of words in the free monoid:

$$\delta(q/\rho, x) = qx/\rho. \tag{3.12}$$

Theorem 3.8. (Right Congruences Theorem) ([180], Proposition 6.1.6) *Let ϱ be an equivalence relation on the free monoid X^*. Then the following conditions are equivalent:*

(i) *ϱ defines an automaton $(X^*/\varrho, X, \delta)$, where $\delta(w/\varrho, x) = wx/\varrho$ for all $x \in X, w \in X^*$;*

(ii) *ϱ is a right congruence on X^*.*

Theorem 3.9. (Minimal Automaton Theorem) ([180], Proposition 6.1.10) *The minimal automaton of a language L over the alphabet X is the automaton*

$$(X^*/\eta_L, X, \delta, 1/\eta_L, L/\eta_L)$$

given by the Nerode equivalence η_L, where the next-state function is defined by $\delta(w/\varrho, x) = wx/\varrho$ for all $x \in X, w \in X^*$.

Exercise 3.16. Use Nerode equivalence to find the minimal automata recognizing the following languages:

(i) $L_1 = xy^*$;

(ii) $L_1 = x(x+y)^+y$;

(iii) $L_1 = xy^* + yx^*$;

(iv) $L_1 = (xy)^*$;

(v) $L_1 = (xy)^* + (yx)^*$.

3.9 Krohn-Rhodes Decomposition Theorem

A finite automaton may have output associated with the states (in a Moore machine) or with the transitions (in a Mealy machine). A *finite automaton* or *Mealy machine* is a 6-tuple $(Q, X, Y, \delta, \lambda, q_0)$, where

- Q is a finite set of states;
- X is a finite set called the input alphabet;
- Y is a finite set called the output alphabet;
- $\delta : Q \times X \to Q$ is the next-state function;

- $\lambda : Q \times X \to Y$ is the output function;
- $q_0 \in Q$ is the start state.

If the output function $\lambda(q, x)$ depends only on the state q, then the machine is called a *Moore machine.*

The *output* of a Moore automaton $M = (Q, X, Y, \delta, \lambda, q_0)$ in response to input $x_1, x_2, \ldots, x_n$ is $\lambda(x_1), \lambda(x_2), \ldots, \lambda(x_n)$. Note that any Moore machine gives output $\lambda(q_0)$ in response to input ε.

The *output* of a Mealy machine $M = (Q, X, Y, \delta, \lambda, q_0)$ in response to input x_1, x_2, ..., x_n is $\lambda(q_0, x_1)$, $\lambda(q_1, x_2)$, ..., $\lambda(q_{n-1}x_n)$, where q_0, q_1, ..., q_n is the sequence of states such that $\delta(q_{i-1}, x_i) = q_i$ for $i = 1, \ldots, n$. Note that this sequence has length n rather than length $n + 1$ as for Moore machine, and on input ε a Mealy machine gives output ε.

The *output table* shows all the values of the output function. On the transition graph the output values are written next to the inputs, right after them.

For a machine M and an input w denote by $T_M(w)$ the output produced by M on w. A Mealy machine M and a Moore machine M' are *equivalent* if, for all inputs w, $yT_M(w) = T_{M'}(w)$, where y is the output of M' for its initial state.

Example 3.18. (Identity-Reset automaton or IR Flip-Flop) $Q = \{q_0, q_1\}$, $X = \{e, 0, 1\}$, $Y = \{0, 1\}$, and

δ	e	0	1
q_0	q_0	q_0	q_1
q_1	q_1	q_0	q_1

λ	e	0	1
q_0	q_0	q_0	q_0
q_1	q_1	q_1	q_1

Transition diagram of the IR flip-flop is given in Figure 3.36.

Exercise 3.17. Find the transformation monoid of the IR Flip-Flop.

Solution. $M_{\mathcal{A}} = \{f_\Lambda, f_0, f_1\}$. □

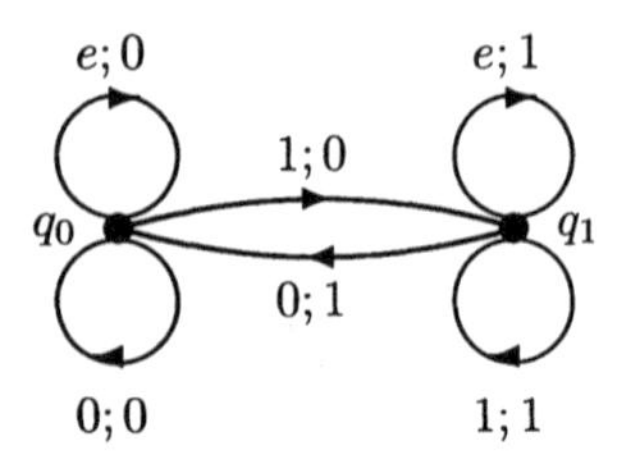

Figure 3.36: IR Flip-Flop

Theorem 3.10. ([255], §2.4) *For every Moore machine there exists an equivalent Mealy machine.*

A *finite transducer* is a 6-tuple $(Q, X, Y, \delta, q_0, T)$, where

- Q is a finite set of states;
- X is a finite set called input alphabet;
- Y is a finite set called output alphabet;
- δ is the transition-and-output function from $Q \times X^*$ to finite subsets of $Q \times Y^*$;
- $q_0 \in Q$ is the start state;
- $T \subseteq Q$ is the set of final states.

A finite transducer is called a *generalized sequential machine* if δ is a function from $Q \times X$ to finite subsets of $Q \times Y^*$, i.e., if the transducer reads exactly one symbol with each transition.

Definition 3.8. An automaton $\mathcal{A}' = (Q', X', Y', \delta', \lambda')$ is called a subautomaton of the automaton $\mathcal{A} = (Q, X, Y, \delta, \lambda)$ if $Q' \subseteq Q$ and δ', λ' are the restrictions on $Q' \times X$ of δ, λ, respectively. In this case we write $\mathcal{A}' \subseteq \mathcal{A}$.

Definition 3.9. Let $\mathcal{A}_1 = (Q_1, X_1, Y_1, \delta_1, \lambda_1)$ and $\mathcal{A}_2 = (Q_2, X_2, Y_2, \delta_2, \lambda_2)$ be automata. We say that F is a *homomorphism* from $\mathcal{A}_1$ to $\mathcal{A}_\in$ and write $F : \mathcal{A}_1 \to \mathcal{A}_2$, if F is a triplet $F = (\xi, \alpha, \beta)$.

Let $\mathcal{A} = (Q_1, X_1, Y_1, \delta_1, \lambda_1, q_1)$ and and $\mathcal{A} = (Q_2, X_2, Y_2, \delta_2, \lambda_2, q_2)$ be two automata. The *parallel connection* of $\mathcal{A}_1$ and $\mathcal{A}_2$ is the automaton

$$\mathcal{A}_1 \times \mathcal{A}_2 = (Q_1 \times Q_2, X_1 \times X_2, Y_1 \times Y_2, (\delta_1, \delta_2), (\lambda_1, \lambda_2), (q_1, q_2)) \quad (3.13)$$

with the transition diagram illustrated in Figure 3.37. Here the functions

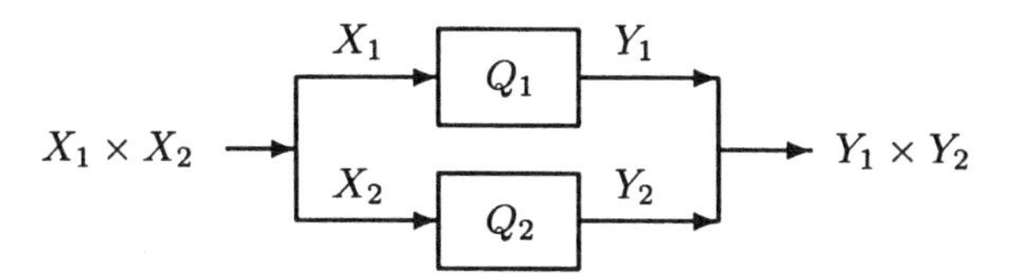

Figure 3.37: Parallel connection of automata.

are defined by the rules

$$\begin{aligned} (\delta_1, \delta_2)((r_1, r_2), (x_1, x_2)) &= (\delta_1(r_1, x_1), \delta_2((r_2, x_2)), \\ (\lambda_1, \lambda_2)((r_1, r_2), (x_1, x_2)) &= (\lambda_1(r_1, x_1), \lambda_2((r_2, x_2)), \end{aligned}$$

for all $r_1 \in Q_1$, $r_2 \in Q_2$, $x_1 \in X_1$, $x_2 \in X_2$.

If $Y_1 \subseteq X_2$, then the *series connection* of the automata $\mathcal{A}_1 = (Q_1, X_1, Y_1, \delta_1, \lambda_1)$ and $\mathcal{A}_2 = (Q_2, X_2, Y_2, \delta_2, \lambda_2)$ is the automaton

$$\mathcal{A}_1 \mapsto\!\!\dashv \mathcal{A}_2 = (Q_1 \times Q_2, X_1, Y_2, (\delta_1, \delta_2\delta_1), \lambda_2(\lambda_1)). \quad (3.14)$$

illustrated in Figure 3.38.

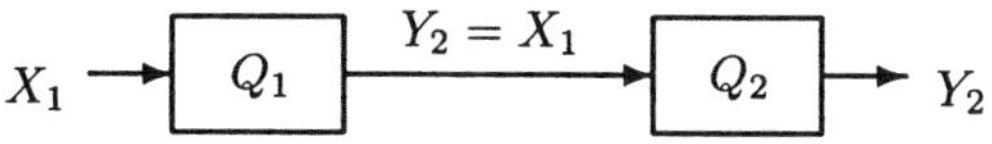

Figure 3.38: Series connection of automata.

We say that a semigroup S_1 *divides* a semigroup S_2, and write $S_1 | S_2$, if S_1 is a homomorphic image of a subsemigroup of S_2.

Theorem 3.11. (Krohn-Rhodes Decomposition Theorem) ([190], 29.16) *Each finite automaton $\mathcal{A}$ can be simulated by series/parallel connections of the following two types of automata:*

(1) IR flip-flops;

(2) automata whose transformation monoids are simple groups dividing the transformation monoid of $\mathcal{A}$.

If the transformation monoid of an automaton is a group, then the automaton is called a *group automaton.*

Exercise 3.18. Prove that a complete deterministic finite state automaton is a group automaton if and only if every two transition edges with equal labels have distinct endpoints.

3.10 Grammars and Rewriting Systems

A *grammar* is a quadruple $G = (N, T, S, P)$, where N and T are disjoint alphabets, $S \in N$, and P is a finite relation from V^*NV^* into V^*, where $V = N \cup T$, and

- the elements of N are called *nonterminal symbols* or *grammar symbols* of G;
- the elements of G are called *terminal symbols* of G;
- $V = V_G = N \cup T$ is the *complete vocabulary* of $\mathcal{G}$;
- S is the *start symbol* or the *axiom* of G;
- every element (x, y) in P is also written in the form $x \to y$ or $x \to_G y$ and is called a *production* or *rewriting rule.*

Let $x, y \in V^*$. The notation $x \to y$ and $x \to_G y$ means that $x = x_1x'x_2, y = y_1y'y_2$ and $x' \to y' \in P$, where $x_1, x_2 \in V^*$. Then we say that x *directly derives* y with respect to G. We say that x derives y and write $x \Rightarrow y$ or $x \Rightarrow_G y$, if there exist $x = w_1, w_2, \ldots, w_n = y$ such that $w_i \to w_{i+1} \in P$ for $i = 1, \ldots, n-1$.

The *language generated by* G is defined by

$$L(G) = \{x \in T^* \mid S \Rightarrow^* x\}.$$

Two grammars G_1 and G_2 are said to be *equivalent* if $L(G_1) = L(G_2)$.

A grammar is said to be *monotonous* or *length-increasing* if $|x| \leq |y|$ for all $x \to y \in P$.

A grammar is said to be *context-sensitive* if each production $x \to y \in P$ has the form $x = x_1 A x_2$, $y = x_1 z x_2$ for $x_1, x_2 \in V_G^*$, $A \in N$, $z \in V_G^+$. Production of the form $S \to \varepsilon$ is allowed in monotonous and context-sensitive grammars, provided that S does not occur in th right hand side members of rules in P.

A grammar is *linear* if each production $x \to y \in P$ has $x \in N$ and $y \in T^* \cup T^* N T^*$.

A grammar is *right linear* if each production $x \to y \in P$ has $x \in N$ and $y \in T^* \cup T^* N$. A grammar is *left linear* if each production $x \to y \in P$ has $x \in N$ and $y \in T^* \cup N T^*$.

A grammar is *regular* if each production $x \to y \in P$ has $x \in N$ and $y \in T^* \cup N T^*$.

Theorem 3.12. ([195], §3.1) *The family of languages defined by monotonous grammars coincides with the family of languages generated by context-sensitive grammars.*

Theorem 3.13. (Ginsburg's Theorem) ([195], §3.1) *For every language L, the following conditions are equivalent:*

(i) *L is generated by a right linear grammar;*

(ii) *L is generated by a left linear grammar;*

(iii) *L is accepted by a finite state automaton.*

The main types of exercises in this topic are listed below.

Exercise 3.19. Given a right linear grammar, find an automaton that recognizes the same language.

Exercise 3.20. Find grammars that generate the following languages:

(i) $L = x + y + z$;

(ii) $L = xy + yx$;

(iii) $L = xy^*$;

(iv) $L = x^*y$;

(v) $L = xy^* + yx^*$;

(vi) $L = x(x + y)x$;

(vii) $L = x(x + y)y^+$;

(viii) $L = x(x + y)^+x$;

Exercise 3.21. For all finite state automata with transition diagrams listed below, find right linear grammars that generate languages accepted by these automata.

(i) $\mathcal{A}_1$:

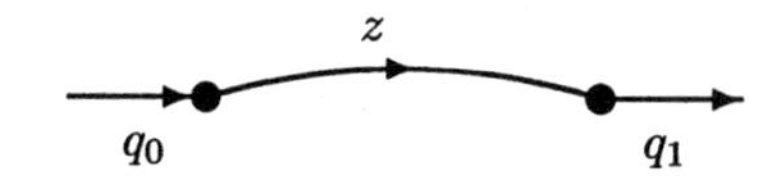

(ii) $\mathcal{A}_2$:

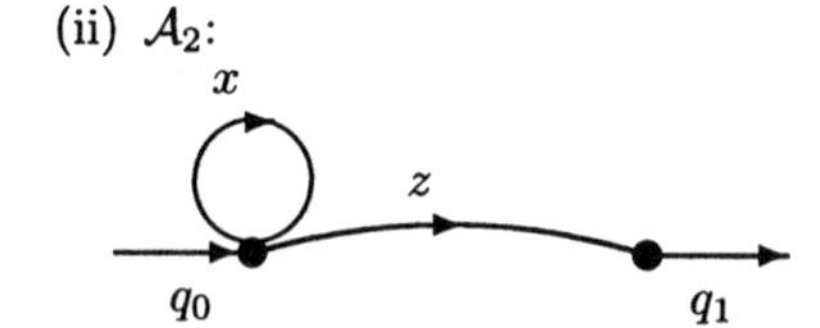

(iii) $\mathcal{A}_3$:

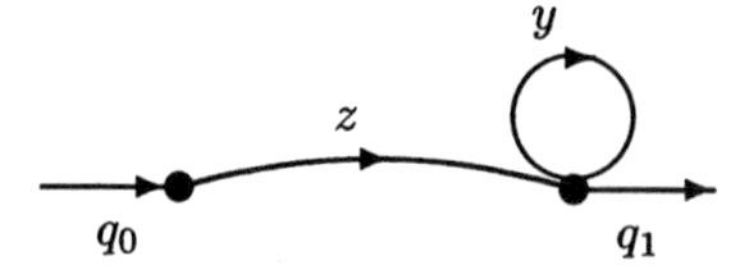

(iv) $\mathcal{A}_4$:

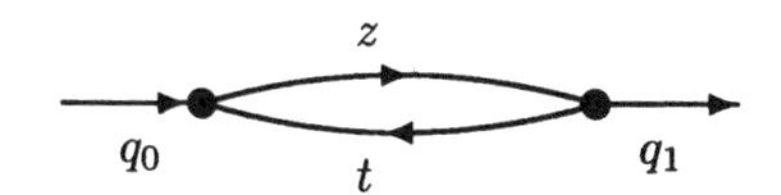

(v) $\mathcal{A}_5$:

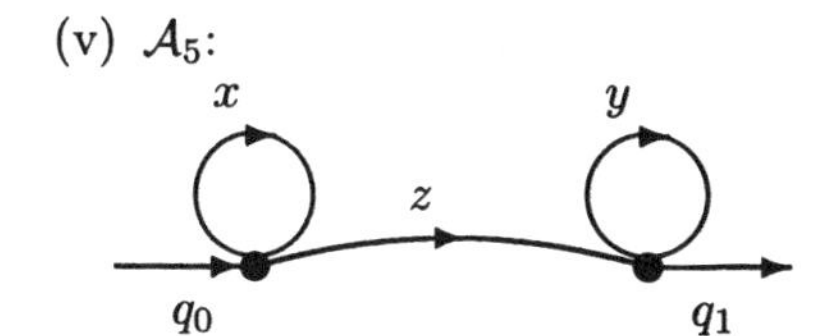

(vi) $\mathcal{A}_6$:

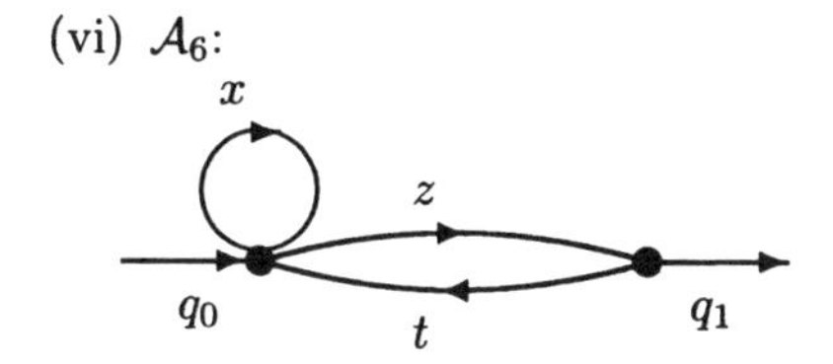

(vii) $\mathcal{A}_7$:

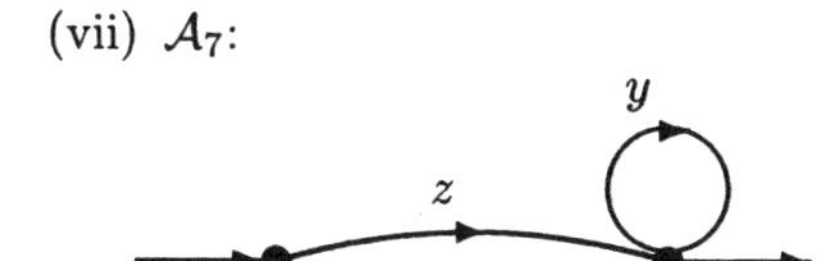

(viii) $\mathcal{A}_8$:

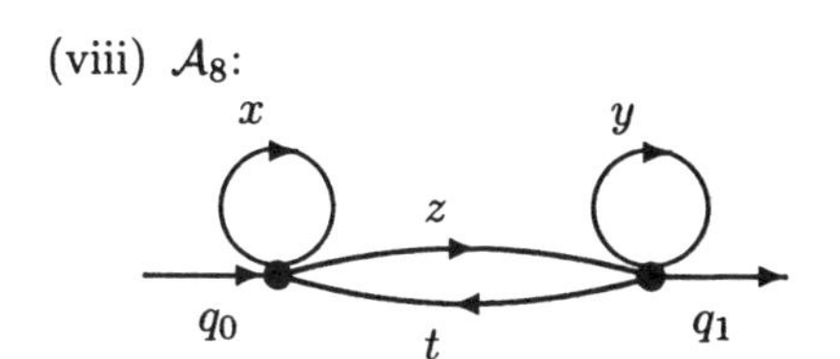

A grammar is *context-free* if each production $x \to y \in P$ has $x \in N$. A language is *context-free* if it is generated by a context-free grammar.

A *pushdown machine* is a quadruple $\mathcal{M} = (Q, X, Z, \delta)$, where

- Q is the set of *states*;
- X is the *input alphabet*;
- Z is the *stack alphabet*;

- δ is a finite subset of $(A \cup \{\varepsilon\} \times Q \times Z \times Z^* \times Q$, called the set of *transition rules*.

Every element (x, q, z, h, q') of δ is called a *rule*, and if $x = \varepsilon$, then it is called an ε-rule. The first three components are pre-conditions in the behaviour of the pushdown machine, and the last two components are post-conditions. The set δ can be regarded as a function from $(A \cup \{\varepsilon\} \times Q \times Z$ to $Z^* \times Q$.

An *internal configuration* of a pushdown machine is a pair $(q, h) \in Q \times Z^*$, where q is the current state and h is a word in Z^* that is in the stack of the machine, the first letter of h being on the bottom of the stack. A *configuration* is a triple (x, q, h), where $x \in X^*$ is the input word to be read, and (q, h) is an internal configuration.

In order to recognize words with pushdown machines we have to define the start configuration and accepting configurations. The convention is that there is only one starting internal configuration $i = (q, z)$, where the state q is the initial state, and the letter z is the initial stack symbol. The set K of internal accepting configurations usually is of the form $T = \cup_{q \in Q}\{q\} \times T_q$ where T_q is a regular language in Z^*.

A *pushdown automaton* is a 5-tuple $\mathcal{A} = (Q, X, Z, \delta, i, T)$, where

- (Q, X, Z, δ) is a pushdown machine;
- $i \in Q \times Z^*$ is the initial internal configuration;
- T is the set of terminal internal configurations.

Theorem 3.14. ([178], §6) *Every context-free language can be recognized by a pushdown automaton $\mathcal{A} = (Q, X, Z, \delta, i, T)$ with the set of terminal internal configurations in each of the following forms:*

(i) $T = F \times Z^*$ *where F is a subset of Q;*

(ii) $T = Q \times \{\varepsilon\}$;

(iii) $T = F \times \{\varepsilon\}$ *where F is a subset of Q;*

(iv) $T = Q \times Z^* Z'$ *where* Z' *is a subset of* Z.

Theorem 3.15. (Sakarovitch's Theorem) *Let* $\mathcal{A} = (Q, X, Z, \delta, i, T)$ *be a pushdown automaton. If* $T = \cup_{q \in Q}\{q\} \times T_q$ *where all* T_q *are context free languages in* Z^*, *then the language recognized by* $\mathcal{A}$ *is context-free.*

Exercise 3.22. Given a context-free grammar, find a pushdown automaton that accepts the language generated by the grammar.

Exercise 3.23. Given a pushdown automaton, find a context-free grammar that generates the language accepted by the automaton.

3.11 Other Classes of Automata

A digitized image in the finite resolution $m \times n$ consists of $m \times n$ pixels each of which takes a Boolean value (0 for no ore deposit, 1 for ore deposit), or a real value (say the percentage of useful minerals). In this section we consider square images of resolution $2^n \times 2^n$. The first task is to encode a 2-dimensional image as a language.

There are several methods for encoding of 2-dimensional images. Let us assume that the whole image lies in an $m \times n$ square and is represented by unit subsquares. To illustrate the first method let us draw the subsquare with address xy^2 (Figure 3.39).

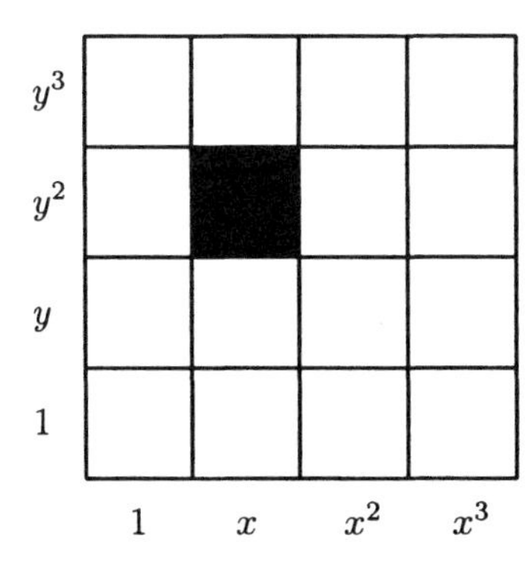

Figure 3.39: Square labeled by xy^2.

An image representing 2-dimensional data can be encoded as a subset of the group $\mathbb{Z}_m \times \mathbb{Z}_n$ (see Figure 3.40). The whole image can be recorded as

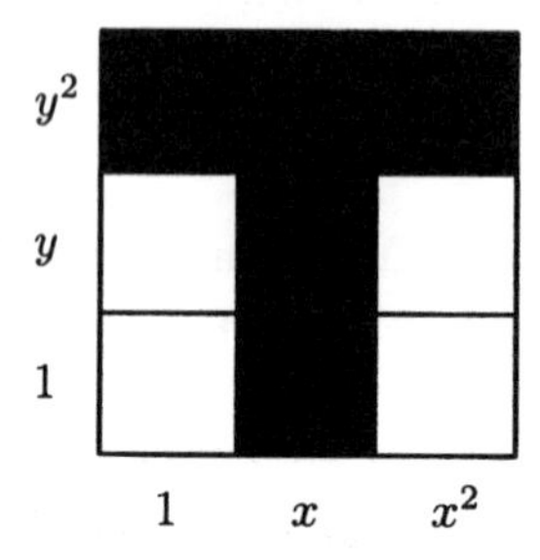

Figure 3.40: $x + xy + xy^2 + x^2y^2 + y^2$.

the polynomial $x + xy + xy^2 + x^2y^2 + y^2$. We can define a linear automaton that accepts the image as in Figure 3.41.

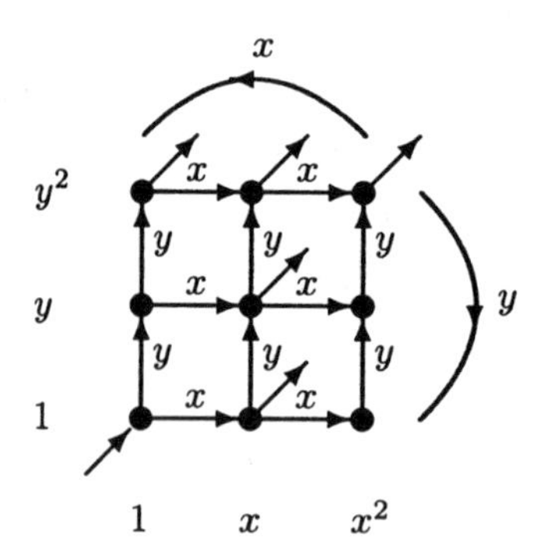

Figure 3.41: x, xy, xy^2, x^2y^2, y^2.

An automaton $(Q, X, Y, \delta, \lambda, q_0, T)$ is called a *linear automaton* if Q and Y are linear spaces and the mappings $\delta_x : q \mapsto \delta(q, x)$ and $\lambda_x : q \mapsto \lambda(q, x)$ are linear mappings, for every $x \in X$. We can also assume that X is a linear space too. Moreover, we may assume that it is a ring by taking the monoid ring $F[X^*]$. Thus, a linear automaton is a right module over $F[X^*]$.

Next, we assume that the whole image lies in a unit square and is represented by subsquares of size 2^{-n}. We use the empty word ε as the address of the whole unit square. Its quadrants are addressed by single letters from the alphabet $X = \{x, y, z, t\}$ as shown in Figure 3.42. Then

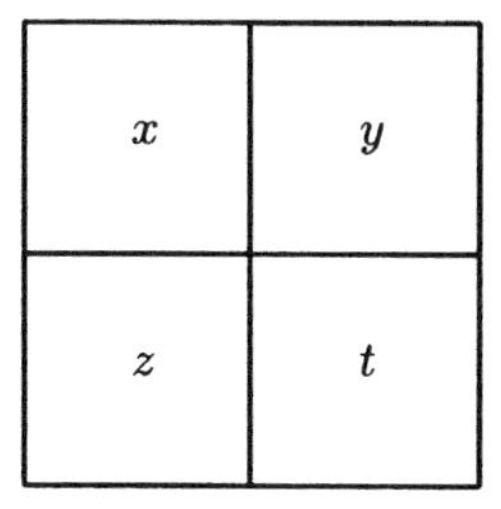

Figure 3.42: Labeling of quadrants

inductively the 4 subsquares of the square with address w are labeled as wx, wy, wz, wt as in Figure 3.43.

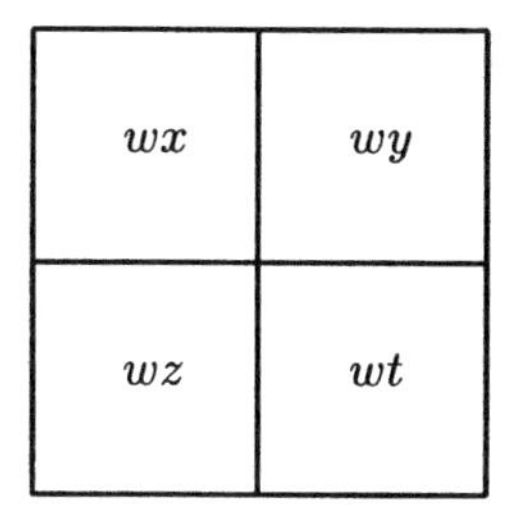

Figure 3.43: Labeling of subsquares of square w

To illustrate let us draw the subsquare with address $xtzx$ (Figure 3.44).

An image representing 2 dimensional data is a language in X^*. A real-valued image is a function $f : A^m \to \mathbb{R}$ or $f : X^* \to \mathbb{R}$.

Another method is to consider the free monoid X^{**}. Elements of $\{0,1\}^{**}$ are sequences of words of $\{0,1\}^*$. Here are two examples, the first

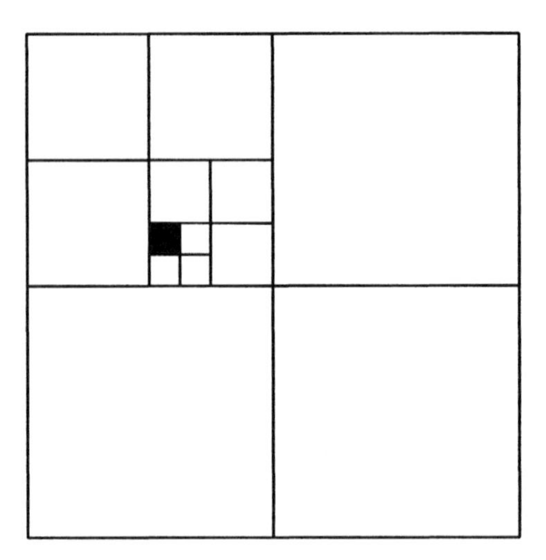

Figure 3.44: Square labeled by $xtzx$

of which happens to be "rectangular".

010 0
111 11111
010 010
101 11

Weighted finite automata are used for analysis of real-valued 2 dimensional data. A *weighted finite automaton* A over alphabet X is defined by

- a row vector $I^A \in \mathbb{R}^{1\times m}$ (called the initial distribution),
- a column vector $F^A \in \mathbb{R}^{m\times 1}$ (called the terminal distribution),
- weight matrices $W_a^A \in \mathbb{R}^{m\times m}$, for all $x \in X$ (transition matrices).

The weighted finite automaton determines a *multiresolution transition function* f_A over X^* by

$$f_A(x_1 x_2 \dots x_k) = I^A \cdot W_{x_1}^A \cdot W_{x_2}^A \cdot \dots \cdot W_{x_k}^A \cdot F^A.$$

We display weighted finite automata using diagrams that are similar to those used for finite state automata. States are represented by circles. If $(W_a)_{i,j} \neq 0$, then there is an edge from state i to state j labeled by $x : (W_a)_{i,j}$.

Let's look at the weighted finite automaton over the alphabet $X = \{x, y, z, t\}$ with the initial distribution $I = (1, 0)$, the final distribution $F = \begin{bmatrix} \frac{1}{2} \\ 1 \end{bmatrix}$, and the weight matrices

$$W_x = \begin{bmatrix} \frac{1}{2} & 0 \\ 0 & 1 \end{bmatrix}, W_y = \begin{bmatrix} \frac{1}{2} & \frac{1}{4} \\ 0 & 1 \end{bmatrix}, W_z = \begin{bmatrix} \frac{1}{2} & \frac{1}{4} \\ 0 & 1 \end{bmatrix}, W_t = \begin{bmatrix} \frac{1}{2} & \frac{1}{2} \\ 0 & 1 \end{bmatrix}$$

The diagram of A is shown in Figure 3.45.

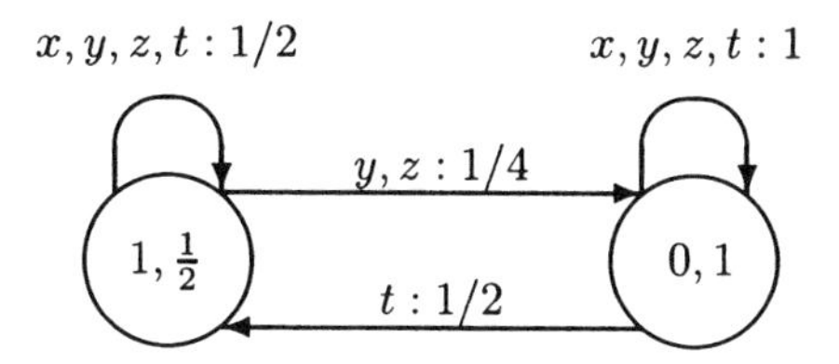

Figure 3.45: Weighted finite automaton A

To find values of f_A, we take a unit square again, and represent subsquares of size 2^{-n} by words over the alphabet x, y, z, t. Suppose that w is a word labeling a subsquare. To find the value $f_A(w)$, we look at all walks in the diagram of the weighted finite automaton such that the letters labeling edges of the walk form the word w. The weight of a walk is obtained by multiplying together the weights of all the transitions on the walk, the initial distribution value of the first node, and the final distribution node of the last node on the walk. Then $f_A(w)$ is the sum of the weights of all walks whose labels form the word w.

For example, $f_A(xt)$ = sum of weights of three walks labeled by $xt = \frac{1}{8} + \frac{1}{4} + 0 = \frac{3}{8}$. Alternatively, $f_A(xt) = IW_xW_tF = \frac{3}{8}$.

A weighted automaton is said to be *deterministic* if the underlying finite automaton obtained by omitting the weights is deterministic. This means that there do not exist two distinct edges beginning in the same state and labeled by the same letter. Nondeterministic weighted finite automata are not equivalent to deterministic ones.

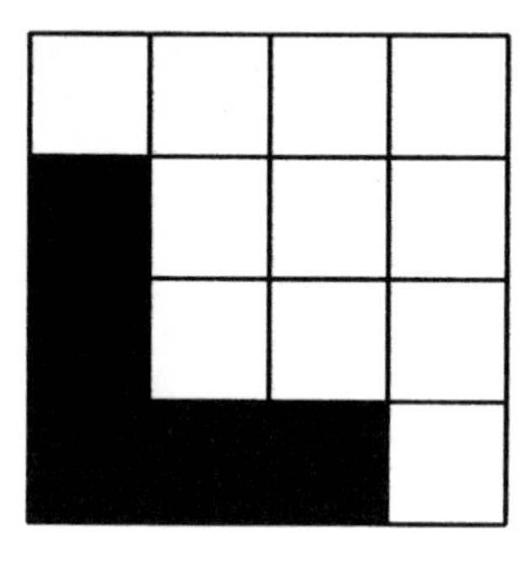

Figure 3.46: $L = \{xy, zx, zz, zt, tz\}$.

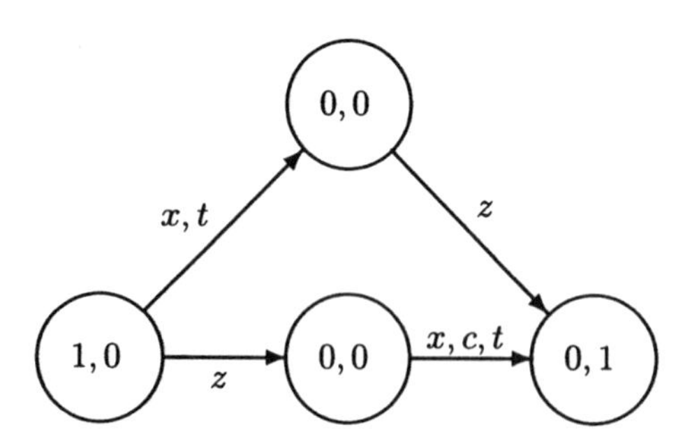

Figure 3.47: WFA for image in Figure 3.46

Exercise 3.24. Find a weighted finite automaton that defines the image in Figure 3.46.

Solution. The image is given by words xz, zx, zz, zt, tz. □

Exercise 3.25. Find a weighted finite automaton that defines the image in Figure 3.48.

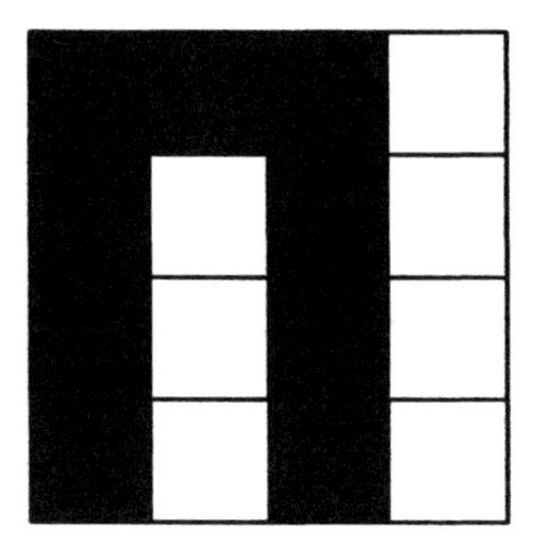

Figure 3.48: $L = \{xx, xy, xz, yx, yz, zx, zz, tx, tz\}$.

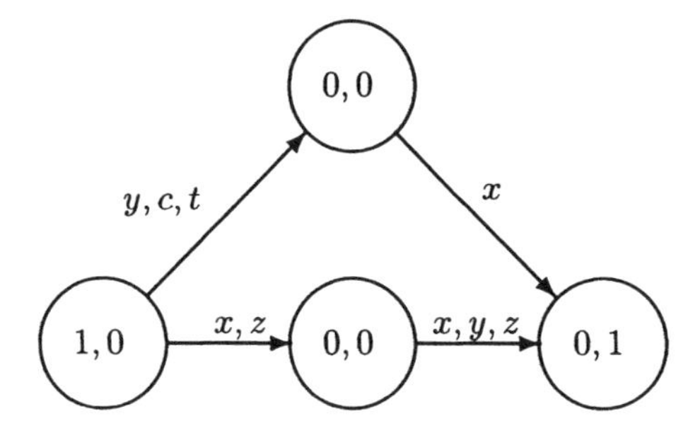

Figure 3.49: WFA for image in Figure 3.48

Solution. The image is encoded as the language

$$L = \{xx, xy, xz, yx, yz, zx, zz, tx, tz\}.$$

□

Example 3.19. The chess board in Figure 3.50 is encoded as the language $(x + y + z + t)^2(y + z)(x + y + z + t)^*$. It is recognized by the automaton with transition diagram in Figure 3.51.

The following small automaton encodes an infinite set of squares.

Figure 3.50: Chess board

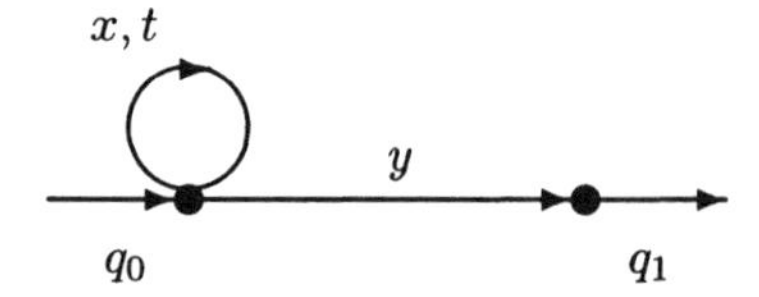

The language $L = L(\mathcal{A})$ accepted by this automaton is given by the regular expression $(x+t)^*y$. The only word of length 1 in L is y. It encodes the second quadrant.

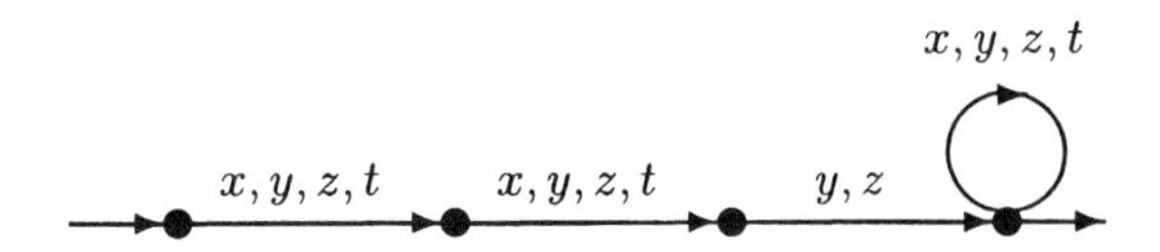

Figure 3.51: FSA encoding the chess board

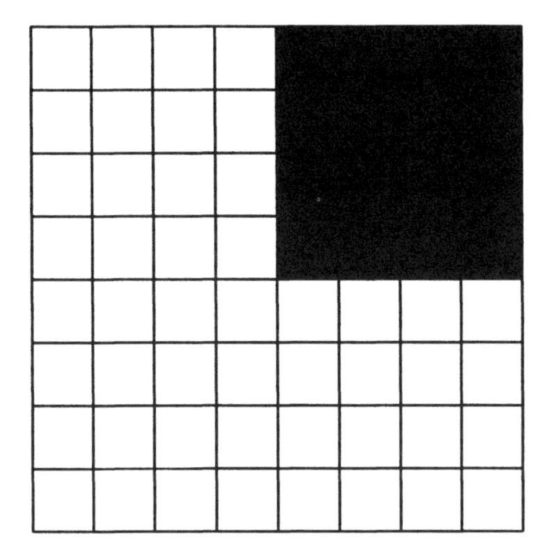

There are two words of length 2 in L: xy, ty. They add two more squares and the image turns into the following one.

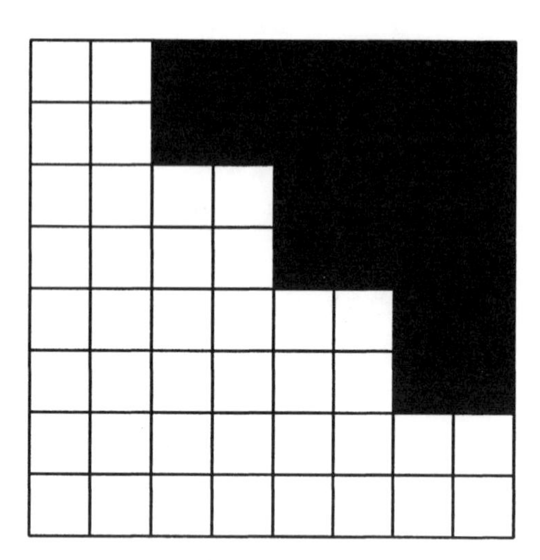

The language L contains 4 words of length 3, namely, xxy, xty, txy, tty. They encode 4 small squares and the image is sketched on the next diagram.

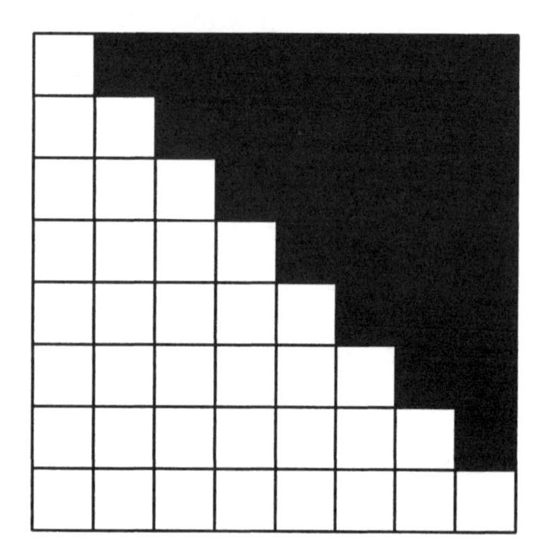

There are 8 words of length 4 in the language L: $xxxy$, $xxty$, $xtxy$, $xtty$, $txxy$, $txty$, $ttxy$, $ttty$.

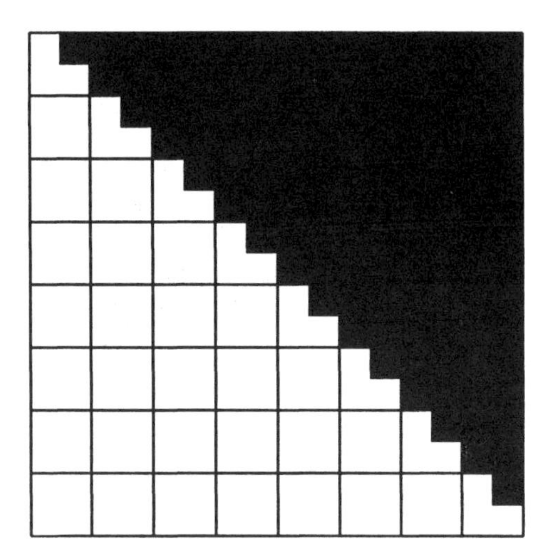

Here is the image encoded by all words of length at most 5 in L.

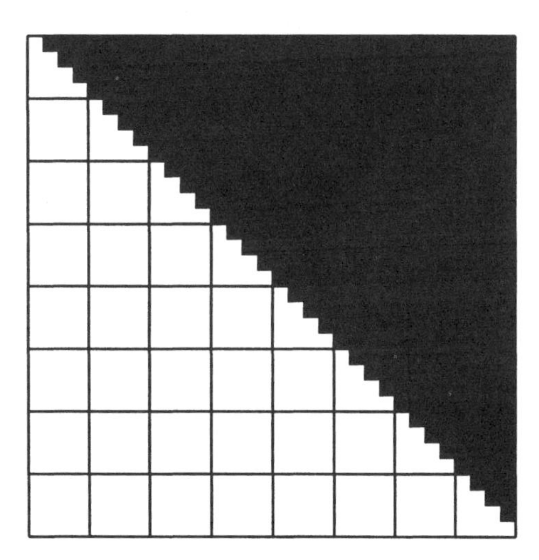

The image encoded by words of length 6 in L is this.

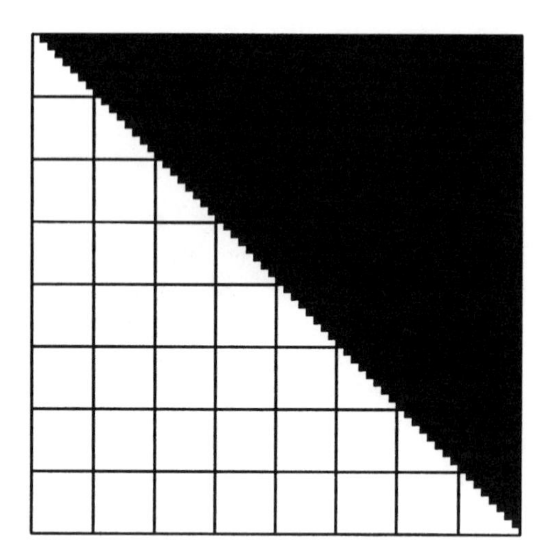

The automaton $\mathcal{A}$ encodes the whole triangle displayed on the next diagram.

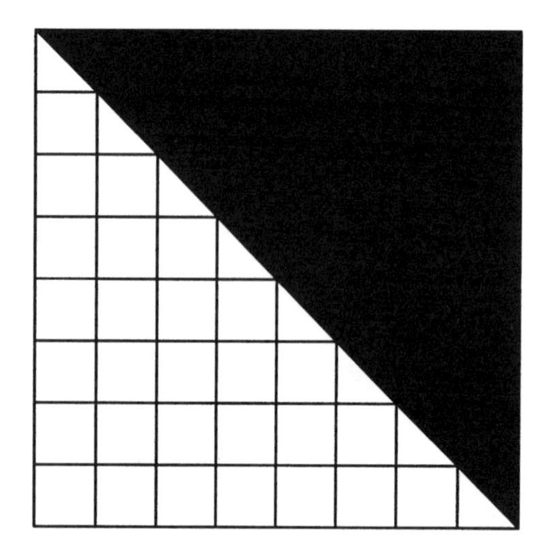

A *finite linear automaton* is a 6-tuple $(Q, X, Y, \delta, \lambda, q_0)$, where

- the set of states Q is a finite dimensional linear space over a finite field $GF(p^r)$;

- X is a finite set called the input alphabet;
- Y is a linear space over $GF(p^r)$ called the output alphabet;
- $\delta : Q \times X \to Q$ is the next-state function such that $\delta(q, x)$ is a linear mapping on Q for every $x \in X$;
- $\lambda : Q \times X \to Y$ is the output function such that $\lambda(q, x)$ is a linear mapping on Q for every $x \in X$;
- $q_0 \in Q$ is the start state.

Example 3.20. Let Q and Y be the spaces of n-dimensional and m-dimensional rows over the field $GF(p^r)$. Suppose that a pair of matrices (Q_x, Y_x) and a pair of vectors (q'_x, y'_x) are defined for each $x \in X$, where X_x is an $n \times n$ matrix, Y_x is an $n \times m$ matrix, $q'_x \in Q$ and $y'_x \in Y$. Define the functions δ and λ by the rule

$$\begin{aligned} \delta(q, x) &= qQ_x. \\ \lambda(q, x) &= qQ_x. \end{aligned}$$

Then $(Q, X, Y, \delta, \lambda)$ is a linear automaton.

We use the signs $\cdot, \circ$ as an alternative notation for δ, λ:

$$\delta(q, x) = q \cdot x, \quad \lambda(q, x) = q \circ x.$$

The *eventual-state function* is defined by

$$q \cdot (x_1 x_2 \cdots x_n) = \delta(\ldots(\delta(\delta(q, x_1), x_2))\ldots, x_n),$$

for $q \in Q$ and $x_1, x_2, \ldots, x_n \in X$. Similarly, the *eventual-output function* is defined by

$$q \circ (x_1 x_2 \cdots x_n) = \lambda(\ldots(\lambda(\lambda(q, x_1), x_2))\ldots, x_n),$$

for $q \in Q$ and $x_1, x_2, \ldots, x_n \in X$.

It is always possible to assume that X and Y are also linear spaces over $F = GF(p^r)$. First, by extending δ to the eventual-state function and λ to the eventual-output function, we may assume that X and Y are monoids. After that, we can extend δ to the whole monoid ring $F[X]$ by the distributive law, and λ to the monoid ring $F[Y]$. Thus we may assume that from the very beginning X and Y are linear spaces over F.

Example 3.21. Let Q and Y be the spaces of n-dimensional and m-dimensional rows over the field $GF(p^r)$. Suppose that a pair of matrices (Q_x, Y_x) is defined for each $x \in X$, where X_x is an $n \times n$ matrix and Y_x is an $n \times m$ matrix. Define the functions δ and λ by the rules

$$\begin{aligned}\delta(q,x) &= qQ_x,\\ \lambda(q,x) &= qY_x.\end{aligned}$$

Then $(Q, X, Y, \delta, \lambda)$ is a linear automaton. It is called a *matrix automaton.*

Every finite automaton can be linearized by considering the semigroup ring on the output monoid.

Definition 3.10. Let Q and Y be the spaces of n-dimensional and m-dimensional rows over the field $GF(p^r)$. Suppose that a pair of matrices (Q_x, Y_x) and a pair of vectors (q'_x, y'_x) are defined for each $x \in X$, where X_x is an $n \times n$ matrix, Y_x is an $n \times m$ matrix, $q'_x \in Q$ and $y'_x \in Y$. Define the functions δ and λ by the rules

$$\begin{aligned}\delta(q,x) &= qQ_x + q'_x,\\ \lambda(q,x) &= qY_x + y'_x.\end{aligned}$$

Then $(Q, X, Y, \delta, \lambda)$ is called an affine automaton. In this case we say that $\delta(q,x)$ and $\lambda(q,x)$ are affine transformations of Q for every $x \in X$.

A *finite group automaton* is an algebraic system $\mathcal{A} = (Q, G, \delta)$, where

- Q is a finite set of *states*;
- G is a finite group of *input symbols*;

- $\delta : Q \times G \to Q$ is a *transition function* satisfying the equality

$$\delta(q, gh) = \delta(\delta(q, g), h) \tag{3.15}$$

for all $q \in Q, g, h \in G$ (see [219]).

Notation $qg = q \cdot g$ is also used for $\delta(q, g)$, where $q \in Q$, $g \in G$. Group automata occur in the Krohn-Rhodes Decomposition Theorem where words of the free monoid on the input alphabet are identified with their images in the transition group. The following example of a group automaton with transition diagram displayed in Figure 3.52 illustrates this concept.

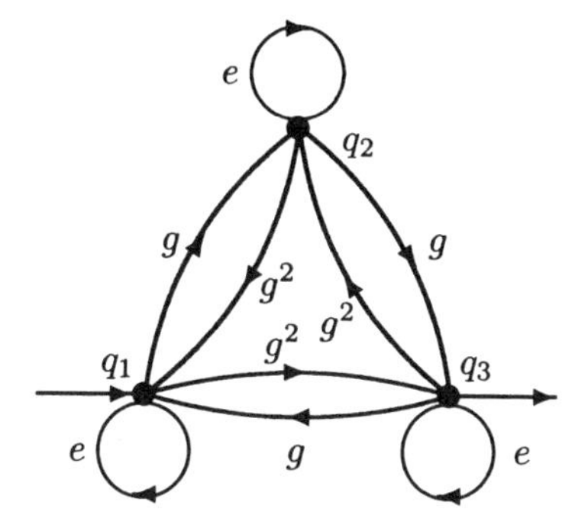

Figure 3.52: A group automaton $\mathcal{A} = (\{1, 2, 3\}, \{e, g, g^2\}, \delta)$.

3.12 Tree Languages

Terms of a free algebra $T_X(\Omega)$ can be used to encode trees. For a nonnegative integer n, denote by Ω_n the set of all n-ary operations in Ω, so that $\Omega = \bigcup_{i=1}^{\infty} \Omega_n$. The trees of terms are defined inductively by the rules:

(i) if x is a letter of X or a nullary operation in Ω, then x corresponds to one vertex labeled by x, and this vertex is also called the *root* of the tree;

(ii) if $n \geq 0$, $\gamma \in \Omega_m$, and $t_2, \ldots, t_m$ are terms with known trees, then the tree of the term $\gamma(t_1, \ldots, t_n)$ is obtained by connecting the roots of the trees of $t_1, \ldots, t_n$ to a new root labeled by γ as illustrated in Figure 3.53.

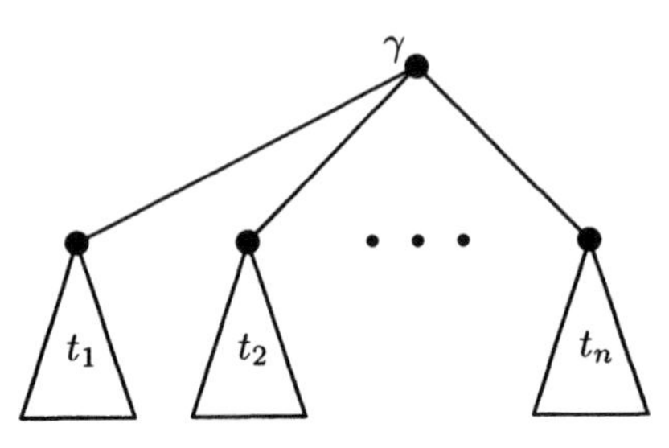

Figure 3.53: $\gamma(t_1, t_2, \ldots, t_n))$.

Example 3.22. As in Example 2.29, let $X = \{x, y, z\}$ and $\Omega = \Omega_0 \cup \Omega_1 \cup \Omega_2 \cup \Omega_3$, where $\Omega_0 = \{\alpha\}$, $\Omega_1 = \{\beta\}$, $\Omega_2 = \{\gamma\}$, $\Omega_3 = \{\delta, \mu\}$. Then the trees in Figures 3.54 and 3.55 represent terms in $F_X(\Omega)$.

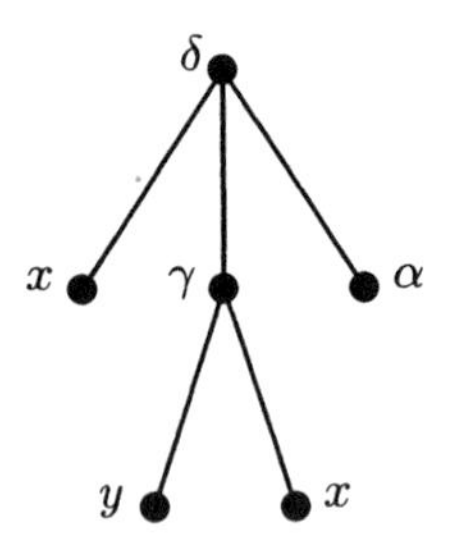

Figure 3.54: $\delta(x, \gamma(y, x), \alpha)$.

The set $\mathrm{Tr}(A)$ of translations of a Σ-algebra A is the smallest set of mappings $A \to A$ which is closed under composition, and contains the identity mapping 1 of A and all elementary translations

$$\tau(\xi) = \sigma^A(a_1, \ldots, a_{i-1}, \xi, a_{i+1}, \ldots, a_m) \qquad (m \geq 0, \sigma \in \Sigma_m, a_j \in A).$$

Let G be an algebra with a subset Q, and let $x \in G$. The *context* of x with respect to Q is defined by

$$\mathrm{Cont}_Q(x) = \{\tau \in \mathrm{Tr}(G) \mid \tau(x) \in Q\}.$$

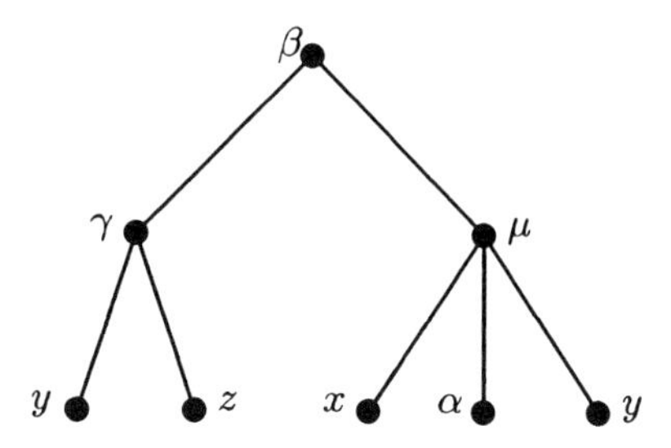

Figure 3.55: $\beta(\gamma(y,z),\mu(x,\alpha,y))$.

The *syntactic algebra* $\mathrm{Syn}(L)$ of a language $L \subseteq G$ is the quotient of G by the *Myhill congruence*

$$\mu_L = \{(w_1, w_2) \in G \times G \mid \mathrm{Cont}_L(w_1) = \mathrm{Cont}_L(w_2)\}.$$

A subset Q of an algebra A is said to be *disjunctive* if every two distinct elements of A have different contexts with respect to Q.

Theorem 3.16. ([242], Corollary 3.10) *A tree language is recognizable if and only if its syntactic algebra is finite.*

Lemma 3.5. ([242], Proposition 3.6) *An algebra A is syntactic if and only if it has a disjunctive subset.*

Lemma 3.6. ([242], Proposition 3.7) *Every subdirectly irreducible algebra is syntactic.*

Denote by $FG(X)$ be the free groupoid over X. If $t \in FG(X)$, then we denote by $\ell(t)$ the leftmost variable of t. For each term $t \in FG(X)$, denote by $G[t] = (V(t), E(t))$ the rooted graph with the root $\ell(t)$, where the vertex set $V(t)$ is the set of variables occurring in t, and the set $E(t)$ of edges is defined inductively by

- $E(x) = \emptyset$ if $x \in X$ is a variable,
- $E(ts) = E(t) \cup E(s) \cup \{(\ell(t), \ell(s))\}$.

Each translation τ of a graph algebra $\mathrm{Alg}(D)$ can be written in the form:

$$\tau = u_1^{f_1} \cdots u_n^{f_n},$$

where $f_i \in \{r, l\}$ and $u_i^r(x) = xu_i$, $u_i^l(x) = u_i x$, $n \geq 1$. If $n = 0$, then $\tau = 1$. A translation of a graph algebra is said to be *trivial* if 0 occurs in it. As it was done for terms, for any translation τ and variable x, the element $\tau(x)$ can be represented by a rooted graph $G[\tau(x)] = (V(\tau(x)), E(\tau(x)))$ with the root $\ell(\tau(x))$.

Example 3.23. ([164]) Let $u_1, \ldots, u_7, x$ be pairwise distinct vertices of a graph D. Then $u_7^l u_6^r u_5^r u_4^l u_3^l u_2^r u_1^r(x) = u_7(((u_4(u_3(xu_1)u_2)))u_5)u_6)$ is represented by the rooted graph shown in Figure 3.56.

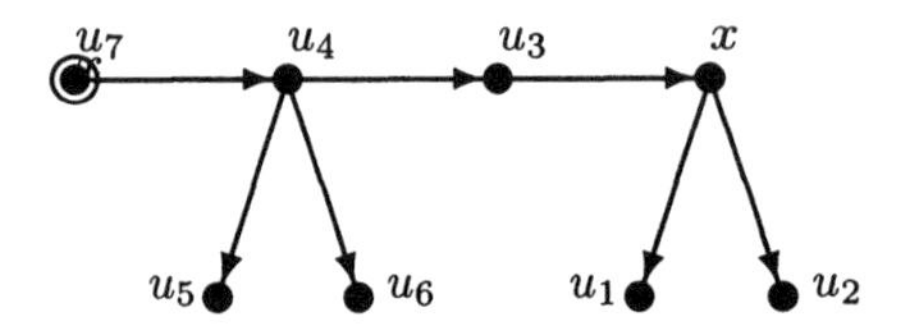

Figure 3.56: $G[u_7^l u_6^r u_5^r u_4^l u_3^l u_2^r u_1^r(x)]$

3.13 Exercises

1 **(a)** Draw a transition diagram and compute $\delta(p, xxyx)$ for the automaton with the set of states $Q = \{p, q, r, s\}$, input alphabet $X = \{x, y\}$, start state s, final states $T = \{r\}$, and transition function δ given by the table

states	alphabet	
	x	y
p	p	r
q	q	s
r	r	q
s	q	p

(b) Draw a transition diagram and compute $\delta(s, xy)$ for the nondeterministic automaton with the set of states $Q = \{q, r, s\}$, input alphabet $X = \{x, y\}$, start state s, final states $T = \{r\}$, and transition function δ given by the following table

states	alphabet	
	x	y
q	{q}	{s}
r	{r,q}	{r,s}
s	{q,s}	{r}

(c) For the automaton in (b), draw a transition diagram for an equivalent deterministic automaton recognizing the same language.

2 **(a)** Draw a transition diagram and compute $\delta(p, yxyy)$ for the automaton with the set of states $Q = \{p, q, r, s\}$, input alphabet $X = \{x, y\}$, start state s, final states $T = \{r\}$, and transition function δ given by the table

states	alphabet	
	x	y
p	r	p
q	p	r
r	r	q
s	q	q

(b) For the automaton in 2(a), find the Nerode equivalence

$$\eta = \{(s, t) \mid sw \in T \Leftrightarrow tw \in T \text{ for all } w \in X^*\}.$$

(c) Find the transformation monoid of the automaton in 2(a).

3 **(a)** Draw a transition diagram and compute $\delta(s, xy)$ for the nondeterministic automaton with the set of states $Q = \{q, r, s\}$, input alphabet $X = \{x, y\}$, start state s, final states $T = \{r\}$, and transition function δ given by the following table

states	alphabet	
	x	y
q	{q,r}	{r,s}
r	{r}	{r,q}
s	{q,s}	{r}

(b) For the automaton in (a), find the Nerode equivalence

$$\eta = \{(s,t) \mid sw \in T \Leftrightarrow tw \in T \text{ for all } w \in X^*\}.$$

(c) Find the transformation monoid of the automaton in (a).

(d) For the automaton in (a), draw a transition diagram for an equivalent deterministic automaton recognizing the same language.

4 **(a)** Let $\mathcal{A}_1$ be the automaton with the set of states $Q = \{p, q, r, s\}$, input alphabet $X = \{x, y\}$, start state s, terminal states $F = \{r\}$, and transition function δ defined by the table on the right. Draw a transition diagram of this automaton.

states	alphabet	
	x	y
p	r	s
q	s	r
r	r	q
s	q	p

(b) Use Algorithm 1 to find a regular expression for $L(\mathcal{A}_1)$.

(c) Use Algorithm 2 to find a regular expression for $L(\mathcal{A}_1)$.

5 **(a)** Let $\mathcal{A}_2$ be the automaton with the set of states $Q = \{p, q, r, s\}$, input alphabet $X = \{x, y\}$, start state s, terminal states $F = \{r\}$, and transition function δ defined by the table on the right. Draw a transition diagram of this automaton.

states	alphabet	
	x	y
p	q	r
q	p	q
r	r	p
s	s	r

(b) Use Algorithm 1 to find a regular expression for $L(\mathcal{A}_2)$.

(c) Use Algorithm 2 to find a regular expression for $L(\mathcal{A}_2)$.

6 Design an automaton that reads strings of 0's and 1's, and determines if the sum of all digits read so far is divisible by 3.

7 Design an automaton that reads strings of 0's and 1's, and checks if there are no more than three consecutive 1's. Thus the automaton accepts an input string unless it contains a substring 1111.

8 **(a)** Let $\mathcal{A}_1$ be the automaton with the set of states $Q = \{q_1, q_2, q_3, q_4, q_5\}$, input alphabet $X = \{a, b\}$, start state q_1, terminal states $F = \{q_4\}$, and transition function δ defined by the table on the right. Draw a transition diagram of this automaton.

states	alphabet	
	a	b
q_1	q_2	q_3
q_2	q_5	q_3
q_3	q_4	q_5
q_4	q_5	q_5
q_5	q_5	q_5

(b) Use Algorithm 1 to find a regular expression for the language accepted by $\mathcal{A}_1$.

(c) Use Algorithm 2 to find a regular expression for $L(\mathcal{A}_1)$.

9 (a) Let $\mathcal{A}_2$ be the automaton with the set of states $Q = \{1, 2, 3, 4\}$, input alphabet $X = \{x, y\}$, start state 1, the set of terminal states $F = \{3\}$, and transition function δ defined by the table on the right. Draw a transition diagram of this automaton.

states	alphabet	
	x	y
1	2	4
2	4	3
3	3	4
4	4	4

(b) Use Algorithm 1 to find a regular expression for the language accepted by $\mathcal{A}_2$.

(c) Use Algorithm 2 to find a regular expression for $L(\mathcal{A}_2)$.

10 Use Algorithm 2 to find a regular expression for the language accepted by the automaton defined by the following transition diagram.

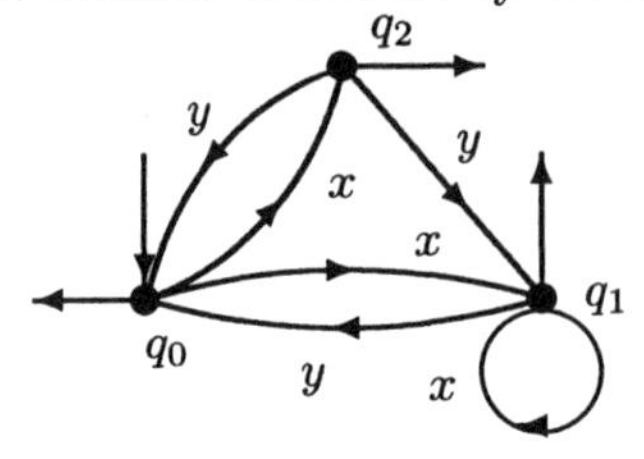

11 (a) Use Algorithm 2 to find the language accepted by the following automaton.

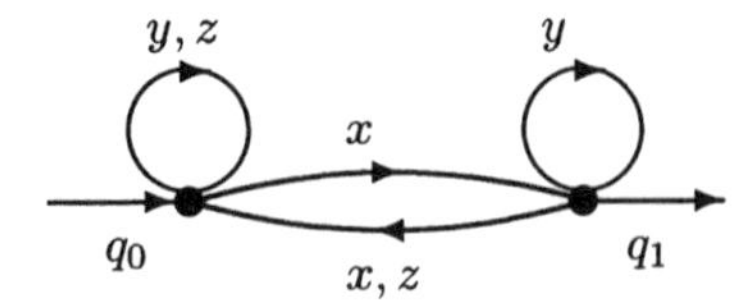

(b) Find the transformation monoid of the automaton given in (a).

(c) State the theorem on the transformation monoid of a minimal automaton of a language. Use it to find the syntactic monoid of the language

$$L = (y + z + xy^*(x + z))^*xy^*.$$

12 (a) Find the language accepted by the automaton $\mathcal{A}_1$ in Figure 3.57.

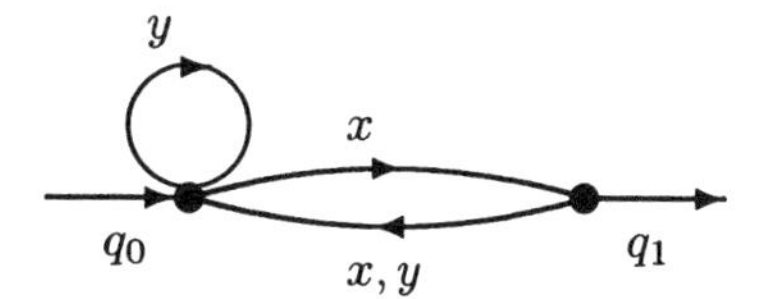

Figure 3.57: Automaton $\mathcal{A}_1$.

(b) Find the transformation monoid of the automaton $\mathcal{A}_1$ in Figure 3.57.

(c) State the theorem on the transformation monoid of a minimal automaton of a language. Use it to find the syntactic monoid of the language $L = [y + x(x + y)]^*x$.

13 (a) Explain how the syntactic monoid of L can be expressed in terms of the Myhill congruence.

(b) Find the syntactic monoid of $L = a + b^*$.

14 (a) Explain how the minimal automaton recognizing L can be defined using the Nerode equivalence.

(b) Find the minimal automaton of $L = a + b^*$.

15 (a) Compute successive approximations of the Nerode equivalence on the following automaton.

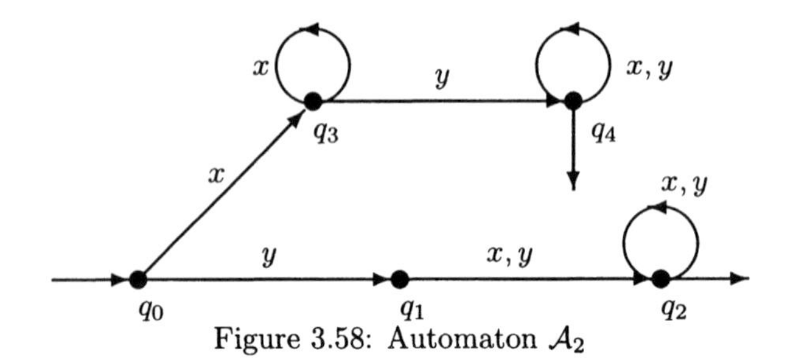

Figure 3.58: Automaton $\mathcal{A}_2$

(b) Find a minimal automaton recognizing the same language as the automaton $\mathcal{A}_2$ in (b).

Chapter 4

Syntactic Monoids of Automata

This chapter contains a complete description of all graphs with algebras isomorphic to syntactic semigroups of languages. To simplify notation here we consider languages which are subsets of the free semigroup X^+ over an alphabet X.

Theorem 4.1. ([161]) *Let $D = (V, E)$ be a graph. Then the following conditions are equivalent:*

(1) *the graph algebra $Alg(D)$ is a syntactic semigroup of a language;*

(2) *the graph D satisfies the following properties:*

   (i) *D has at most one isolated vertex;*

   (ii) *D does not contain complete graphs with three vertices;*

   (iii) *D has no connected component with more than one vertex of zero in-degree;*

   (iv) *in each connected component C of D all vertices of nonzero in-degree induce a complete subgraph;*

   (v) *every vertex of zero in-degree is adjacent to all vertices of nonzero in-degree in its connected component.*

(3) *D has at most one isolated vertex, and every other connected component is isomorphic to one of the graphs* K_1, K_2, K_1^1, K_2^1.

It easily follows from Myhill-Nerode Theorem (Theorem 3.7) that there exist non-regular languages with syntactic semigroups isomorphic to graph algebras.

Proposition 4.1. ([159]) *For any graph* $D = (V, E)$*, the following conditions are equivalent:*

(1) *the graph algebra* $Alg(D)$ *is associative;*

(2) *for all* $(x, y) \in E$ *and* $z \in V$,

$$(x, z) \in E \Leftrightarrow (y, z) \in E;$$

(3) *each connected component of D is isomorphic to* N_1*, or a complete graph, or a direct sum of a null graph and a complete graph.*

Proof. (1) $\Rightarrow$ (2): Consider any $(x, y) \in E$ and $z \in V$. Since $xy = x$, and $(xy)z = x(yz)$ by (1), we get

$$(x, z) \in E \Leftrightarrow (xy)z = x \Leftrightarrow x(yz) = x \Leftrightarrow (y, z) \in E.$$

(2) $\Rightarrow$ (1): Take any three elements x, y and $z \in V$. If $(x, y) \notin E$, then $(xy)z = 0 = x(yz)$. If, however, $(x, y) \in E$, then $(xy)z = xz$, and so by (2) we get

$$(xy)z = x \Leftrightarrow (x, z) \in E \Leftrightarrow (y, z) \in E \Leftrightarrow x(yz) = x.$$

Thus (1) holds in both the cases.

(2) $\Rightarrow$ (3): Let C be a connected component of D with at least two vertices. Denote by K the subgraph induced by all vertices of C with nonzero in-degrees.

Pick any vertex $v \in V(K)$. There exists an edge (u, v), where $u \in V(C)$. By (2),

$$(u, v) \in E \Leftrightarrow (v, v) \in E.$$

Therefore all vertices in K have loops.

Consider an edge (u, v), where $u, v \in V(K)$. In view of (2),

$$(u, u) \in E \Leftrightarrow (v, u).$$

Since u has a loop, $(v, u) \in E$. Thus the subgraph K is undirected.

In order to prove (3) it suffices to verify that C contains edges (x, y) for all vertices $x \in V(C)$ and $y \in V(K)$.

If $x = y$, then $(x, y) \in E$, because K contains all loops. Let y, z_1, z_2, ..., z_{n-1}, x be the shortest path from y to x, where $n > 0$.

By the definition of a path $(z_1, y) \in E$ or $(y, z_1) \in E$. If $(y, z_1) \in E$, then $z_1 \in K$; whence $(z_1, y) \in E$, because K is undirected. Therefore (z_1, y) is an edge in any case.

Suppose that $n > 1$. If $(z_2, z_1) \in E$, then $(z_2, y) \in E$, because (2) shows that E is a transitive relation. If, however, $(z_1, z_2) \in E$, then $(z_2, y) \in E$ by (2), again. Thus $y, z_2, \ldots, z_{n-1}, x$ is a shorter path from y to x. This contradicts the minimality of n and shows that $n = 1$.

Therefore $x = z_1$, and so $(x, y) = (z_1, y) \in E$, as claimed. It follows that K is complete, and moreover (3) holds.

(3) $\Rightarrow$ (2): Take any $(x, y) \in E$ and $z \in V$. Let C by the connected component of D containing x and y. Clearly, we may assume that $C \neq N_1$. Denote by K the comlete subgraph of C mentioned in (3). Since K contains all vertices of nonzero in-degree, we get

$$(x, z) \in E \Leftrightarrow z \in K \Leftrightarrow (y, z) \in E,$$

that is (2) holds. □

Proposition 4.1 immediately gives us the following corollary:

Corollary 4.1. *If the graph algebra of a graph D is associative, then for all $x, y, z \in V$, the following conditions hold:*

(i) $(x, x), (x, y) \in E \Rightarrow (y, x) \in E$;

(ii) $(x, y) \in E \Rightarrow (y, y) \in E$.

Proof of Theorem 4.1. (1) $\Rightarrow$ (2): Suppose that the graph algebra $\mathrm{Alg}(D)$ of D is a syntactic semigroup. Then $\mathrm{Alg}(D)$ is associate, and by Proposition 4.1 every connected component C of D with more than one vertex has a complete subgraph K_C such that each vertex of $C \setminus K$ has in-degree zero and is adjacent to all vertices in K_C. This establishes conditions (iv) and (v).

Further, by Lemma 3.3 there exists $T \subseteq \mathrm{Alg}(D)$ such that $\mathrm{Cont}_T(x)$ is distinct from $\mathrm{Cont}_T(y)$, for all $x \neq y \in \mathrm{Alg}(D)$. Note that the set $\mathrm{Alg}(D) \setminus T$ also satisfies this condition. Therefore we may always assume that $0 \notin T$.

First, we prove the auxiliary fact that all vertices of in-degree zero belong to T. Indeed, since $0 \notin T$, the context of each vertex with in-degree zero which is not in T, equals $\emptyset = \mathrm{Cont}_T(0)$, and so all these vertices should belong to T.

Second, we verify that D has not more than one isolated vertex. All isolated vertices belong to T, as we have seen above. Therefore the context of every isolated vertex consists of one and the same pair $(1, 1)$. Therefore condition (i) follows.

Next, we show that the maximal complete subgraph of each connected component of D has at most two vertices. Take a connected component C of D with more than one vertex, and let K_C be the complete subgraph of C induced by vertices of nonzero in-degree.

Since $0 \notin T$, all elements of $K_C \cap T$ have the same context

$$(V(C) \cap T)^1 \times V(K_C)^1;$$

whence $|V(K_C) \cap T| \leq 1$. Similarly, the context of every element in $V(K_C) \setminus T$ equals $(V(C) \cap T) \times V(C)^1$, and hence $|V(K_C) \setminus T| \leq 1$. Thus K_C has at most two vertices, that is, condition (ii) holds.

Finally, consider any connected component C of D with more than one vertex, and denote by K_C the complete subgraph induced in C by all vertices of nonzero in-degree.

Since every vertex of C with in-degree zero belongs to T, its context is equal to $\{1\} \times V(K_C)^1$. Therefore each connected component of D has at most one vertex with in-degree zero. Thus condition (iii) holds, too.

(2) $\Rightarrow$ (3): Let us take a connected component C of D. If C has one vertex, then it is isomorphic to N_1 or K_1, and so by (i) we may assume that C has more than one vertex. Denote by K_C the subgraph induced in C by all vertices of nonzero in-degree. By (iv), K_C is a complete graph. It follows from (ii) that K_C is isomorphic to K_1 or K_2. In view of (iii), we have $|V(C \setminus K_C)| \leq 1$. If $|V(C \setminus K_C)| = 1$, then by (v), the only vertex of $C \setminus C_K$ is adjacent to all vertices of K_C. Therefore C is isomorphic to K_1, or K_2, or K_1^1, or K_2^1.

(3) $\Rightarrow$ (1): Suppose that D has at most one isolated vertex and every other connected component is isomorphic to K_1, K_2, K_1^1, or K_2^1. Then the graph algebra $\mathrm{Alg}(D)$ is a semigroup by Proposition 4.1, and so we can use Lemma 3.3.

Let us define a subset T of $\mathrm{Alg}(D)$. We include all vertices of zero in-degree in T. For each connected component C of D, we choose one vertex in the maximal connected subgraph K_C of C, and put it in T.

Consider any two distinct elements x, y of $\mathrm{Alg}(D)$. We are going to show that $\mathrm{Cont}_T(x) \neq \mathrm{Cont}_T(y)$.

If D has an isolated vertex v, then v is the only element of $\mathrm{Alg}(D)$ with $\mathrm{Cont}_T(v) = \{(1,1)\}$. Besides, 0 is the only element of $\mathrm{Alg}(D)$ with empty context with respect to T. Therefore we may assume that neither x nor y is an isolated vertex of D, and $x, y \neq 0$.

Suppose that x and y belong to connected components C and C' of D, respectively. Let K_C and $K_{C'}$ be the maximal complete subgraphs of these components. Denote by ℓ the element of T chosen in K_C. The following four cases are possible:

Case 1: $x, y \in K_C$. Then only one of them, say x, belongs to T. Hence $(1,1) \in \text{Cont}_T(x) \setminus \text{Cont}_T(y)$.

Case 2: $x \in K_C, y \notin K_C$. Since K_C is a complete graph, we get $(\ell, x) \in E$. Hence $\ell x = \ell \in T$. Since $y \notin K_C$, there is no edge (ℓ, y), and so $\ell y = 0 \notin T$. Therefore $(\ell, 1) \in Cont_T(x) \setminus \text{Cont}_T(y)$.

Case 3: $x \in C \setminus K_C, y \in K_{C'}$. This case is similar to Case 2.

Case 4: $x \in C \setminus K_C, y \in C' \setminus K_{C'}$. Then x is adjacent to all vertices of K_C, and so $(x, \ell) \in E$. It follows that $(1, \ell) \in Cont_T(x) \setminus \text{Cont}_T(y)$.

Thus $\text{Cont}_T(x) \neq \text{Cont}_T(y)$ in all cases. By Lemma 3.3, the graph algebra of D is a syntactic semigroup. □

Corollary 4.2. ([159]) *Let $D = (V, E)$ be an undirected graph. Then the graph algebra Alg(D) is a syntactic semigroup if and only if D has at most one isolated vertex and all other connected components of D are complete graphs with not more than two vertices.*

Lemma 4.1. ([159]) *A 0-direct union of syntactic semigroups is a syntactic semigroup, too.*

Proof. Let $S = \cup_{i \in I} S_i$ be a 0-direct union of semigroups S_i, where $i \in I$. Suppose that all the semigroups S_i are syntactic. Then by Lemma 3.3 every semigroup S_i contains a subset T_i such that every two distinct elements of S_i have different contexts with respect to T_i. Note that we can always assume that $0 \notin T$ for all $i \in I$. (Indeed, if $0 \in T_i$ for some $i \in I$, we can take the subset $S_i \setminus T_i$ instead of T.) Put $T = \cup_{i \in I} S_i$. The context of 0 with respect to T is empty, as well as w.r.t. all T_i, and for any element $x \in S_i, x \neq 0$, $\text{Cont}_{T_i}(x) = \text{Cont}_T(x)$. Therefore $\text{Cont}_T(x) \neq \text{Cont}_T(y)$ whenever $x \neq y$. □

Lemma 4.2. ([197]) *The graph algebra of an undirected graph D is subdirectly irreducible if and only if D is connected and no more than one pair of distinct vertices have identical neighbourhoods.*

Proof. Suppose that D is disconnected. Choose one of the connected components of D and denote it by $D_1 = (V_1, E_1)$. Let β be the equivalence relation on Alg(D) with equivalence classes $V_1 \cup \{0\}$ and $\{v\}$ for all $v \in V \setminus V_1$. Denote by γ the equivalence relation on Alg(D) with classes $(V \setminus V_1) \cup \{0\}$ and $\{v\}$ for all $v \in V_1$. It is straighforward to check that β and γ are congruences on Alg(D) and $\beta \cap \gamma = \iota$. Therefore Alg($D$) is not subdirectly irreducible by Lemma 9.1. Further, assume that D is connected.

Consider any vertex $v \in V$. We claim that the congruence $\Theta(v,0)$ generated by the pair $(v,0)$ is the complete relation. Indeed, for each $u \in V$ there exists a walk from v to u, say $v = v_0, v_1, \dots, v_m = u$, in D. Hence

$$u = u(v_{m-1}(v_{m-2} \cdots (v_2(v_1 v)) \dots)),$$

and so

$$u \cdot (v,0) = (uv, u0) = (u,0) \in \Theta(v,0).$$

For any $v \in V$, denote by N_v the set of all vertices adjacent to v. If $N_u = N_v$, then the congruence $\Theta(u,v)$ is equal to the equivalence relation with classes $\{u,v\}$ and $\{w\}$ for all $w \in V \setminus \{u,v\}$. If $N_u = N_v$, then $\Theta(u,v)$ is the complete relation. Indeed, we may assume that there exists $w \in N_u \setminus N_v$, and then $w = wu\Theta(u,v)wv = 0$, i.e., $(w,0) \in \Theta(u,v)$. □

Proposition 4.2. ([159]) *The graph algebra of an undirected graph is a syntactic semigroup if and only if it is isomorphic to a* 0*-direct union of subdirectly irreducible semigroups.*

Proof. It is easily seen that Alg(D) is a 0-direct union of the graph algebras of all connected components of the graph D.

Suppose that Alg(D) is a syntactic semigroup. Theorem 4.1 implies that D has at most one isolated vertex, and every other connected component is isomorphic to one of the graphs K_1, K_2. Obviously, in each of these components at most one pair of distinct vertices have the same neighbourhoods. By Lemma 4.2, the graph algebra of an undirected graph is subdirectly irreducible if and only if it is connected and not more than one pair of distinct vertices have identical neighbourhoods. Hence all connected components of D are subdirectly irreducible. Therefore $\mathrm{Alg}(D)$ is isomorphic to a 0-direct union of subdirectly irreducible semigroups.

Conversely, suppose that $\mathrm{Alg}(D)$ is isomorphic to a 0-direct union of subdirectly irreducible semigroups. It is known and easy to verify that every subdirectly irreducible semigroup is a syntactic semigroup of some language (because $\bigcap_{s \in S} \sigma_{\{s\}}$ is the identity relation on S). Therefore all these subsemigroups are syntactic. According to Lemma 4.1, a 0-direct union of syntactic semigroups is syntactic, as well. □

The following example shows that Proposition 4.2 does not generalise from undirected graphs to all graphs.

Example 4.1. ([159]) The graph algebra of the graph K_2^1 is a syntactic semigroup by Theorem 4.1, but it cannot be represented as a 0-direct union of subdirectly irreducible semigroups. Indeed, first note that $\mathrm{Alg}(K_2^1)$ is indecomposable into 0-direct unions of proper subsemigroups. Second, consider the following two principal congruences $\Theta(b,c)$ and $\Theta(a,0)$ on $\mathrm{Alg}(K_2^1)$ with equivalence classes

$$\begin{aligned} \Theta(b,c) &= \{\{b,c\},\{a\},\{0\}\}, \\ \Theta(a,0) &= \{\{a,0\},\{b\},\{c\}\}, \end{aligned}$$

where a is the vertex of in-degree zero in $\mathrm{Alg}(K_2^1)$, and b, c are the other vertices of $\mathrm{Alg}(K_2^1)$. Since $\Theta(b,c) \cap \Theta(a,0)$ is the equality congruence, we see that $\mathrm{Alg}(K_2^1)$ is subdirectly reducible.

Chapter 5

Congruences on Automata

Let $G = (V, E)$ be a graph, and let $\mathrm{Alg}(G)^1 = \mathrm{Alg}(G) \cup \{1\}$ denote its graph algebra with the identity element 1 adjoined. We take a subset T of $\mathrm{Alg}(G) \cup \{1\}$, an arbitrary mapping $f: X \to \mathrm{Alg}(G) \cup \{1\}$, and define the following *right automaton* $\mathrm{Atm}_r(G, T)$:

- the set of states of $\mathrm{Atm}_r(G, T)$ is $V(G) \cup \{1\}$;
- 1 is the initial state;
- T is the set of terminal states;
- the next-state function is given by $a \cdot x = af(x)$, for all $a \in V(G) \cup \{1\}$, $x \in X$.

Hence the eventual-state function is defined by

$$a \cdot (x_1 x_2 \cdots x_n) = (\ldots((af(x_1))f(x_2))\ldots)f(x_n),$$

for $a \in \mathrm{Alg}(G)^1$ and $x_1, x_2, \ldots, x_n \in X$. The transition diagram of this automaton is a labeled Cayley graph of $\mathrm{Alg}(G)$.

A congruence ρ on $\mathrm{Atm}_r(G, T)$ is said to be *essential* if, for every congruence δ on $\mathrm{Atm}_r(G, T)$, the equality $\rho \bigwedge \delta = \iota$ implies $\delta = \iota$. A congruence ρ on $\mathrm{Atm}_r(G, T)$ is said to be a *direct summand* if there exists a congruence δ on $\mathrm{Atm}_r(G,T)$ such that $\rho \bigwedge \delta = \iota$ and $\rho \vee \delta = \eta_T$. Then δ is called a *complement* of ρ. The following theorem shows that automata defined by graph algebras satisfy all three semisimplicity properties introduced in [209].

Theorem 5.1. ([161]) *Let G be a finite graph, and let $Atm_r(G,T)$ be an automaton of the graph algebra of G. Then the following conditions hold:*

(O1) *$Atm_r(G,T)$ has no proper essential congruences;*

(O2) *the Nerode equivalence η_T is a join of minimal congruences;*

(O3) *every congruence on $Atm_r(G,T)$ is a direct summand.*

The next theorem describes all languages recognized by automata of graph algebras in terms of regular expressions.

Theorem 5.2. ([160]) *A nontrivial languge $L \subseteq X^*$ is recognized by an automaton of a graph algebra of some graph if and only if the letters of the alphabet can be reordered and denoted by $x_1, x_2, \dots, x_n$ so that there exist nonnegative integers $k \le \ell \le m \le n$ and subsets $Y_1, Y_2, \dots, Y_\ell$ of $\{x_1, \dots, x_m\}$ such that $\{x_{\ell+1}, \dots, x_m\} \subseteq Y_1 \cap \dots \cap Y_\ell$ and L is defined by one of the following regular expressions:*

(R1) $(x_{\ell+1} + \dots + x_m)^*(x_1 Y_1^* + x_2 Y_2^* + \dots + x_k Y_k^*)$,

(R2) $(x_{\ell+1} + \dots + x_m)^*(x_1 Y_1^* + x_2 Y_2^* + \dots + x_k Y_k^* + 1)$,

(R3) $(x_{\ell+1} + \dots + x_m)^*((x_1 + \dots + x_k)X^* + (x_{k+1}X^*(\{x_1, \dots, x_\ell\} \setminus Y_{k+1}) + \dots + x_\ell X^*(\{x_1, \dots, x_\ell\} \setminus Y_\ell))X^*) + X^*(x_{m+1} + \dots + x_n)X^*$,

(R4) $(x_{\ell+1} + \dots + x_m)^*((x_1 + \dots + x_k)X^* + (x_{k+1}X^*(\{x_1, \dots, x_\ell\} \setminus Y_{k+1}) + \dots + x_\ell X^*(\{x_1, \dots, x_\ell\} \setminus Y_\ell))X^* + 1) + X^*(x_{m+1} + \dots + x_n)X^*$.

The following definitions are required for our proofs. Let $G = (V, E)$ be a finite graph. For any vertex $v \in V$, put

$$\mathrm{Out}(v) = \{w \in V \cap f(X) \mid (v, w) \in E\}.$$

Define an equivalence relation γ_{out} on $\text{Alg}(G) \setminus \{0\}$ as the collection of all pairs (a, b) such that $\text{Out}(a) = \text{Out}(b)$ and either $a, b \in T$ or $a, b \in \text{Alg}(G) \setminus T$. In order to consider the cases where $0 \in T$ and $0 \notin T$ simultaneously, we define the following sets

$$T_0 = \begin{cases} T \setminus \{1\} & \text{if } 0 \in T, \\ \text{Alg}(G) \setminus T & \text{otherwise,} \end{cases}$$
$$\overline{T}_0 = \text{Alg}(G) \setminus T_0.$$

First, we describe all congruences on the automaton $\text{Atm}_r(G, T)$. The equivalence class of a congruence ρ containing x is denoted by x/ρ.

Lemma 5.1. *Let G be a finite graph, and let $Atm_r(G, T)$ be an automaton defined by the graph algebra of G. An equivalence relation ρ on $Atm_r(G, T)$ is a congruence if and only if the following conditions hold:*

(i) *the class $0/\rho$ is contained in T_0;*

(ii) *the restriction of ρ on the set $Alg(G) \setminus (0/\rho)$ is contained in γ_{out}.*

Proof. The 'if' part: Let ρ be an equivalence relation on $\text{Atm}_r(G, T)$ satisfying (i) and (ii). Since γ_{out} satisfies condition (C2) by definition, it follows that the same can be said of ρ.

In order to verify (C1), consider any pair $(a, b) \in \rho$ and a vertex $c \in f(X)$. First, if $a, b \in 0/\rho$, then $ac, bc \in \{a, b, 0\} \subseteq 0/\rho$, and so $(ac, bc) \in \rho$. Second, consider the case where $a, b \notin 0/\rho$. Clearly, (ii) implies $\text{Out}(a) = \text{Out}(b)$. If $c \in \text{Out}(a)$, then $(ac, bc) = (a, b) \in \rho$. If $c \notin \text{Out}(a)$, we get $(ac, bc) = (0, 0) \in \rho$, too.

We have proved that (C1) holds. Thus ρ is a congruence on the automaton $\text{Atm}_r(G, T)$.

The 'only if' part: Let ρ be a congruence on $\text{Atm}_r(G, T)$. Condition (i) easily follows from (C2).

In order to prove (ii), consider any pair $(a, b) \in \rho$ such that $a, b \notin 0/\rho$. Suppose to the contrary that $(a, b) \notin \gamma_{\text{out}}$. Then $\text{Out}(a) \neq \text{Out}(b)$, and we may assume that there exists an element c in $\text{Out}(a) \setminus \text{Out}(b)$. By condition (C1), we get $(a, 0) = (ac, bc) \in \rho$, a contradiction with the choice of a. Thus ρ satisfies (i) and (ii). □

In particular, we see that γ_{out} is a congruence. Since the Nerode equivalence is the largest congruence on $\text{Atm}_r(G, T)$, we get the following

Corollary 5.1. *Let G be a finite graph, and let $Atm_r(G,T)$ be an automaton defined by the graph algebra of G. Then the Nerode equivalence on $Atm_r(G,T)$ is given by*

$$\eta_T = (T_0 \times T_0) \cup (\gamma_{\text{out}} \cap (\overline{T}_0 \times \overline{T}_0)).$$

For any two elements a, b of $\text{Alg}(G)$, consider the equivalence relation

$$\Theta(a, b) = \iota \cup (\{a, b\} \times \{a, b\}).$$

Lemma 5.1 easily gives us the following

Lemma 5.2. *For any $a, b \in Alg(G)$, the equivalence relation $\Theta(a, b)$ is a congruence if and only if one of the following conditions holds:*

(i) $a, b \in \overline{T}_0$ *and* $Out(a) = Out(b)$*;*

(ii) $0 \neq a, b \in T_0$ *and* $Out(a) = Out(b)$*;*

(iii) $a, b \in T_0$ *and* $0 \in \{a, b\}$*.*

Proof of Theorem 5.1. Let ρ be an essential congruence on $\text{Atm}_r(G, T)$. Suppose there exists an element $c \in T_0$ such that $(c, 0) \notin \rho$. Lemma 5.2 shows that $\Theta(c, 0)$ is a congruence. Obviously, $\Theta(c, 0) \bigwedge \rho = \iota$, a contradiction with the fact that ρ is essential. Therefore $0/\rho = T_0$. Suppose now that there exist two vertices $a, b \notin T_0$ such that $(a, b) \in \eta_T$ but $(a, b) \notin \rho$. Lemma 5.2 and Corollary 5.1 show that $\Theta(c, 0)$ is a congruence. We get $\Theta(a, b) \bigwedge \rho = \iota$ again. Therefore $\rho = \eta_T$, and so ρ is not proper. Thus condition (O1) holds.

Corollary 5.1 tells us that $0/\eta_T = T_0$ and the restrictions of η_T and γ_{out} on the set $\text{Alg}(G) \setminus T_0$ coincide. For any $a, b \notin T_0, a \neq b$, such that $(a,b) \in \gamma_{\text{out}}$ and for any vertex $c \in T_0$, the equivalence relations $\Theta(a,b)$ and $\Theta(c,0)$ are congruences by Lemma 5.2. Evidently, these congruences are minimal. We claim that η_T coincides with the congruence

$$\mu = \left(\bigvee_{\substack{a \neq b \notin T_0, \\ (a,b) \in \gamma_{\text{out}}}} \Theta(a,b) \right) \bigvee \left(\bigvee_{0 \neq c \in T_0} \Theta(c,0) \right)$$

Indeed, the inclusion $\mu \subseteq \eta_T$ is obvious. For the reversed inclusion, consider any $(d,e) \in \eta_T$, where $d \neq e$. If $d, e \notin T_0$, then $(d,e) \in \gamma_{\text{out}}$ by Corollary 5.1, and so $\Theta(d,e)$ occurs in the definition of μ. If $d \in T_0$ and $e = 0$, then $(d,e) \in \Theta(d,0) \subseteq \mu$. If $e \in T_0$ and $d = 0$, then $(d,e) \in \mu$, too. If $d, e \in T_0 \setminus \{0\}$, then $(d,e) \in \Theta(d,0) \circ \Theta(e,0) \subseteq \mu$. Thus $\eta_T = \mu$, and so condition (O2) holds.

Finally, take any congruence ρ on $\text{Atm}_r(G,T)$. We define a congruence δ in the following way. Choose a single element in each equivalence class of ρ contained in T_0, except the class $0/\rho$, and put these elements together with 0 in the class $0/\delta$. Further, consider each class K of γ_{out}, contained in $\overline{T}_0$, choose a single element in every class of δ contained in K, and put them together in one new class of δ corresponding to K. All the remaining elements of T_0 will form singleton classes of δ.

By Lemma 5.1, δ is a congruence on $\text{Atm}_r(G,T)$. Clearly, $\rho \bigwedge \delta = \iota$. For any $a, b \in T_0$, by the definition of $0/\delta$ there exist $a', b' \in 0/\delta$ such that $(a,a'), (b,b') \in \rho$. Hence $(a,b) \in \rho \vee \delta$. Similarly, for each class $K \subseteq \overline{T}_0$ of γ_{out}, the join of the restrictions of ρ and δ on K equals $K \times K$. It follows from Corollary 5.1 that $\rho \vee \delta = \eta_T$. Therefore ρ is a direct summand. Thus condition (O3) holds, as well. □

Proof of Theorem 5.2. The 'if' part. Define a graph $G = (V, E)$ as follows: the set V of vertices is $\{x_1, x_2, \ldots, x_\ell\}$, and the set E of edges contains all pairs (x_i, y) for $i = 1, \ldots, \ell$ and $y \in Y_i \cap V$. Let f be the mapping from X

to $\mathrm{Alg}(G)^1$ defined by

$$f(x_i) = \begin{cases} x_i & \text{if } 1 \le i \le \ell, \\ 1 & \text{if } \ell+1 \le i \le m, \\ 0 & \text{if } m+1 \le i \le n. \end{cases}$$

Put $I = (x_{\ell+1} + \cdots + x_m)^*$ and $Z_i = \{x_1, \ldots, x_\ell\} \setminus Y_i$ for $i = 1, \ldots, \ell$.

First, suppose that the language L is given by the regular expression (R1). Put $T_1 = \{x_1, \ldots, x_k\}$, and consider the automaton $\mathrm{Atm}_r(G, T_1)$ on the graph algebra $\mathrm{Alg}(G)$ with the set T_1 of terminal states and mapping f. Take any word w in L. If $w \in Ix_iY_i^*$ for some $1 \le i \le k$, then $1 \cdot w = x_i \in T_1$, because $x_iY_i = x_i$ in $\mathrm{Alg}(G)$; and so w is accepted by $\mathrm{Atm}_r(G, T_1)$.

Conversely, suppose that a word $w = v_1 \ldots v_s$ is accepted by the automaton $\mathrm{Atm}_r(G, T_1)$, i.e., $1 \cdot w \in T_1$ in $\mathrm{Atm}_r(G, T_1)$. It follows that $v_1, \ldots, v_{a-1} \in I$, and $v_a = x_i$ for some $1 \le a \le s$, $1 \le i \le k$, and $v_{a+1}, \ldots, v_s \in Y_i$. Therefore $w \in Ix_iY_i^*$, and so w belongs to L. Thus $\mathrm{Atm}_r(G, T_1)$ recognizes the language given by (R1).

Second, suppose that the language L is defined by the regular expression (R2). Consider the automaton $\mathrm{Atm}_r(G, T_2)$ on the same graph algebra $\mathrm{Alg}(G)$ with the new set $T_2 = \{x_1, \ldots, x_k, 1\}$ of terminal states and the same mapping f. As above, we can verify that $\mathrm{Atm}_r(G, T_2)$ recognizes L. In particular, the empty word belongs to the language defined by $(x_{\ell+1} + \cdots + x_m)^*$, and $\mathrm{Atm}_r(G, T_2)$ accepts it, too.

Third, suppose that L is defined by the regular expression (R3). Consider the automaton $\mathrm{Atm}_r(G, T_3)$ on the graph algebra $\mathrm{Alg}(G)$ with the same mapping f and the new set $T_3 = \{x_1, \ldots, x_k, 0\}$ of terminal states.

Take any word w in L. If $w \in Ix_iX^*$ for some $1 \le i \le k$, then $1 \cdot w \in \{x_i, 0\} \in T_3$. If $w \in Ix_iX^*Z_iX^*$ for $k+1 \le i \le \ell$, then $1 \cdot w = 0 \in T_3$. Finally, if $w \in X^*x_iX^*$ for some $m+1 \le i \le n$, then $1 \cdot w = 0 \in T_3$, again. Thus all words of L are recognized by $\mathrm{Atm}_r(G, T_3)$.

Consider any word $w = v_1 \ldots v_s$ accepted by $\mathrm{Atm}_r(G, T_3)$. If at least one of the letters $v_1, \ldots, v_s$ belongs to the set $\{x_{m+1}, \ldots, x_n\}$, then

$1 \cdot w = 0$, and so w lies in L, as required. Therefore we may assume that w is a word over the alphabet $\{x_1, \ldots, x_m\}$. Two cases are possible.

Case 1: The state $1 \cdot w$ is equal to 0 in $\text{Atm}_r(G, T_3)$. Then there exists $1 \le a \le b < s$ such that $v_1, \ldots, v_{a-1} \in I$, $v_a, v_b \in V$ and (v_a, v_b) is not an edge of G. Taking i such that $v_a = x_i$, we get $v_b \in Z_i$, and so $w \in Ix_iX^*$ if $1 \le i \le k$, and $w \in Ix_iX^*Z_iX^*$ if $k+1 \le i \le \ell$. Therefore $w \in L$.

Case 2: The state $1 \cdot w$ coincides with x_i in $\text{Alg}(G)$, for some $1 \le i \le k$. Then it follows that $w \in Ix_iX^*$. Therefore w belongs to L again. Thus we have proved that $\text{Atm}_r(G, T_3)$ recognizes the language defined by (R3).

Fourth, suppose that L is defined by the regular expression (R4). Consider the automaton $\text{Atm}_r(G, T_4)$ on the graph algebra $\text{Alg}(G)$ with the same mapping f and the new set $T_4 = \{x_1, \ldots, x_k, 1, 0\}$ of terminal states. As above, we can verify that L is recognized by $\text{Atm}_r(G, T_4)$.

The 'only if' part. Suppose that L is recognized by the automaton $\text{Atm}_r(G, T)$ of a graph algebra $\text{Alg}(G)$ of some graph $G = (V, E)$ with a function $f : X \to V \cup \{1, 0\}$. We may assume that

$$f(x_i) \in \begin{cases} V \cap T & \text{if } 1 \le i \le k, \\ V \setminus T & \text{if } k+1 \le i \le \ell, \\ \{1\} & \text{if } \ell+1 \le i \le m, \\ \{0\} & \text{if } m+1 \le i \le n. \end{cases}$$

For $i = 1, \ldots, \ell$, denote by Y_i the union of $\{x_{\ell+1}, \ldots, x_m\}$ and set of all letters x_j such that $f(x_j) \notin \{1, 0\}$ and $(f(x_i), f(x_j))$ is an edge of G. Clearly,

$$1 \cdot x_i y = \begin{cases} 0 & \text{if } y \in X \setminus Y_i, \\ x_i & \text{otherwise.} \end{cases}$$

If $0 \notin T$ and $1 \notin T$, then it is routine to verify that the language L accepted by $\text{Atm}_r(G, T)$ is described by the regular expression (R1). If $0 \notin T$ and $1 \in T$, then we see that L is determined by the regular expression (R2). If $0 \in T$ and $1 \notin T$, then we can check that L is defined by the expression (R3). Finally, if $0 \in T$ and $1 \in T$, then we can show that L is given by (R4). This completes the proof. □

Let $D = (V, E)$ be a graph, $\text{Alg}(D)^1$ the graph algebra with identity 1 adjoined, T a subset of $\text{Alg}(D)^1$, and let $f\colon X \to \text{Alg}(D)^1$ be an arbitrary mapping. Now we consider the *left automaton* $\text{Atm}_\ell(D, T)$, where

- the set of states is $\text{Alg}(D)^1$;
- 1 is the initial state;
- T is the set of terminal states;
- the next-state function is defined by left multiplications of elements of the graph algebra, i.e., $a \cdot x = f(x)a$, for $a \in \text{Alg}(D)^1$, $x \in X$.

An easy illustrating example is given in Figure 5.1.

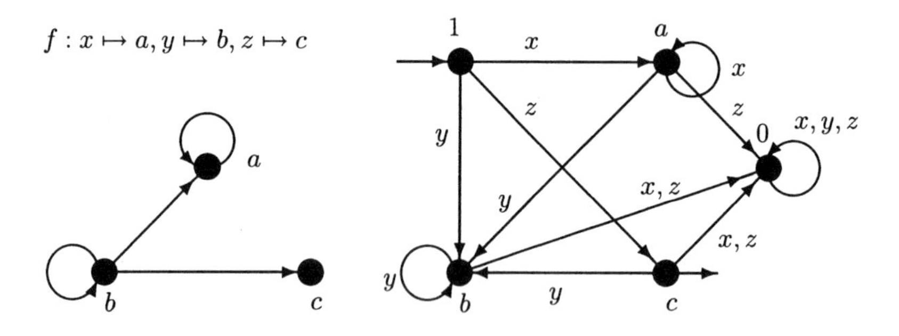

Figure 5.1: Graph and its automaton

The *language recognized* or *accepted* by $\text{Atm}_\ell(D, T)$ is

$$\{u \in X^* \mid 1 \cdot u \in T\}$$

To illustrate let us mention that if D is a null graph, then all products in $\text{Alg}(D)$ are zero, and in this case the language recognized by $\text{Atm}_\ell(D, T)$ is $f^{-1}(1)^* f^{-1}(T)$ if $0 \notin T$, and $X^* f^{-1}(T) \cup X^*(X - f^{-1}(1))X^*(X - f^{-1}(1))X^*$ if $0 \in T$.

Let D' be the subgraph induced in D by the set $V' = V \cap f(X)$ of vertices. Consider the automaton $\mathrm{Atm}_\ell(D', T)$, defined by the graph D' and the same function f. It is easily seen that $\mathrm{Atm}_\ell(D', T)$ recognizes the same language as the original automaton $\mathrm{Atm}_\ell(D, T)$. Thus further we may assume that $V \subseteq f(X)$.

The next main theorem describes all left algebra automata satisfying the conditions (O1), (O2), and (O3) for congruences, and shows that in this case all three properties are equivalent (see Theorem 5.3). Besides, for each automaton $\mathrm{Atm}_\ell(D, T)$, descriptions of all congruences and the Nerode equivalence are included (Lemma 5.3 and Lemma 5.4).

If the class of an equivalence relation ϱ containing 1 is a singleton, then we say that ϱ is an *equivalence relation of the automaton* $\mathrm{Atm}_\ell(D, T)$. Thus, now an equivalence relation ρ on the set of states is called a *congruence* of $\mathrm{Atm}_\ell(D, T)$ if and only if it satisfies the following three conditions:

(C1) $(a, b) \in \rho$ implies $(a \cdot x, b \cdot x) \in \rho$ for all $a, b \in \mathrm{Alg}(D), x \in X$;

(C2) if $(a, b) \in \rho$ and $a \in T$, then $b \in T$;

(C3) the class containing 1 is a singleton.

The largest congruence on $\mathrm{Atm}_\ell(D, T)$ or the *Nerode equivalence* η_T is now described by

$$\begin{aligned} \eta_T \quad = \quad & \{(a, b) \in \mathrm{Alg}(D) \times \mathrm{Alg}(D) \mid a \cdot u \in T \text{ iff } b \cdot u \in T \\ & \text{for all } u \in X^*\} \cup \{(1, 1)\}. \end{aligned} \tag{5.1}$$

For any subset S of $\mathrm{Alg}(D)^1$, denote by $\overline{S}$ the set $\mathrm{Alg}(D) \setminus S$. Note that $\overline{S}$ never contains 1. Clearly, an equivalence relation on $\mathrm{Alg}(D)^1$, with 1 in a separate class, saturates a subset $S \subseteq \mathrm{Alg}(D)$ if and only if it saturates $\overline{S}$. In order to consider the cases where $0 \in T$ and $0 \notin T$ simultaneously, define the following set:

$$T_0 = \begin{cases} T \setminus \{1\} & \text{if } 0 \in T, \\ \overline{T} & \text{otherwise.} \end{cases}$$

We say that a subset $S \subseteq \mathrm{Alg}(D)$ is *in-closed* if $\mathrm{In}(S) \subseteq S$, where $\mathrm{In}(S) = \bigcup_{s \in S} \mathrm{In}(s)$. Putting $\mathrm{In}(0) = \emptyset$ we see that $\{0\}$ is in-closed.

Let C_0 be the set of all elements $c \in T_0$ such that there does not exist any vertex $v \in \overline{T}_0$ with a walk from v to c. Obviously, C_0 is the largest in-closed subset of T_0, and it always contains 0.

Theorem 5.3. ([166]) *Let $D = (V, E)$ be a finite graph, and let $T \subseteq V \cup \{0, 1\}$. Then the following conditions are equivalent:*

(O1) *the automaton $Atm_\ell(D, T)$ has no proper essential congruences;*

(O2) *the Nerode equivalence η_T on $Atm_\ell(D, T)$ is a join of minimal congruences;*

(O3) *every proper congruence on $Atm_\ell(D, T)$ is a direct summand;*

(O4) *there exists an in-closed subset M of $C_0 \setminus \{0\}$ satisfying the following four properties:*

(i) *all connected components of the subgraph induced by M in D are strongly connected;*

(ii) *all vertices in M with nonzero in-degrees have pairwise distinct in-neighbourhoods;*

(iii) *for each $a \in C_0 \setminus \{0\}$, there exists $a' \in M$ such that $In(a) = In(a')$;*

(iv) *for all $a, b \in T_0 \setminus C_0$ or $a, b \in \overline{T}_0$, the equality $In(a) \cap \overline{C}_0 = In(b) \cap \overline{C}_0$ implies $In(a) = In(b)$.*

For any subset S of T_0, consider three auxiliary relations

$$\begin{aligned}
\mu_{S,S} &= (S \cup \{0\}) \times (S \cup \{0\}), \\
\mu_S^{T_0} &= \{(a, b) \mid \mathrm{In}(a) \cap \overline{S} = \mathrm{In}(b) \cap \overline{S} \text{ and } a, b \in T_0 \setminus (S \cup \{0\})\}, \\
\mu_S^{\overline{T}_0} &= \{(a, b) \mid \mathrm{In}(a) \cap \overline{S} = \mathrm{In}(b) \cap \overline{S} \text{ and } a, b \in \overline{T}_0\}.
\end{aligned}$$

We introduce the relation

$$\mu_S = \{(1, 1)\} \cup \mu_{S,S} \cup \mu_S^{T_0} \cup \mu_S^{\overline{T}_0}. \tag{5.2}$$

Clearly, μ_S is an equivalence relation on $\mathrm{Atm}_\ell(D,T)$, and $\mu_S = \mu_{S\cup\{0\}} = \mu_{S\setminus\{0\}}$. It may happen that $S_1 \subseteq S_2$, but $\mu_{S_1} \not\subseteq \mu_{S_2}$, for example, if D has isolated vertices. The following lemma describes all congruences in $\mathrm{Con}(D,T)$.

Lemma 5.3. ([166]) *Let ρ be an equivalence relation on $Atm_\ell(D,T)$. Denote by S the class of ρ containing 0. Then ρ is a congruence on $Atm_\ell(D,T)$ if and only if S is an in-closed subset of C_0 and $\rho \subseteq \mu_S$. In particular, for every in-closed subset $S \subseteq C_0$, the relation μ_S is a congruence on $Atm_\ell(D,T)$.*

Proof. The 'if' part: Suppose that $\rho \subseteq \mu_S$ and S is an in-closed subset of C_0. Since μ_S satisfies conditions (C2) and (C3), it follows that the same can be said of ρ. In order to verify (C1) for ρ, consider any pair $(a,b) \in \rho$ and $f(x) = c$ where $c \notin \{0,1\}$, i.e. $c \in V$.

First, if $c \in \mathrm{In}(a) \cap \mathrm{In}(b)$, then $(a \cdot x, b \cdot x) = (ca, cb) = (c,c) \in \rho$.

Second, if $c \notin \mathrm{In}(a) \cup \mathrm{In}(b)$, then $(a \cdot x, b \cdot x) = (0,0) \in \rho$, too.

Third, suppose that $c \in \mathrm{In}(a) \setminus \mathrm{In}(b)$. We claim that $c \in S$. Indeed, if $a \in S$, then $c \in \mathrm{In}(a) \subseteq \mathrm{In}(S) \subseteq S$, because S is in-closed. If, however, $a \notin S$, then $\rho \subseteq \mu_S$ implies that $\mathrm{In}(a) \cap \overline{S} = \mathrm{In}(b) \cap \overline{S}$, and $c \in S$ again. It follows that $(a \cdot c, b \cdot c) = (ca, cb) = (c, 0) \in S \times S \subseteq \rho$.

The case where $c \in \mathrm{In}(b) \setminus \mathrm{In}(a)$ is similar, and so we have proved that (C1) holds. Thus ρ is a congruence on $\mathrm{Atm}_\ell(D,T)$.

The 'only if' part: Suppose that ρ is a congruence on the automaton $\mathrm{Atm}_\ell(D,T)$. Clearly, $S \subseteq T_0$, because ρ saturates T. To prove that S is in-closed, take any vertex $a \in S$. Condition (C1) implies that $(b,0) = (ba, b0) \in \rho$, for every $b \in \mathrm{In}(a)$. Therefore $\mathrm{In}(S) \subseteq S$.

In order to show that $\rho \subseteq \mu_S$, pick any pair $(a,b) \in \rho$. If $a, b \in S$, then $(a,b) \in \mu_S$, because $S \cup \{0\}$ is an equivalence class of μ_S.

Furthermore, assume that $a, b \notin S$. Condition (C2) shows that ρ saturates T, and so $a, b \in T_0 \setminus S$ or $a, b \in \overline{T}_0$. If there exists $c \in \mathrm{In}(a) \setminus$

In(b), then (C1) implies $(c, 0) = (ca, cb) \in \rho$, and we get $c \in S$. Hence In$(a) \setminus$ In$(b) \subseteq S$. Similarly, In$(b) \setminus$ In$(a) \subseteq S$, and so In$(a) \cap \overline{S} =$ In$(b) \cap \overline{S}$. By the definition of μ_S we see that $(a, b) \in \mu_S$. Thus $\rho \subseteq \mu_S$. □

Lemma 5.4. ([166]) *The Nerode equivalence η_T coincides with μ_{C_0}.*

Proof. By Lemma 5.3, the class $0/\eta_T$ of the Nerode equivalence is in-closed. Therefore it is contained in the largest in-closed subset C_0 of T_0.

To prove the reversed inclusion, consider any vertex $c \in C_0$ and use the equality (5.1). Since C_0 is in-closed, we see that $c \cdot w \in T_0$ for all $w \in X^*$. Obviously, $0 \cdot w \in T_0$ for all $w \in X^*$, too. It follows from (5.1) that $(c, 0) \in \eta_T$. Thus $C_0 = 0/\eta_T$.

Lemma 5.3 tells us that $\eta_T \subseteq \mu_{C_0}$. However, η_T is the largest congruence. Therefore $\eta_T = \mu_{C_0}$. □

For a subset A of Alg(D), denote by $\Theta(A)$ the equivalence relation $\iota \cup (A \times A)$ on $\mathrm{Atm}_\ell(D, T)$.

Lemma 5.5. ([166]) *For any $A \subseteq$ Alg(D), the equivalence relation $\Theta(A)$ is a congruence if and only if one of the following conditions holds:*

(i) *all vertices of A have the same in-neighbourhoods, and either $A \subseteq \overline{T}_0$ or $A \subseteq T_0 \setminus \{0\}$;*

(ii) *A is an in-closed subset of T_0 and $0 \in A$.*

Proof. The 'if' part: If (i) holds, then $\Theta(A) \subseteq \mu_{\{0\}}$. If (ii) holds, then $\Theta(A) \subseteq \mu_A$. In both cases $\Theta(A)$ is a congruence by Lemma 5.3.

The 'only if' part: Suppose that $\Theta(A)$ is a congruence. If $0 \notin A$, then $0/\Theta(A) = \{0\}$, and so $\Theta(A) \subseteq \mu_{\{0\}}$ by Lemma 5.3. The definition of $\mu_{\{0\}}$ shows that (i) holds. If $0 \in A$, then $0/\Theta(A) = A$, and so A is an in-closed subset of T_0 by Lemma 5.3, i.e., condition (ii) is satisfied. □

Lemma 5.6. ([166]) *Let M be an in-closed subset of the set V of vertices. Then the subgraph of $D = (V, E)$ induced by M is strongly connected if and* *only if M is a minimal nonempty in-closed subset of V.*

Proof. If M is a singleton, then the assertion is trivial, and so we assume that $|M| > 1$. For any vertex $v \in V$, denote by $\mathrm{In}^*(v)$ the set of all vertices $u \in V$ such that there exists a walk from u to v. Obviously, $\mathrm{In}^*(v)$ is the smallest in-closed set containing v. It is easily seen that, for every in-closed subset M of vertices, the subgraph of D induced by M is strongly connected if and only if $M = \mathrm{In}^*(v)$ for every $v \in M$. Let us verify that the latter fact is equivalent to M being a minimal in-closed subset of V.

To prove one implication, suppose to the contrary that $M = \mathrm{In}^*(v)$, for every $v \in M$, but there exists an in-closed proper subset N of M. Then, for each vertex $v \in N$, the set $\mathrm{In}^*(v)$ is contained in the in-closed set N, contradicting $M = \mathrm{In}^*(v)$.

For the converse implication, suppose that M is a minimal in-closed subset of V. Take any $v \in M$. Obviously, $\mathrm{In}^*(v) \subseteq M$ since M is in-closed. Since the set $\mathrm{In}^*(v)$ is in-closed, by the minimality of M we get $M = \mathrm{In}^*(v)$. This completes our proof. □

Lemma 5.7. *Let the Nerode equivalence η_T be the join of minimal congruences α_i on the automaton $Atm_\ell(D,T)$, for $i \in I$. Then $Atm_\ell(D,T)$ has no proper essential congruences.*

Proof. Let ρ be a proper congruence on $\mathrm{Atm}_\ell(D,T)$. For every $i \in I$, either $\alpha_i \subseteq \rho$, or $\alpha_i \bigwedge \rho = \iota$. If all the α_i are contained in ρ, then $\rho = \eta_T$, a contradiction with ρ being proper. Therefore there exists $i \in I$ such that $\alpha_i \bigwedge \rho = \iota$. Thus ρ is not essential. □

Lemma 5.8. *Let all proper congruence of $Atm_\ell(D,T)$ be direct summands. Then the Nerode equivalence η_T on $Atm_\ell(D,T)$ is a join of minimal congruences.*

Proof. Denote by α_i, where $i \in I$, all minimal congruences on $\mathrm{Atm}_\ell(D,T)$. Suppose to the contrary that $\alpha = \bigvee_{i\in I} \alpha_i$ is properly contained in η_T. By

(O3) there exists a congruence $\delta \neq \iota$ such that $\alpha \perp \delta$. Choose a minimal proper congruence $\delta' \subseteq \delta$. Then $\delta' \subseteq \alpha$, and so $\delta' = \alpha \bigwedge \delta' \subseteq \alpha \bigwedge \delta = \iota$. This contradiction shows that $\eta_T = \bigvee_{i \in I} \alpha_i$. □

Lemma 5.9. *Let M be an in-closed subset M of $C_0 \setminus \{0\}$ satisfying the properties (i) to (iv), and let ρ be a proper congruence on $Atm_\ell(D, T)$. Then ρ is a direct summand.*

Proof. First, consider the easy case where $M = \emptyset$. Condition (iii) implies that $C_0 = \{0\}$. Lemma 5.3 shows that an equivalence relation is a congruence if and only if it is contained in $\mu_{\{0\}}$. By Lemma 5.4, $\eta_T = \mu_{\{0\}}$. Denote by K_i, where $i \in I$, all $\mu_{\{0\}}$-classes not equal to $\{0\}$. Evidently, every class K_i is a disjoint union of some ρ-classes: $K_i = \bigcup_{j \in J_i} K_{ij}$ for some J_i. In this notation all classes of the congruence ρ are $\{0\}$ and the K_{ij}, where $i \in I, j \in J_i$. For each $i \in I$, pick one element a_{ij} in each K_{ij}. The equivalence relation $\tau = \bigcup_{i \in I} \Theta(\{a_{ij} \mid j \in J_i\})$ is a congruence, because $\tau \subseteq \mu_{\{0\}}$. It is easily seen that $\rho \perp \tau$. Thus condition (O3) holds.

Second, consider the case where $M \neq \emptyset$. If a vertex v of T_0 has in-degree 0 and does not belong to M, then it is an isolated vertex of the subgraph induced in D by $M \cup \{v\}$, and so we can adjoin it to M. Therefore without loss of generality we may assume that M contains all vertices of T_0 with in-degree zero. We are going to define a congruence δ which is a direct complement to ρ. To this end we have to introduce notation for certain subsets in C_0 and classify the classes of ρ in relation to these sets.

Denote by $M_1, \ldots, M_n$ the sets of vertices of all connected components of the subgraph induced by M in D. By (i) all these components are strongly connected. Lemma 5.6 says that $M_1, \ldots, M_n$ are minimal in-closed subsets of $C_0 \setminus \{0\}$. Take any minimal in-closed subset M' of $C_0 \setminus \{0\}$, which does not consist of a single vertex with in-degree 0. Since the subgraph induced by M' is strongly connected by Lemma 5.6, we get $\text{In}(a) \neq \emptyset$ for any vertex $a \in M'$. Therefore (iii) implies that $\text{In}(a) = \text{In}(b)$, for some $b \in M_i$, where $1 \leq i \leq n$. Since both M' and M_i are in-closed, we get

$\mathrm{In}(a) \subseteq M' \cap M_i$, and so $M' = M_i$, because they are minimal. Thus M is a disjoint union of all minimal in-closed subsets of $C_0 \setminus \{0\}$.

If M_i consists of a single vertex with in-degree zero, then put $N_i = M_i$. Otherwise, denote by N_i the set of all vertices a in C_0 such that $\mathrm{In}(a) = \mathrm{In}(a')$ for some $a' \in M_i$. Condition (iii) implies that $C_0 \setminus \{0\}$ is a disjoint union of the N_i, $i = 1, \ldots, n$.

For any i, put $P_i = N_i \setminus M_i$. Note that $\mathrm{Out}(a) \cap C_0 = \emptyset$, for every vertex a of $\bigcup_{i=1}^{n} P_i$. Indeed, if $(a, b) \in E$, for some $b \in C_0$, then by (iii) there exists $b' \in M$, such that $\mathrm{In}(b) = \mathrm{In}(b')$, and so $(a, b') \in E$, which is impossible, because M is in-closed.

Denote by S the class $0/\rho$. By Lemma 5.3, S is in-closed, and so $S \cap M$ is in-closed, too. Moreover, since M is a disjoint union of all minimal in-closed subsets of C_0, we see that both the sets $M \setminus S$ and $M \cap S$ are disjoint unions of some minimal in-closed subsets M_i of M. Assume that $M \cap S = \bigcup_{i=1}^{k} M_i$ and $M \setminus S = \bigcup_{i=k+1}^{n} M_i$, and that every set $M_1, \ldots, M_s, M_{k+1}, \ldots, M_t$ consists of a single vertex with in-degree zero, where $0 \leq s \leq k \leq t \leq n$. For $i = s+1, \ldots, k$, let $P_i' = P_i \cap S$. The set C_0 and its subsets are illustrated in Figure 5.2. Note that C_0 is a disjoint union of some ρ-classes. The class S is the disjoint union $\{0\} \cup (\bigcup_{i=1}^{k} M_i) \cup \left(\bigcup_{i=s+1}^{k} P_i' \right)$.

We claim that each ρ-class $K \subseteq C_0$ such that $K \neq S$ has one of the following four types:

Type 1: $K \subseteq P_i$, where $i = t+1, \ldots, n$, and all vertices of K have the same in-neighbourhoods.

Type 2: $K \subseteq M_i \cup P_i$, where $i = t+1, \ldots, n$, $|K \cap M_i| = 1$ and all vertices of K have the same in-neighbourhoods.

Type 3: $K \subseteq (P_{s+1} \setminus P_{s+1}') \cup \cdots \cup (P_k \setminus P_k') \cup M_{k+1} \cup \cdots \cup M_t$ and $K \cap M \neq \emptyset$.

Type 4: $K \subseteq (P_{s+1} \setminus P_{s+1}') \cup \cdots \cup (P_k \setminus P_k')$.

N_n	M_n	P_n	
$\vdots$	$\vdots$	$\vdots$	
N_{t+2}	M_{t+2}	P_{t+2}	
N_{t+1}	M_{t+1}	P_{t+1}	
N_t	M_t		
$\vdots$	$\vdots$		
N_{k+2}	M_{k+2}		
N_{k+1}	M_{k+1}		
N_k	$\mathbf{M_k}$	$\mathbf{P'_k}$	$P_k \setminus P'_k$
$\vdots$	$\vdots$	$\vdots$	$\vdots$
N_{s+2}	$\mathbf{M_{s+2}}$	$\mathbf{P'_{s+2}}$	$P_{s+2} \setminus P'_{s+2}$
N_{s+1}	$\mathbf{M_{s+1}}$	$\mathbf{P'_{s+1}}$	$P_{s+1} \setminus P'_{s+1}$
N_s	$\mathbf{M_s}$		
$\vdots$	$\vdots$		
N_2	$\mathbf{M_2}$		
N_1	$\mathbf{M_1}$		
	$\mathbf{0}$		

Figure 5.2: Partition of the set C_0

Indeed, first suppose that K contains a vertex a of N_i, for some i, where $t+1 \le i \le n$. Take any b in K. By the definition of N_i we get $\text{In}(a) = \text{In}(a')$ for $a' \in M_i$. Since the in-neighbourhoods of all vertices in M_i are nonempty, we see that $\emptyset \ne \text{In}(a) \subseteq M_i$. By Lemma 5.3, $\text{In}(b) \cap \overline{S} = \text{In}(a)$. If $b \in N_j$, then $\text{In}(b) \subseteq M_j$, and so $j = i$. Hence $b \in N_i$. It follows that $\text{In}(b) = \text{In}(a)$. Thus $K \subseteq N_i$, and the in-neighbourhoods of all vertices of K are equal to $\text{In}(a)$. Now, if $K \cap M_i = \emptyset$, then K is of Type 1. Otherwise, if K contains a vertex of M_i, then (ii) shows that $|K \cap M_i| = 1$, and so K is of Type 2.

Suppose now that K contains a vertex a in the union

$$(P_{s+1} \setminus P'_{s+1}) \cup \cdots \cup (P_k \setminus P'_k) \cup M_{k+1} \cup \cdots \cup M_t.$$

Take any $b \in K$. Lemma 5.3 shows that $\text{In}(b) \cap \overline{S} = \text{In}(a) \cap \overline{S}$. We have $\text{In}(a) \subseteq M_{s+1} \cup \cdots \cup M_k \subseteq S$, and so $\text{In}(b) \subseteq S$. Observe that

$b \notin \bigcup_{i=t+1}^{n} N_i$, because otherwise $\emptyset \neq \mathrm{In}(b) \subseteq \bigcup_{i=t+1}^{n} M_i$, which is impossible, because $S \cap \bigcup_{t+1}^{n} M_i = \emptyset$. Therefore $b \in \bigcup_{i=1}^{t} N_i$. Since $S \cap K = \emptyset$, we get $K \subseteq (P_{s+1} \setminus P'_{s+1}) \cup \cdots \cup (P_k \setminus P'_k) \cup M_{k+1} \cup \cdots \cup M_t$. Now, if $K \cap (M_{k+1} \cup \cdots \cup M_t) \neq \emptyset$, then K is of Type 3. Otherwise, K is of Type 4.

Let us mention that using Lemma 5.3 it is possible to show that every partition of C_0 such that one class is a disjoint union $\{0\} \cup (\bigcup_{i=1}^{k} M_i) \cup \bigcup_{i=s+1}^{k} P'_i$, and other classes are arbitrary subsets of Types 1, 2, 3, or 4, is the restriction on C_0 of some congruence on $\mathrm{Atm}_\ell(D,T)$.

In order to define the required δ, we first construct its class containing 0. Let Q_1 be the union $M_{t+1} \cup \cdots \cup M_n$. In each ρ-class of Type 1 pick a vertex v belonging to some set P_i, where $t+1 \leq i \leq n$, and denote the set of chosen vertices by Q_2. Choose one vertex with in-degree zero in each ρ-class of Type 3, and denote the set of these vertices by Q_3. We claim that the set $Q = Q_1 \cup Q_2 \cup Q_3$ is in-closed. Indeed, take a vertex $a \in Q$. If $a \in Q_1$, then $\mathrm{In}(a) \subseteq Q_1$, because Q_1 is a union of in-closed sets. If $a \in Q_2$, then a belongs to a set P_i, for some $i \in \{t+1, \ldots, n\}$, and there exists a vertex $a' \in M_i$ such that $\mathrm{In}(a) = \mathrm{In}(a') \subseteq M_i \subseteq Q_1$. If $a \in Q_3$, then $\mathrm{In}(a) = \emptyset$. Thus $\mathrm{In}(a) \subseteq Q$, in all cases. Therefore

$$\delta_1 = \Theta(Q \cup \{0\})$$

is a congruence by Lemma 5.5.

Next, in each ρ-class of Type 4 we pick a vertex u, which belongs to some $P_i \setminus P'_i$, where $s+1 \leq i \leq k$, and denote the set of these vertices u by U. For each $u \in U$, by (iii) there exists a vertex $u' \in M_i$ such that $\mathrm{In}(u) = \mathrm{In}(u')$. Hence $\Theta(u,u')$ is a congruence by Lemma 5.5. Denote by U' a minimal set of all vertices $u' \in M$ such that, for each $u \in U$, there exists $u' \in U'$ with $\mathrm{In}(u) = \mathrm{In}(u')$. Define a congruence

$$\delta_2 = \bigvee_{u \in U} \Theta(u,u').$$

It is routine to verify that the family of all classes of δ_2 consists of all classes

$$H_{u'} = \{u'\} \cup \{u \in U \mid \mathrm{In}(u) = \mathrm{In}(u')\}, \tag{5.3}$$

where u' runs over U'.

Furthermore, let K_i, $i \in I$, be all η_T-classes not equal to C_0. Since η_T is the largest congruence, every class K_i is a disjoint union of some ρ-classes: $K_i = \bigcup\limits_{j \in J_i} K_{ij}$ for some J_i. Therefore all classes of ρ lying outside of C_0 are K_{ij}, where $i \in I, j \in J_i$. For $i \in I$, pick one element a_{ij} in each K_{ij} and consider the equivalence relation $\tau_i = \Theta(\{a_{ij} \mid j \in J_i\})$. For each $i \in I$, all the elements a_{ij}, $j \in J_i$, belong to the same η_T-class K_i. Hence $\mathrm{In}(a_{ij}) \cap \overline{C}_0$ is the same for all $j \in J_i$ in view of Lemma 5.4. Condition (iv) implies that all the elements a_{ij}, where $j \in J_i$, have the same in-neighbourhood. Therefore τ_i is a congruence by Lemma 5.5. Put

$$\delta_3 = \bigcup_{i \in I} \tau_i.$$

For any $i_1, i_2 \in I$, $i_1 \neq i_2$, we see that the only possibly non-singleton class $\{a_{i_1 j} \mid j \in J_{i_1}\}$ of τ_{i_1} does not intersect the class $\{a_{i_2 j} \mid j \in J_{i_2}\}$ of τ_{i_2}. It follows that δ_3 is an equivalence relation. Since all the τ_i are congruences, it follows that δ_3 is a congruence, too.

Finally, define

$$\delta = \delta_1 \bigcup \delta_2 \bigcup \delta_3.$$

Since $(U \cup U') \cap Q = \emptyset$, we see that the relation $\delta_1 \cup \delta_2$ is transitive, and therefore it is a congruence. By construction, all non-singleton classes of $\delta_1 \cup \delta_2$ lie in C_0, while all non-singleton classes of δ_3 lie outside of C_0. It follows that δ is a transitive relation and, moreover, a congruence.

Now, we are going to verify that $\rho \perp \delta$. Take any pair $(c, d) \in \rho \bigwedge \delta$. First, assume that $(c, d) \in \delta_1$. By the definition of δ_1 and because $Q \cap S = \emptyset$, we get either $c = d$ or $c, d \in Q$. In the latter case we see that $\mathrm{In}(c)$ and $\mathrm{In}(d)$ are nonempty since $Q \cap S = \emptyset$. Now $(c, d) \in \rho$ implies $\mathrm{In}(c) = \mathrm{In}(d)$, because Q is in-closed. Therefore $c = d$ by (ii). Second, suppose that

$(c,d) \in \delta_2$. Then $c = d$, or $c, d \in H_{u'}$ for some $u' \in U'$. The definition (5.3) of the class $H_{u'}$ shows that in the latter case $\mathrm{In}(c)$ and $\mathrm{In}(d)$ are nonempty and by (ii) $c = d$, too. If $(c,d) \in \delta_3$, then $(c,d) \in \tau_i$, for some $i \in I$, and the definition of τ_i yields $c = d$. Therefore $\rho \bigwedge \delta = \iota$.

In order to prove that $\rho \vee \delta = \eta_T$, take any pair $(c,d) \in \eta_T$. If $c, d \in T_0 \setminus C_0$ or $c, d \in \overline{T}_0$, then c, d lie in some η_T-class K_i. Let $c \in K_{ij_1}$ and $d \in K_{ij_2}$, for some $j_1, j_2 \in J_i$, and let $c' \in K_{ij_1}$ and $d' \in K_{ij_2}$ be the elements which were chosen in these ρ-classes when we defined τ_i. Then $(c,c'), (d,d') \in \rho$ and $(c',d') \in \tau_i$, which yields $(c,d) \in \rho \circ \tau_i \circ \rho \subseteq \rho \vee \delta$. Finally, assume that $c, d \in C_0$. It is enough to verify that $(c, 0) \in \rho \vee \delta$. To this end, consider the ρ-class K that contains c. Note that the set Q contains one element in each ρ-class of Types 1, 2, and 3 contained in C_0. Therefore if K is of Type 1, 2, or 3, then $(c,0) \in \rho \circ \delta_1$. If K is of Type 4, then there exist $u \in K \cap U$ and the corresponding $u' \in M_{s+1} \cup \cdots \cup M_k$ such that $(u, u') \in \delta_2$. Hence $(c, 0) \in \rho \circ \delta_2 \circ \rho$. Therefore in all these cases $(c, 0) \in \rho \vee \delta$. Thus $\eta_T = \rho \vee \delta$, as required. □

Lemma 5.10. *If the automaton $Atm_\ell(D,T)$ has no proper essential congruences, then conditions* (i) *to* (iv) *are satisfied.*

Proof. Assume that the automaton If $C_0 = \{0\}$, then conditions (i) to (iv) are vacuously true for $M = \emptyset$. Thus we may assume that $C_0 \neq \{0\}$. In particular, $\eta_T \neq \iota$ by Lemma 5.4.

Let $M_1, \ldots, M_n$ be all nonempty minimal in-closed subsets in $C_0 \setminus \{0\}$. Clearly, they are pairwise disjoint, and $n \geq 1$. The union $M = M_1 \cup \cdots \cup M_n$ is in-closed. Besides, it contains all vertices of C_0 with in-degree zero. We are going to verify that M satisfies (i) to (iv).

The connected components of the subgraph induced by M in D are the subgraphs induced by $M_1, \ldots, M_n$. By Lemma 5.6, each subgraph of this sort is strongly connected. Thus (i) is satisfied.

Suppose that (ii) does not hold, i.e., there exist two distinct vertices a, b of M with nonzero in-degree such that $\mathrm{In}(a) = \mathrm{In}(b) \neq \emptyset$. Since all the

M_i are in-closed, it follows that a and b lie in the same set M_i.

Denote by $\mathrm{Out}^*(M_i)$ the set of all vertices $u \in V$ such that there exists a walk from some vertex of M_i to u. Consider the set

$$H = C_0 \setminus \mathrm{Out}^*(M_i).$$

Since $M \setminus M_i$ is in-closed, there do not exist any walks from vertices of M_i to vertices of $M \setminus M_i$. Therefore

$$M \setminus M_i \subseteq H.$$

We are going to prove that μ_H is a proper essential congruence.

In order to verify that μ_H is a congruence, by Lemma 5.3 it suffices to check that the set H is in-closed. To this end, take any $u \in H$, and suppose to the contrary that there exists $v \in \mathrm{In}(u) \setminus H$. Since C_0 is in-closed, we get $v \in C_0$. Then $v \in \mathrm{Out}^*(M_i)$, and so there exists a walk from some vertex $x \in M_i$ to v. Completing this walk by the edge (v, u) we get a walk from x to u; whence $u \in \mathrm{Out}^*(M_i)$. This contradicts the choice of u and establishes that μ_H is a congruence.

By (5.2) and the definition of μ_H, we get $(a, b) \in \mu_H$. Hence $\mu_H \neq \iota$. Besides, $\mu_H \neq \eta_T$, because $\eta_T = \mu_{C_0} \supseteq \{0\} \times M_i$ and $\mu_H \cap (\{0\} \times M_i) = \emptyset$. Thus μ_H is a proper congruence.

It remains to show that μ_H is essential. Take an arbitrary congruence $\delta \neq \iota$. We need to verify that $\mu_H \bigwedge \delta \neq \iota$. Denote by S the equivalence class of δ containing 0. By Lemma 5.3, the set S is in-closed.

First, consider the case where $S \neq \{0\}$. Choose a minimal in-closed subset S' in $S \setminus \{0\}$. We have $S' \subseteq M$. If $S' = M_i$, then $(a, b) \in \mu_H \bigwedge \delta$. If $S' \neq M_i$, then $S' \subseteq H$, and therefore $\Theta(S' \cup \{0\}) \subseteq \mu_H \bigwedge \delta$. Hence $\mu_H \bigwedge \delta \neq \iota$ in this case.

Second, suppose that $S = \{0\}$. Pick any pair $(c, d) \in \delta$ with $c \neq d$. Lemma 5.3 tells us that $(c, d) \in \mu_{\{0\}}$. We get $\mathrm{In}(c) = \mathrm{In}(d)$.

If $\mathrm{In}(c) = \emptyset$, then either $c, d \in \overline{T}_0$ and $(c, d) \in \mu_H$, or $c, d \in M$. Note that neither c nor d belong to M_i, because every vertex of in-degree zero forms a minimal (singleton) in-closed subset in M and $|M_i| \geq 2$. Hence $c, d \in M \setminus M_i \subseteq H$, and so $(c, d) \in \mu_H$.

Further, assume that $\mathrm{In}(c) \neq \emptyset$. It can be easily seen that $c \in \mathrm{Out}^*(M_i)$ if and only if $d \in \mathrm{Out}^*(M_i)$ because $\mathrm{In}(c) = \mathrm{In}(d)$. Now, suppose that one of the vertices, say c, belongs to H. Then $\mathrm{In}(c) \subseteq C_0$, because C_0 is in-closed. Since $(c, d) \in \mu_{\{0\}}$, we get $d \in T_0$. Assume that $d \in T_0 \setminus C_0$. Since C_0 is the largest in-closed subset of T_0, there exists $x \in \mathrm{In}(d) \setminus C_0$, which is impossible because $\mathrm{In}(d) = \mathrm{In}(c) \subseteq C_0$. Hence $c \in H$ if and only if $d \in H$. Therefore $(c, d) \in \mu_H$, again.

Thus, in the second case, we get $\delta \subseteq \mu_H$, and so $(\mu_H) \bigwedge \delta = \delta \neq \iota$, as required. Hence the congruence μ_H is essential. This contradicts (O1) and shows that (ii) holds.

In the case where $H = \emptyset$ the proof remains valid too.

Next, let us verify (iii). If M_i consists of a single vertex with in-degree zero, then put $N_i = M_i$. Otherwise, denote by N_i the set of all vertices a in C_0 such that $\mathrm{In}(a) = \mathrm{In}(a')$ for some $a' \in M_i$. Since the M_i are in-closed, we see that the sets $N_1, \ldots, N_n$ are pairwise disjoint. The set $N = N_1 \cup \cdots \cup N_n$ is in-closed, because $\mathrm{In}(a) \subseteq M \subseteq N$ for every vertex $a \in N$.

We claim that $\mu_{\{0\}} \subseteq \mu_N$. Indeed, take any pair $(a, b) \in \mu_{\{0\}}$. Since the definition of $\mu_{\{0\}}$ is more demanding then the definition of μ_N are the same on $\overline{T}_0$, we only have to consider the case where $a, b \in T_0 \setminus \{0\}$. The definition of $\mu_{\{0\}}$ implies $\mathrm{In}(a) = \mathrm{In}(b)$. If $a, b \in T_0 \setminus N$, then $(a, b) \in \mu_N$. Further, we may assume that one of these vertices, say a, belongs to N_i. First, suppose that a has in-degree zero, and so is the only vertex of some $N_i = M_i$. Then b is also a vertex with in-degree zero, hence $b \in C_0$, and so $b \in M \subseteq N$, too. This means that $(a, b) \in \mu_N$, as required. Second, suppose that $|\mathrm{In}(a)| > 0$. By the definition of N_i, there exists $a' \in M_i$ such that $\mathrm{In}(a) = \mathrm{In}(a')$. Then $\mathrm{In}(b) = \mathrm{In}(a') \subseteq C_0$, hence $b \in C_0$, and so $b \in N_i \subseteq N$, again. Thus in both cases $a, b \in N$, and hence $(a, b) \in \mu_N$, by the definition of μ_N. Therefore $\mu_{\{0\}} \subseteq \mu_N$, as claimed.

Suppose that $N \neq C_0 \setminus \{0\}$. Then μ_N is a proper congruence. By (O1), μ_N is not essential. Hence there exists a congruence $\delta \neq \iota$ such that $\mu_N \bigwedge \delta = \iota$. Denote by K the class of δ containing 0. If $K = \{0\}$, then $\delta \subseteq \mu_{\{0\}} \subseteq \mu_N$. This contradiction shows that $K \neq \{0\}$. By Lemma 5.3, K is an in-closed subset of C_0, and we can choose a minimal in-closed subset M_{n+1} in $K \setminus \{0\}$ that does not intersect N. This contradicts the definition of M and shows that $N = C_0 \setminus \{0\}$. Thus (iii) holds.

It remains to prove (iv). Suppose to the contrary that there exist two elements a, b, which both belong either to the set $T_0 \setminus C_0$ or to the set $\overline{T}_0$, and $\text{In}(a) \cap \overline{C}_0 = \text{In}(b) \cap \overline{C}_0$, but $\text{In}(a) \neq \text{In}(b)$. Consider the equivalence relation ρ on $\text{Atm}_\ell(D, T)$ which has the same classes as η_T with the only exception: we divide a/η_T into two new classes,

$$\{c \in a/\eta_T \mid \text{In}(c) = \text{In}(a)\} \text{ and } \{c \in a/\eta_T \mid \text{In}(c) \neq \text{In}(a)\}.$$

Since $a \notin C_0$, it follows from Lemma 5.3 that ρ is a congruence on $\text{Atm}_\ell(D, T)$.

We prove that ρ is essential. Assume that $\rho \bigwedge \delta = \iota$ for some congruence δ. Then $0/\delta = \{0\}$, because $0/\rho = C_0$. Take any $(c, d) \in \delta$. If $c \notin a/\eta_T$, then $c/\rho = c/\eta_T \supseteq c/\delta$; whence $(c, d) \in \rho$. If $d \notin a/\eta_T$, then $(c, d) \in \rho$ and so further we assume that $c, d \in a/\eta_T$. Since $0/\delta = \{0\}$, Lemma 5.3 yields $\text{In}(c) = \text{In}(d)$. If $\text{In}(c) = \text{In}(a)$, then both c and d lie in a/ρ by the definition of ρ, and therefore $(c, d) \in \rho$. Similarly, if $\text{In}(c) \neq \text{In}(a)$, then c, d lie in b/ρ, and $(c, d) \in \rho$, again. Thus $(c, d) \in \iota$ in all cases. Thus ρ is essential. This contradiction shows that (iv) is satisfied. □

Proof of Theorem 5.3 follows from Lemmas 5.7 through 5.10. □

Chapter 6

Minimal Automata

Let $\ell : X \to \{+,-\}$ and $f : X \to V$ any mappings, and let T be a subset of V. The *two-sided automaton* $\mathrm{Atm}_t(D) = \mathrm{Atm}_t(D,T) = \mathrm{Atm}_t(D,T,f,\ell)$ of the graph D is the (possibly incomplete) finite state acceptor with

(DA1) the set of states $V \cup \{1\}$;

(DA2) the initial state 1;

(DA3) the set of terminal states T;

(DA4) the next-state function given, for a state u and a letter $x \in X$, by the rule

$$u \cdot x = \begin{cases} f(x) & \text{if } \ell(x) = + \text{ and } (u, f(x)) \in E, \text{ or if } u = 1, \\ u & \text{if } \ell(x) = - \text{ and } (f(x), u) \in E. \end{cases}$$

If a vertex $v \in V$ does not belong to $f(X)$, then this state is inaccessible in $\mathrm{Atm}_t(D,T)$, and so without loss of generality we may assume that all vertices are images of letters of the alphabet X.

Let $Y \subseteq S$, and let ϱ be an equivalence relation on Y. The class of ϱ containing x is denoted by x/ϱ. If there is no need to indicate the set Y explicitly, we may call the equivalence relation an *incomplete* equivalence relation on S. Often we omit the word 'incomplete' when there is no ambiguity. The set Y is called the *ground set* of ϱ, and is denoted by G_ϱ.

An incomplete equivalence relation ϱ on the set of states of $\mathrm{Atm}_t(D,T)$ is called an incomplete *congruence* if it defines the quotient automaton recognizing the same language. Defining the quotient automaton modulo an incomplete congruence, as usual, one has to drop all states which do not belong to the ground set of the congruence, and then introduces a new transition function on the set of all equivalence classes of the relation. It follows that ϱ is an incomplete congruence if and only if the following conditions hold, for all $a, b \in V \cup \{1\}$, $x \in X$,

(I1) if $(a,b) \in \varrho$ and $a{\cdot}x$ is defined, then $b{\cdot}x$ is defined too, and $(a{\cdot}x, b{\cdot}x) \in \varrho$;

(I2) if $(a,b) \in \varrho$ and $a \in T$, then $b \in T$;

(I3) $(1,1) \in \varrho$ and the class containing 1 is a singleton;

(I4) if $1 \cdot x_1 \cdots x_n \in T$ for some $x_1, \ldots, x_n \in X$, then

$$x_1, \ldots, x_n, 1 \cdot x_1, 1 \cdot x_1 x_2, \ldots, 1 \cdot x_1 \cdots x_n \in G_\varrho.$$

Let ϱ_1 and ϱ_2 be incomplete relations with ground sets S_1 and S_2, respectively. Then we write $\varrho_1 \leq \varrho_2$ if $\varrho_1 \subseteq \varrho_2$ and $S_1 \supseteq S_2$. The largest incomplete congruence on $\mathrm{Atm}_t(D,T)$ is the *Nerode equivalence* η_T described by

$$\begin{aligned} \eta_T \quad = \quad & \{(a,b) \mid a \cdot u \in T \text{ iff } b \cdot u \in T \text{ for all } u \in X^*; \\ & \text{and } (a \cdot X^*) \cap T \neq \emptyset\} \cup \{(1,1)\}. \end{aligned} \tag{6.1}$$

The following sets of letters and vertices are used in our main theorem and proofs:

$$\begin{aligned} X^{(+)} &= \{x \in X \mid \ell(x) = +\}, \\ X^{(-)} &= \{x \in X \mid \ell(x) = -\}, \\ V^{(+)} &= \{v \in V \mid \exists x \in X^+, f(x) = v\}, \\ V^{(-)} &= \{v \in V \mid \exists x \in X^{(-)}, f(x) = v\}. \end{aligned}$$

Note that the intersection $V^{(+)} \cap V^{(-)}$ may be nonempty in general. If $v \in V$ and $S \subseteq V$, then put

$$\begin{aligned} \mathrm{In}^-(v) &= \{w \in V^{(-)} \mid (w,v) \in E\}, \\ \mathrm{Out}^+(v) &= \{w \in V^{(+)} \mid (v,w) \in E\}, \\ \mathrm{In}^-(S) &= \cup_{s \in S} \mathrm{In}^-(s), \\ \mathrm{Out}^+(S) &= \cup_{s \in S} \mathrm{Out}^+(s). \end{aligned}$$

For a subset S of V, define new equivalence relations

$$\alpha_S = \{(1,1)\} \cup \{(a,b) \mid a,b \in V \setminus S,\ \mathrm{In}^-(a) = \mathrm{In}^-(b)\}, \quad (6.2)$$

$$\Theta(T) = \{(1,1)\} \cup (T \times T) \cup ((V \setminus T) \times (V \setminus T)), \quad (6.3)$$

and consider auxiliary sets

$$\begin{aligned} \beta_S^T &= \{(a,b) \mid \mathrm{Out}^+(a) \setminus S = \mathrm{Out}^+(b) \setminus S \text{ and } a,b \in T\}, \\ \beta_S^{V \setminus T} &= \{(a,b) \mid \mathrm{Out}^+(a) \setminus S = \mathrm{Out}^+(b) \setminus S \text{ and } a,b \in V \setminus T\}. \end{aligned}$$

We introduce new relation β_S as the following disjoint union

$$\beta_S = \{(1,1)\} \cup \beta_S^T \cup \beta_S^{V \setminus T}. \quad (6.4)$$

Clearly, β_S is an equivalence relation on the set of states of $\mathrm{Atm}_t(D,T)$. Our main theorem describes all incomplete congruences on the automaton $\mathrm{Atm}_t(G,T,f,\ell)$.

Denote by T_+ the set of all elements $v \in V$ such that either $v \in T$ or there exist a vertex $t \in T \cap V^{(+)}$ and a walk $v = v_0, v_1, \ldots, v_n = t$, from v to t with $n \geq 1$ and all vertices $v_1, \ldots, v_n$ in $V^{(+)}$. Let $C = C_D$ be the set of all vertices $c \in V$ such that $c \notin T_+$ and if $c \in V^{(-)}$ then $(v,c) \notin E$ for all $v \in T_+$.

Theorem 6.1. *The automaton $Atm_t(D,T,f,\ell)$ is minimal if and only if $C_D = \emptyset$ and $\alpha_\emptyset \cap \beta_\emptyset \cap \Theta(T) = \iota$.*

Theorem 6.2. *Let ϱ be an incomplete equivalence relation on $Atm_t(D,T,f,\ell)$, and let $S = V \setminus G_\varrho$. Then ϱ is a congruence of this automaton if and only if S is a subset of C_D and*

$$\varrho \subseteq \alpha_S \cap \beta_S \cap \Theta(T). \quad (6.5)$$

Proof of Theorem 6.2. The 'only if' part. Take any incomplete congruence ϱ of the automaton $\mathrm{Atm}_t(D, T, f, \ell)$.

Let us begin by showing that condition (I4) implies that $S = V \setminus G_\varrho$ is a subset of $C = C_D$. To this end consider any vertex v which does not belong to C. All we have to verify is that v lies in G_ϱ. In view of the definitions of C the following cases may occur.

Case 1. $v \in T$. Then $1 \cdot v = v \in T$, and so $1 \cdot v \in G_\varrho$ by condition $(I4)$.

Case 2. $v \notin T$ and $v \in T_+$. Then the definition of T_+ means that there exist a vertex $t \in T \cap V^{(+)}$ and a walk $v = v_0, v_1, \ldots, v_n = t$, from v to t with $n \geq 1$ and all vertices $v_1, \ldots, v_n$ in $V^{(+)}$. By the definition of $\mathrm{Atm}_t(D, T, f, \ell)$ we get

$$1 \cdot v_0 v_1 \ldots v_n = t \in T.$$

Hence condition (I4) yields $v = 1 \cdot v_0 \in G_\varrho$.

Case 3. $c \in V^{(-)}$ and $(v, c) \in E$ for some $v \in T$. By (DA4), $1 \cdot vc = v \in T$. Therefore $v = 1 \cdot v \in G_\varrho$ in view of (I4).

Case 4. $c \in V^{(-)}$ and $(v, c) \in E$ for some $v \in T_+$, $v \notin T$. Then there exist $t \in T \cap V^{(+)}$ and a walk $v = v_0, v_1, \ldots, v_n = t$, from v to t with $n \geq 1$ and all vertices $v_1, \ldots, v_n$ in $V^{(+)}$ such that $(v, c) \in E$. It follows from (DA4) that

$$1 \cdot vcv_1v_2 \cdots v_n = t \in T.$$

Hence (I4) implies $v \in G_\varrho$ again.

Next take (a, b) in ϱ. We check that (a, b) is in all three equivalence relations in the right hand side of (6.5). Since $(1, 1)$ belongs to all of them, by (I3) we may assume $a, b \neq 1$. Condition (I2) shows that (a, b) always lies in $\Theta(T)$. Therefore it remains to prove that (a, b) belongs to $\alpha_S \cap \beta_S$.

Suppose to the contrary that $(a, b) \notin \alpha_S$. Since $S = V \setminus G_\varrho$ and $\varrho \subseteq G_\varrho \times G_\varrho$, we have $a, b \notin S$. Therefore it follows from the definition of α_S that $\mathrm{In}^-(a) \neq \mathrm{In}^-(b)$. We may assume that there exists $u \in \mathrm{In}^-(a) \setminus \mathrm{In}^-(b)$. Choose x in $X^{(-)}$ such that $f(x) = u$. Then $ax = a$ and bx is undefined.

This contradicts condition (I1) and shows that $(a, b) \in \alpha_S$.

If there exists an element u in $\mathrm{Out}^+(a) \cap \overline{S} \setminus \mathrm{Out}^+(b) \cap \overline{S}$, then $u = f(x)$ for some $x \in X^{(+)}$; whence $ax = x$ and bx is undefined, a contradiction to (I1). Therefore $\mathrm{Out}^+(a) \cap \overline{S} \subseteq \mathrm{Out}^+(b) \cap \overline{S}$. The reversed inclusion is proven in exactly the same way; whence

$$\mathrm{Out}^+(a) = \mathrm{Out}^+(b). \tag{6.6}$$

In proving that $(a, b) \in \beta$ first note that if $a, b \in T$, then $\mathrm{Out}^+(a) = \mathrm{Out}^+(b)$ implies $(a, b) \in \beta^T$. If, however, a or b is not in T, then $a, b \in V \setminus T$ as indicated above. Hence $(a, b) \in \beta^{V \setminus T}$ again, and we get $(a, b) \in \beta$. Therefore $(a, b) \in \beta$ in both cases. Thus (6.5) is satisfied.

The 'if' part. Let ϱ be an incomplete equivalence relation such that $S = V \setminus G_\varrho$ is a subset of C and the inclusion (6.5) holds. We claim that ϱ is a congruence.

Indeed, since $\varrho \subseteq \Theta(T)$, conditions (I2) and (I3) are obvious. In order to verify (I1), choose an arbitrary pair $(a, b) \in \varrho$ and $x \in X$ such that ax is defined. Note that $a, b \notin S$ by the definition of G_ϱ. The following cases are possible.

Case 1: $\ell(x) = -$. Since $(a, b) \in \alpha_S$, we get $\mathrm{In}^-(a) = \mathrm{In}^-(b)$. If $f(x) \notin \mathrm{In}^-(a)$, then ax and bx are undefined. This contradiction shows that $f(x) \in \mathrm{In}^-(a)$. Therefore $ax = a$, $bx = b$, and so $(ax, bx) \in \varrho$.

Case 2: $\ell(x) = +$. Since $(a, b) \in \beta$, we get $\mathrm{Out}^+(a) \setminus S = \mathrm{Out}^+(b) \setminus S$. If $f(x) \notin \mathrm{Out}^+(a)$, then ax and bx are undefined, a contradiction. Therefore $f(x) \in \mathrm{Out}^+(a) \setminus S \subseteq \mathrm{Out}^+(b)$, and so bx is defined too. It follows that $(ax, bx) = (f(x), f(x)) \in \varrho$, because $\ell(x) = +$.

Thus, if ax is defined, then (ax, bx) always belongs to ϱ, i.e., (I1) holds.

It remains to prove condition (I4). Choose any elements $x_1, \ldots, x_n \in X$ such that $1 \cdot x_1 \cdots x_n \in T$. All we have to verify is that $x_1, \ldots, x_n$, $1 \cdot x_1$, $1 \cdot x_1 x_2, \ldots, 1 \cdot x_1 \cdots x_n$ are not in S.

Denote by $i_1, i_2, \ldots, i_m$ all integers such that $x_{i_1}, \ldots, x_{i_m} \in X^{(+)}$ and $i_1 \leq i_2 \leq \cdots \leq i_m$. Clearly, all $x_{i_1}, \ldots, x_{i_m}$ are in T_+. Consider any k such that $1 \leq k \leq n$.

First, suppose that $\ell(x_k) = +$. Then $k = \ell_q$ for some q, and $x_q = 1 \cdot x_1 \cdots x_n$. If $q = m$, then it follows from (DA4) that $x_q = 1 \cdot x_1 \cdots x_n \in T$. Hence $1 \cdot x_1 \cdots x_q = x_q \notin C$. If, however, $q < m$, then (DA4) implies

$$1 \cdot x_q \cdots x_n = 1 \cdot x_1 \cdots x_n \in T$$

Hence $x_q \in T^+$, and so $1 \cdot x_1 \cdots x_q = x_q \notin C$ again.

Second, consider the case where $\ell(x_k) = -$. If $i_1 > k$, then $1 \cdot x_1 \cdots x_q = x_1 \in T_+$. This equality immediately implies that $1 \cdot x_1 \cdots x_q \notin C$. Besides, the same equality together with the definition of C via T_+ yield that $x_q \notin C$. If, however, $i_1 \leq k$, then denote by r the maximum integer with $i_r \leq k$. We get $1 \cdot x_1 \cdots x_q = x_{i_r} \in T_+$. This equality immediately shows that $1 \cdot x_1 \cdots x_q \notin C$. Besides, the same equality together with the definition of C also yield that $x_q \notin C$.

Thus we see that x_q, $1 \cdot x_1 \cdots x_q$ are not in C. Therefore they do not belong to $S \subseteq C$. This means that (I4) is satisfied, which completes the proof. □

Proof of Theorem 6.1. An automaton is minimal if and only if its Nerode equivalence is the identity relation. Since the intersection in the right hand side of (6.5) is an incomplete equivalence relation, we see that it coincides with the Nerode equivalence. Hence the proof follows from Theorem 6.2. □

Proof of Theorem 6.1. An automaton is minimal if and only if its Nerode equivalence is the identity relation. Since the intersection in the right hand side of (6.5) is an equivalence relation, we see that it coincides with the Nerode equivalence. □

Chapter 7

Languages

Trivial languages are recognized by graph algebras. The first theorem of this chapter deals with nontrivial regular languages, and gives necessary and sufficient conditions for there to exist a graph algebra recognizing a regular language.

Theorem 7.1. ([160]) *A nontrivial language $L \subseteq X^+$ is recognized by the graph algebra of some undirected graph if and only if the letters of X can be reordered and denoted by*

$$\begin{aligned} X &= \{x_{11}, \ldots, x_{1k_1}, \ldots, x_{\ell 1}, \ldots, x_{\ell k_\ell}, y_{11}, \ldots, y_{1n_1}, \ldots, \\ &\quad y_{\ell 1}, \ldots, y_{\ell n_\ell}, z_1, \ldots, z_p, w_1, \ldots, w_t\}, \end{aligned}$$

so that either L or $X^+ \setminus L$ has the following regular expression

$$\begin{aligned} & (z_1 + \cdots + z_p) + \\ & + \quad (x_{11} + \cdots + x_{1k_1})(x_{11} + \cdots + x_{1k_1} + y_{11} + \cdots + y_{1n_1})^* \\ & + \quad \cdots \qquad\qquad (7.1) \\ & + \quad (x_{\ell 1} + \cdots + x_{\ell k_\ell})(x_{\ell 1} + \cdots + x_{\ell k_\ell} + y_{\ell 1}, \ldots, y_{\ell n_\ell})^* \end{aligned}$$

where $p, \ell, t, n_1, \ldots, n_\ell \geq 0; \ell + p \geq 1; k_1, \ldots, k_\ell \geq 1$.

First, suppose that a nontrivial language L is defined by the regular expression (7.1). Then it recognized by the automaton in Figure 7.1. Indeed, we can see immediately that this automaton accepts L. As explained earlier, an automaton $\mathcal{A} = (Q, X, \cdot, i, T)$ is minimal if and only if the Nerode congruence is equal to the identity relation. Using this it is routine to verify that the automaton in Figure 7.1 is minimal. Note that in the special cases where some sets labelling edges of this diagram are empty, we assume that these edges are not included in the diagram, and if there is not any edge from the start state to the rightmost state (which happens if and only if $\ell = 1$ and $p = t = 0$), we remove this inaccessible state too, because every minimal automaton is accessible.

Second, suppose that the complement $X^+ \setminus L$ of a nontrivial language L is defined by (7.1). If $p = 0$, $\ell = 1$ and $n_1 = 0$, then $L = \emptyset$, a contradiction to our assumption that L is nontrivial. In all other cases the shown in Figure 7.2 accepts L, and again it is routine to verify that the Nerode congruence on this automaton is trivial, and so it is minimal.

The following algorithm verifies whether a regular language is recognized by a graph algebra.

Algorithm

// Input: A regular expression R defining a language L.

// Output: Minimal graph algebra recognizing this language, if it exists.

Step 1. Find an automaton $\mathcal{A}$ recognizing L.

Step 2. Reduce $\mathcal{A}$ and find an equivalent minimal automaton $\mathcal{M}$.

Step 3. If $\mathcal{M}$ recognizes a nontrivial language, then check whether $\mathcal{M}$ is of the form shown in Figures 7.1 or 7.2 up to notation of letters of the alphabet.

Step 4. If $\mathcal{M}$ is of the form in Figures 7.1 or 7.2, then the language L is recognized by the graph algebra of an undirected graph D which consists of r copies of K_2, $\ell - r$ copies of K_1 (if $n_1, \ldots, n_r \geq 1$ and $n_{r+1}, \ldots, n_\ell = 0$), and an isolated vertex c_0 (if $p > 0$).

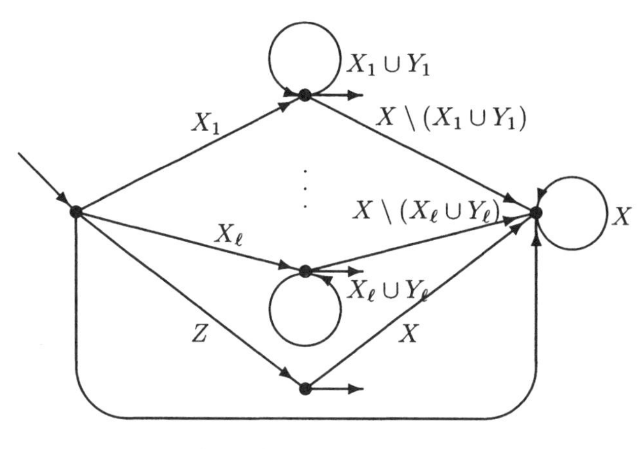

Figure 7.1: $X = Y_1 \cup \cdots \cup Y_\ell \cup Z \cup W$, where $Z = \{z_1, \ldots, z_p\}$, $W = \{w_1, \ldots, w_t\}$, $X_i = \{x_{i1}, \ldots, x_{ik_i}\}$, $Y_i = \{y_{i1}, \ldots, y_{in_i}\}$, $p, \ell, t, n_1, \ldots, n_\ell \geq 0$; $\ell + p \geq 1$; $k_1, \ldots, k_\ell \geq 1$.

Our proof of Theorem 7.1 uses syntactic semigroups.

Lemma 7.1. *For any undirected graph $D = (V, E)$, the following conditions are equivalent:*

(i) *the graph algebra Alg(D) is associative;*

(ii) *for all $(x, y) \in E$ and $z \in V$,*

$$(x, z) \in E \Leftrightarrow (y, z) \in E;$$

(iii) *each connected component of D is either an isolated vertex or a complete graph with all loops.*

Proof easily follows from Proposition 4.1. □

Lemma 7.2. ([160]) *Let $D = (V, E)$ be an undirected graph. Then the graph algebra Alg(D) is a syntactic semigroup of a language if and only if*

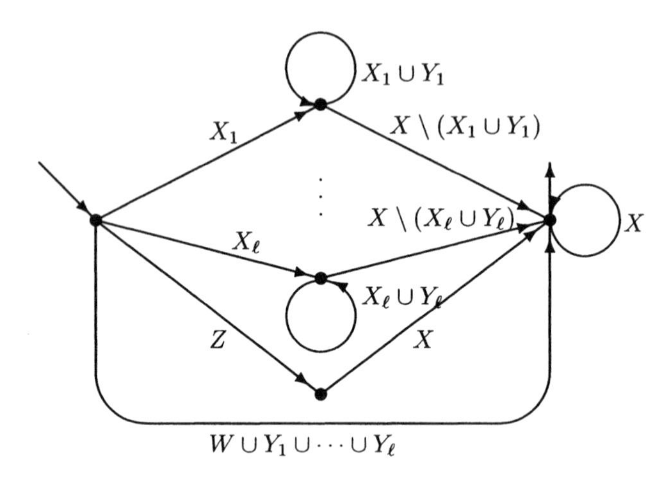

Figure 7.2: Alphabet and all sets of letters as in Fig. 7.1.

D has at most one isolated vertex, and all other connected components of D are complete graphs with no more than two vertices.

Proof. The 'if' part: Suppose that D has at most one isolated vertex and every other connected component is isomorphic to K_1 or K_2. Then the graph algebra $\mathrm{Alg}(D)$ is a semigroup by Lemma 7.1, and so we can use Lemma 3.3.

Let us define a subset T of $\mathrm{Alg}(D)$. We include the only isolated vertex, if it exists, in T. For each connected component of D, we choose one vertex and put it in T.

Consider any two distinct elements x, y of $\mathrm{Alg}(D)$. We are going to show that $\mathrm{Cont}_T(x) \neq \mathrm{Cont}_T(y)$.

If D has an isolated vertex v, then v is the only element of $\mathrm{Alg}(D)$ with $\mathrm{Cont}_T(v) = \{(1,1)\}$. Besides, 0 is the only element of $\mathrm{Alg}(D)$ with empty context w.r.t. T. Therefore we may assume that neither x nor y is an isolated vertex of D, and $x, y \neq 0$.

The following two cases are the only possible ones.

Case 1: x and y belong to the same connected components C. Then precisely one of them, say x, belongs to T. Hence

$$(1,1) \in \mathrm{Cont}_T(x) \setminus \mathrm{Cont}_T(y).$$

Case 2: x and y belong to different connected components C and C' of D, respectively. Denote by ℓ the element of T chosen in C. Since C is a complete graph, we get $(\ell, x) \in E$. Hence $\ell x = \ell \in T$. Since $y \notin C$, there is no edge (ℓ, y), and so $\ell y = 0 \notin T$. Therefore $(\ell, 1) \in \mathrm{Cont}_T(x) \setminus \mathrm{Cont}_T(y)$.

Thus $\mathrm{Cont}_T(x) \neq \mathrm{Cont}_T(y)$ in both cases. By Lemma 3.3, the graph algebra of D is a syntactic semigroup.

The 'only if' part: Suppose that the graph algebra $\mathrm{Alg}(D)$ of D is a syntactic semigroup. Then $\mathrm{Alg}(D)$ is associative, and by Lemma 7.1 every connected component C of D with more than one vertex is a complete graph.

By Lemma 3.3, there exists $T \subseteq \mathrm{Alg}(D)$ such that $\mathrm{Cont}_T(x) \neq \mathrm{Cont}_T(y)$, for all $x \neq y \in \mathrm{Alg}(D)$. Note that the set $\mathrm{Alg}(D) \setminus T$ also satisfies this condition. Therefore we may always assume that $0 \notin T$.

First, we verify that D has no more than one isolated vertex. Indeed, since $0 \notin T$, the context of each isolated vertex, which is not in T, is equal to $\emptyset = \mathrm{Cont}_T(0)$, and so all these vertices should belong to T. Therefore the context of every isolated vertex consists of one and the same pair $(1,1)$. Thus there is at most one isolated vertex in D.

Next, we show that each connected component of D has at most two vertices. Take a connected component C of D with more than one vertex.

Since $0 \notin T$, all elements of $C \cap T$ have the same context $(V(C) \cap T)^1 \times V(C)^1$; whence $|V(C) \cap T| \leq 1$. Similarly, the context of every element in $V(C) \setminus T$ equals $(V(C) \cap T) \times V(C)^1$, and hence $|V(C) \setminus T| \leq 1$. Therefore C has at most two vertices.

Thus each connected component of D, except the only possible isolated vertex, is isomorphic to K_1 or K_2. □

For $k \geq 1$ and $u \in X^+$ such that $|u| \geq k$, denote by $pre_k(u)$ and $suf_k(u)$, respectively, the prefix and the suffix of length k of u, and by $int_k(u)$ the sets of all proper subwords of length k of u. A language $L \subseteq X^+$ is *k-testable* if and only if, for any words $u, v \in X^+$, the conditions $pre_k(u) = pre_k(v), suf_k(u) = suf_k(v)$, and $int_k(u) = int_k(v)$ imply that $u \in L$ iff $v \in L$. A language is said to be *locally testable* if it is k-testable for some integer $k \geq 1$.

Corollary 7.1. ([160]) *All languages recognized by graph algebras of undirected graphs are* 1*-testable.*

Proof of Theorem 7.1. The 'if' part: Suppose that L is represented by the regular expression (7.1). Assume that $n_1, \ldots, n_r \geq 1$ and $n_{r+1}, \ldots, n_\ell = 0$. Consider the graph D which consists of an isolated vertex c_0 (if $p > 0$), r copies of K_2, and $\ell - r$ copies of K_1. By Lemma 7.1, its graph algebra Alg(D) is associative. Let us construct a subset $T \subseteq$ Alg(D) in the following way: we include the only isolated vertex c_0, if it exists, in T and for each other connected component of D we choose one vertex and put it in T. Consider the homomorphism $\phi\colon X^+ \to$ Alg(D). Let ϕ map the letters $z_1, \ldots, z_p$ to c_0, the letters $w_1, \ldots, w_t$ to 0. Further, for $i = r+1, \ldots, \ell$, the function ϕ maps the letters $x_{i1}, \ldots, x_{ik_i}$ to the vertex of the i-th copy of K_1 and, for $i = 1, \ldots, r$, to the vertex of the i-th copy of K_2 which belongs to T, and finally the letters $y_{i1}, \ldots, y_{in_i}$ to another vertex of the i-th copy of K_2.

We claim that $L = \phi^{-1}(T)$. Indeed, take any $u \in L$. Suppose, for example, that u is of the form $u = x_{11}a_1 \cdots a_s$, with $a_1, \ldots, a_s \in \{x_{11}, \ldots, x_{1k_1}, y_{11}, \ldots, y_{1n_1}\}$. Then $\phi(x_{11}), \phi(a_1), \ldots, \phi(a_s)$ lie in the first copy of K_2, and so

$$\phi(u) = \phi(x_1 1)\phi(a_1) \cdots \phi(a_s) = \phi(x_{11}) \in T.$$

On the other hand, let $\phi(u) \in T$ for some $u = a_1 \cdots a_s \in X^+$.

Since $\phi(u) \neq 0$, we see that all the $a_1, \ldots, a_s$ are in the same connected component, say in the i-th copy of K_2 (other cases are similar). Then $a = \phi(u) = \phi(a_1) \cdots \phi(a_n) = \phi(a_1)$, and so $y_1 = x_{ij}$ for some $j = 1, \ldots, k_i$ and $a_2, \ldots, a_s \in \{x_{i1}, \ldots, x_{ik_i}, y_{i1}, \ldots, y_{in_i}\}$. Therefore $u \in L$. Thus the language L is recognized by the finite graph algebra $\mathrm{Alg}(D)$.

Suppose that $X^+ \setminus L$ is represented by the regular expression (7.1). As we have just verified, it follows that $X^+ \setminus L$ is recognized by a finite graph algebra $\mathrm{Alg}(D)$ as $f^{-1}(T)$ for some f, T. Therefore is recognized by the same $\mathrm{Alg}(D)$ as $f^{-1}(\overline{T})$, too.

The 'only if' part: Let $L \subseteq X^+$ be recognizable by a graph algebra $\mathrm{Alg}(D)$ of some undirected graph D. Then there exists a morphism $\phi \colon X^+ \to \mathrm{Alg}(D)$ and a subset T of $\mathrm{Alg}(D)$ such that $L = \phi^{-1}(T)$. The image $\phi(X^+)$ is a subsemigroup of $\mathrm{Alg}(D)$.

We claim that any subalgebra of $\mathrm{Alg}(D)$ is either a graph algebra of some subgraph of D or a left-zero semigroup. Indeed, if a subalgebra A of $\mathrm{Alg}(D)$ is generated by elements that are vertices of the same connected component of D, then A is a left-zero semigroup. If generators are taken from different connected components of D or one of them is 0, then A is isomorphic to the graph algebra of D', where D' is a subgraph of D induced by the elements of $A \setminus \{0\}$.

If we adjoin 0 to a left-zero semigroup, then we get a graph algebra, too. Thus we may assume that the whole $\mathrm{Alg}(D)$ is associative. Lemma 7.1 tells us that each connected component of D is either an isolated vertex or a complete graph with all loops.

Every epimorphic image of a graph algebra (respectively, a left-zero semigroup) is a graph algebra (a left-zero semigroup), again. Therefore, if $L \subseteq X^+$ is recognizable by a graph algebra, then by Lemma 3.2 the following two cases may occur.

Case 1: $\mathrm{Syn}(L)$ is a graph algebra $\mathrm{Alg}(G)$ of a graph G. Then Lemma 7.2 shows that G has at most one isolated vertex c_0, and all other connected components $C_1, \ldots, C_r$ and $C_{r+1}, \ldots, C_\ell$ of G are complete graphs with two or one vertices, respectively.

Then there exists a natural surjection $\phi\colon X^+ \to \mathrm{Alg}(G)$. Put $T = \phi(L)$. It is easily seen that $\mathrm{Cont}_T(x) \neq \mathrm{Cont}_T(y)$, for all $x \neq y \in \mathrm{Alg}(G)$.

First, assume that $0 \notin T$. In the proof of Lemma 7.2 it was shown that in this case T consists of the only isolated vertex c_0, if it exists, and just one vertex from each other connected component of G.

Denote by $z_1, \ldots, z_p$ the elements of X that are mapped to c_0 by ϕ, denote by $w_1, \ldots, w_t$ the elements that are mapped to 0, and by $x_{i1}, \ldots, x_{ik_i}$ the elements that go to the only vertex of C_i for $i = r+1, \ldots, \ell$. Further, for $i = 1, \ldots, r$, let $x_{i1}, \ldots, x_{ik_i}$ be the elements of X that are taken to the vertex of C_i choosen in T, and let $y_{i1}, \ldots, y_{in_i}$ be the elements which are taken to another vertex of C_i.

It is routine to verify that then L is represented by the regular expression of the form (7.1).

Second, assume that $0 \in T$. Then T should contain just one vertex from each connected component isomorphic to K_2. That is, redenoting letters of the alphabet we obtain the complement of the set T considered above. Therefore in this situation we see that $X^+ \setminus L$ is represented by the regular expression (7.1).

Case 2: $\mathrm{Syn}(L)$ is a left-zero semigroup. If a left-zero semigroup has more than two elements, then it is easily seen that, for each subset, there exist two elements with the same contexts w.r.t. this subset. This contradicts Lemma 3.3 and shows that $|\,\mathrm{Syn}(L)| \leq 2$. There exists a natural surjection $\phi\colon X^+ \to \mathrm{Syn}(L)$ such that $L = \phi^{-1}(a)$ for some $a \in \mathrm{Syn}(L)$. If $|\,\mathrm{Syn}(L)| = 1$, then $X^+ = \phi^{-1}(a)$, and so $L = X^+$. If, however, $|\,\mathrm{Syn}(L)| = 2$, then we denote by $x_1, \ldots, x_s$ the elements of X that are mapped to $\{a\}$ by ϕ, where $s \geq 1$. Then L is represented by

$$(x_1 + \cdots + x_s)X^*,$$

which is a special case of (7.1) when $p = 0, \ell = 1, k_1 = s, n_1 = n - s$. $\square$

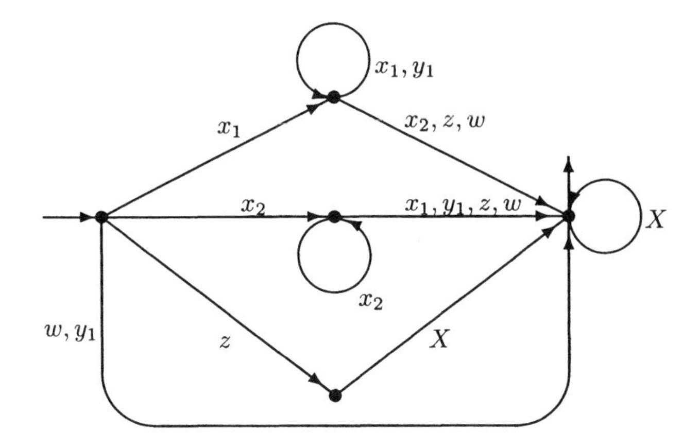

Figure 7.3: $G = K_2 \cup K_1 \cup N_1$

Example 7.1. ([160]) Let L be a language over the alphabet $X = \{x_1, x_2, y_1, z, w\}$ with the complement $\overline{L}$ given by the regular expression

$$x_1(x_1 + y_1)^* + x_2^+ + z.$$

The minimal automaton of L is shown in Figure 7.3. It follows that L is recognized by the graph algebra of the graph $G = K_2 \cup K_1 \cup N_1$, where

$$V(K_2) = \{x_1, y_1\}, V(K_1) = \{x_2\}, V(N_1) = \{z\}.$$

Further, we include a description of languages recognized by left automata in terms of combinatorial properties of words which belong to these languages.

Theorem 7.2. ([162]) *A language L over an alphabet X is recognized by a left automaton if and only if at least one of the following two conditions is satisfied for all $x, y \in X$, and $u, v \in X^*$:*

(i) *$xyu \in L$ implies $yu \in L$, and*

$xu, yxv \in L$ implies $yxu \in L$;

(ii) *$yu \in L$ implies $xyu \in L$, and*

$yxu \in L$ implies either $xu \in L$ or $yxv \in L$.

Proof of Theorem 7.2. The 'if' part: First, suppose that the language L satisfies (i). Introduce a graph $G = (V, E)$ with the set $V = X$ of vertices, and the set E of edges consisting of all pairs (x, y) such that $yxu \in L$ for some $u \in X^*$. Let f be the mapping from X to $\mathrm{Alg}(D)^1$ defined by $f(x) = x$. Put $T = (\{1\} \cup X) \cap L$.

Take an arbitrary word $u = u_1 \cdots u_n$ in L. If $u = 1$, i.e. $n = 0$, then $1 \in T$, and the automaton $\mathrm{Atm}_\ell(D, T)$ accepts u. Further, assume that $n > 0$. The first implication of condition (i) shows that $u_k \cdots u_n \in L$ for all $k = 1, \ldots, n$. It follows that $(u_2, u_1), (u_3, u_2), \ldots, (u_n, u_{n-1}) \in E$ by the definition of E, and $u_n \in T$ by the definition of T. We get $1 \cdot u = f(u_n)(\ldots f(u_1)) = f(u_n) = u_n \in T$, and so the automaton $\mathrm{Atm}_\ell(D, T)$ recognizes u.

Consider any word $u = u_1 \cdots u_n$ accepted by $\mathrm{Atm}_\ell(D, T)$. If $u = 1$, then $1 \in T$, and so $u = 1 \in L$. Further, assume that $n > 0$. As above, $1 \cdot u \in T$ means that $u_n, \ldots, u_1$ is a walk in D and $u_n \in T$. By reversed induction on k we show that $u_k \cdots u_n \in L$ for all $k = 1, \ldots, n$. If $k = n$, then $u_n \in T \subseteq L$ by definition. Assume that $u_k u_{k+1} \cdots u_n \in L$ for some $1 < k \leq n$. Since $(u_{k-1}, u_k) \in E$, it follows that $u_{k-1} u_k v \in L$, for some $v \in X^*$. Then the second implication of (i) yields $u_{k-1} u_k \cdots u_n \in L$, as required. Therefore $u = u_1 \cdots u_n \in L$. Thus L is the language recognized by $\mathrm{Atm}_\ell(D, T)$.

Second, suppose that L satisfies condition (ii). It follows that $xu \in \overline{L}$ implies $u \in \overline{L}$, and $xu, yxv \in \overline{L}$ implies $yxu \in \overline{L}$. Therefore the complement $\overline{L}$ of L satisfies (i). As we have proved above, $\overline{L}$ is recognized by some automaton $\mathrm{Atm}_\ell(D, T)$. Hence L is recognized by the automaton $\mathrm{Atm}_\ell(D, \overline{T})$, where $\overline{T} = \mathrm{Alg}(D)^1 \setminus T$.

The 'only if' part: Suppose that L is recognized by the automaton $\mathrm{Atm}_\ell(D, T)$ of the graph algebra of some graph $D = (V, E)$. Consider two cases.

Case 1: $0 \notin T$. Take a word xyu in L, where $u = u_1 \cdots u_n \in X^*$ and $x, y, u_1, \ldots, u_n \in X$. If $n = 0$, then $1 \cdot xy = f(y)f(x) = f(y) \in T$, and so

$y \in L$. Further, assume that $n > 0$. Since

$$1 \cdot xyu = f(u_n)(\dots(f(u_1)(f(y)f(x)))) = f(u_n) \in T,$$

we see that $f(u_n)$, $f(u_{n-1})$, ..., $f(u_1)$, $f(y)$, $f(x)$ is a walk in D. Therefore

$$1 \cdot yu = f(u_n)(\dots f(u_1)f(y)) = f(u_n) \in T,$$

and so $yu \in L$. We see that the first implication of (i) holds.

Now, pick xu and yxv in L, where $u = u_1 \cdots u_n$, $v = v_1 \cdots v_k \in X^*$ and $x, y \in X$. Clearly, $f(u_n) \in T$, and besides $f(u_n), f(u_{n-1}), \dots, f(u_1), f(x)$ and $f(v_k), \dots, f(v_1), f(x), f(y)$ are walks in D. It follows that $f(u_n), \dots, f(u_1), f(x), f(y)$ is a walk in D, too. Hence $yxu \in L$. Thus the whole condition (i) holds.

Case 2: $0 \in T$. Note that $\overline{L}$ is recognized by the automaton $\mathrm{Atm}_\ell(D, \overline{T})$. Since $0 \notin \overline{T}$, as we have proved in Case 1, the language $\overline{L}$ satisfies condition (i), i.e., $xyu \in \overline{L}$ implies $yu \in \overline{L}$, and $xu, yxv \in \overline{L}$ implies $yxu \in \overline{L}$ for all $x, y \in X$, $u, v \in X^*$. The contrapositive statements show that (ii) holds. □

The next theorem describes all languages recognized by two-sided automata of graphs in terms of combinatorial properties satisfied by these languages and regular expressions for their complements.

Theorem 7.3. ([147]) *For every language L over an alphabet X, the following conditions are equivalent:*

(i) *there exists a graph D such that L is recognized by a two-sided automaton of D;*

(ii) *there exist two disjoint subsets X_- and X_+ of X such that $X = X_- \dot\cup X_+$ and, for all $x \in X_+$, $y \in X_-$, $z \in X$ and $u, v \in X^*$, the following implications hold:*

  (a) *$zxu \in L$ implies $xu \in L$,*

  (b) *$xu, zxv \in L$ implies $zxu \in L$,*

(c) $zyu \in L$ *implies* $zy^*u \in L$,

(d) $zv, zyu \in L$ *implies* $zyv \in L$,

(e) $xyu \in L$ *if and only if* $yxu \in L$;

(iii) *there exist a subset* X_T *of* X, *disjoint subsets* X_- *and* X_+ *of* X, *and relations* $G_1 \subseteq X_+ \times X_+$, $G_2 \subseteq X_- \times X_-$, *and* $G_3 \subseteq X_- \times X_+$ *such that* $X = X_- \dot{\cup} X_+$ *and the language* $X^+ \setminus L$ *has the following regular expression:*

$$(X_N \cap X_-)X_-^* + X^*(X_N \cap X_+)X_-^* + \tag{7.2}$$
$$\sum_{(x_i,x_j)\in G_1\cup G_3^{-1}} X^* x_i X_-^* x_j X^* + \sum_{(x_i,x_j)\in G_2\cup G_3} x_i X_-^* x_j X^*,$$

where X_N *stands for* $X \setminus X_T$.

Note that if one of the sets X_+ or X_- is empty, then the condition (iii) remains in force, and all summands of (7.2) involving empty sets vanish. It follows from Theorem 7.3 that there exists an algorithm deciding whether a regular language is recognizable by two-sided automata of graphs.

A routine verification shows that the next-state function of the automaton $\mathrm{Atm}_t(D, T, f, \ell)$ can be defined by the graph algebra $\mathrm{Alg}(D')$, where $D' = (V, E^{-1})$, if we use the following condition equivalent to (DA4):

(DA5) for $a \in \mathrm{Alg}(D) \cup \{1\}$, $x \in X$,

$$a \cdot x = \begin{cases} f(x)a & \text{if } \ell(x) = + \\ af(x) & \text{if } \ell(x) = -. \end{cases}$$

This means that x acts as a left multiplication by the element $f(x)$ in $\mathrm{Alg}(D')$ if $\ell(x) = +$, and as a right multiplication by $f(x)$ if $\ell(x) = -$.

Proof of Theorem 7.3. First, note that a language L is accepted by $\mathrm{Atm}_t(D, T, f, \ell)$ if and only if $L \setminus \{1\}$ is accepted by

$$\mathrm{Atm}_t(D, T \setminus \{1\}, f, \ell).$$

Since conditions (ii) and (iii) remain unchanged if we replace L by $L \setminus \{1\}$, in the proof we may assume that $L \subseteq X^+$.

(i)$\Rightarrow$(ii): Suppose that the language L is recognized by the automaton $\mathrm{Atm}_t(D, T, f, \ell)$ of some graph $D = (V, E)$. Define

$$X_+ = \{x \in X \mid \ell(x) = +\},$$
$$X_- = \{x \in X \mid \ell(x) = -\},$$

and consider arbitrary elements $x, y \in X$ with $\ell(x) = +$ and $\ell(y) = -$.

To verify the implication (a), take any word zxu in L, where $u \in X^*$ and $z \in X$. Since $1 \cdot zxu$ is defined and belongs to T, it follows that $(f(z), f(x)) \in E$. Besides, $f(z) \cdot f(x) = f(x)$, because $\ell(x) = +$. Therefore $1 \cdot xu = f(x) \cdot u = (f(z) \cdot f(x)) \cdot u = 1 \cdot zxu \in T$. Hence $xu \in L$, as required.

Now, pick arbitrary elements xu and zxv in L such that $u, v \in X^*$ and $z \in X$. Since $1 \cdot zxv$ is defined, we get $(f(z), f(x)) \in E$. Therefore $1 \cdot zxu = (f(z) \cdot f(x)) \cdot u = f(x) \cdot u = 1 \cdot xu \in T$, and so $zxu \in L$. Thus the implication (b) holds, too.

Next, we verify the implication (c). To this end, take a word zyu in L, where $u \in X^*$ and $z \in X$. Since $1 \cdot zyu$ is defined and $\ell(y) = -$, we get $(f(y), f(z)) \in E$. Hence $1 \cdot zy^k u = f(z) \cdot u = (f(z) \cdot f(y)) \cdot u = 1 \cdot zyu \in T$, for all $k \in \mathbf{N}_0$. Therefore $zy^*u \in L$, as required.

Consider any elements zv and zyu in L, where $u, v \in X^*$ and $z \in X$. Clearly, $zyu \in L$ implies $(f(y), f(z)) \in E$, because $\ell(y) = -$. Hence $1 \cdot zyv = (f(z)f(y)) \cdot v = f(z) \cdot v = 1 \cdot zv \in T$, and so $zyv \in L$. We see that (d) holds.

Finally, let us prove the equivalence (e). Take an arbitrary $u \in X^*$ such that $xyu \in L$. It follows that $(f(y), f(x)) \in E$. The implication (c) yields us $xu \in L$. Therefore $1 \cdot yxu = (f(y) \cdot f(x)) \cdot u = f(x) \cdot u = 1 \cdot xu \in T$; whence $yxu \in L$. To verify the converse implication, suppose that $yxu \in L$. Hence $(f(y), f(x)) \in E$. Besides, $xu \in L$ in view of (a). Therefore $1 \cdot xyu = (f(x) \cdot f(y)) \cdot u = f(x) \cdot u = 1 \cdot xu \in T$. This means that $yxu \in L$. Thus the whole condition (ii) holds.

(ii)$\Rightarrow$(i): Condition (ii) involves the sets X_+ and X_-. Let us label x by $\ell(x) = +$ if $x \in X_+$, and by $\ell(x) = -$ otherwise. Introduce a graph $G = (V, E)$ with the set $V = X$ of vertices, and the set $E = E_1 \cup E_2$ of edges where

$$\begin{aligned} E_1 &= \{(x, y) \in X \times X_+ \mid xyw \in L \text{ for some } w \in X^*\}, \\ E_2 &= \{(x, y) \in X_- \times X \mid yxw \in L \text{ for some } w \in X^*\}. \end{aligned}$$

Let f be the mapping from X to V defined by $f(x) = x$. Put $T = X \cap L$.

By induction on the length n of a word u we show that $u \in L$ if and only if u is recognized by the automaton $\text{Atm}_t(D, T, f, \ell)$. If $n = 1$, then this follows immediately from the definition of T. Suppose that the claim is true for all words v of length less than n, where $n > 1$. Take an arbitrary word $u = x_1 \cdots x_n$ in L. We claim that then $\text{Atm}_t(D, T, f, \ell)$ accepts u.

First, consider the case where $x_2 \in X_+$. Implication (a) shows that $x_2 \cdots x_n \in L$, and by the induction hypothesis $x_2 \cdots x_n$ is accepted by $\text{Atm}_t(D)$.

If $\ell(x_1) = +$, then $(f(x_1), f(x_2)) \in E_1$ by the definition of $E_1 \subseteq E$. If, however, $\ell(x_1) = -$, then the equivalence (e) yields $x_2 x_1 x_3 \cdots x_n \in L$, and hence $(f(x_1), f(x_2)) \in E_2$. Thus we always get $(f(x_1), f(x_2)) \in E$. It follows that $1 \cdot u = f(x_1) \cdot x_2 \cdots x_n = f(x_2) \cdots f(x_n) = 1 \cdot x_2 \cdots x_n \in T$. This means that the $\text{Atm}_t(D)$ recognizes u.

Second, consider the case where $x_2 \in X_-$. Then (c) shows that $x_1 x_3 \cdots x_n \in L$. By the induction hypothesis $x_1 x_3 \cdots x_n$ is recognized by $\text{Atm}_t(D)$.

If $\ell(x_1) = -$, then $(f(x_2), f(x_1)) \in E_2$, by the definition of E_2. If $\ell(x_1) = +$, then (e) ensures that $x_2 x_1 x_3 \cdots x_n \in L$; whence $(f(x_2), f(x_1))$ belongs to E_1. In both cases $(f(x_2), f(x_1))$ is an edge of D, and we get $1 \cdot u = f(x_1) \cdot x_2 \cdots x_n = f(x_1) \cdot x_3 \cdots f(x_n) = 1 \cdot x_1 x_3 \cdots x_n \in T$. Therefore the $\text{Atm}_t(D)$ recognizes u, in all cases. Thus the language accepted by $\text{Atm}_t(D)$ contains L.

Conversely, suppose that our word $u = x_1 \cdots x_n$ is accepted by the automaton $\mathrm{Atm}_t(D, T, f, \ell)$. We are going to show that u belongs to L.

First, consider the case where $\ell(x_2) = +$. Clearly, $(f(x_1), f(x_2)) \in E$. Hence we get

$$\begin{aligned} 1 \cdot x_2 \cdots x_n &= (f(x_2) \cdot f(x_3)) \cdots f(x_n) \\ &= (f(x_1) \cdot f(x_2)) \cdots f(x_n) \\ &= 1 \cdot u \in T. \end{aligned}$$

By the induction hypothesis $x_2 \cdots x_n \in L$. Since $x_2 \in X_+$, we get

$$(f(x_1), f(x_2)) \in E_1 \cup E_2.$$

If $(f(x_1), f(x_2)) \in E_1$, then $x_1 x_2 w \in L$, for some $w \in X^*$. Therefore implication (b) gives us $x_1 x_2 \cdots x_n \in L$.

If, however, $(f(x_1), f(x_2)) \in E_2$, then $x_1 \in X_-$ and $x_2 x_1 w \in L$ for some $w \in X^*$. By the equivalence (e) we get $x_1 x_2 w \in L$, and (d) implies $x_1 x_2 \cdots x_n \in L$, again.

Second, suppose that $\ell(x_2) = -$. Then $(f(x_2), f(x_1)) \in E = E_1 \cup E_2$. We get $1 \cdot x_1 x_3 \cdots x_n = (f(x_1) \cdot f(x_3)) \cdots f(x_n) = (f(x_1) \cdot f(x_2)) \cdots f(x_n) = 1 \cdot u \in T$. By the induction hypothesis $x_1 x_3 \cdots x_n \in L$.

If $(f(x_2), f(x_1)) \in E_2$, then $x_1 x_2 w \in L$, for some $w \in X^*$. Hence implication (b) yields $x_1 x_2 \cdots x_n \in L$.

If, however, $(f(x_2), f(x_1)) \notin E_2$, then $(f(x_2), f(x_1)) \in E_1$. Therefore $x_1 \in X_-$ and we get $x_2 x_1 w \in L$ for some $w \in X^*$. Hence $x_1 x_2 w \in L$, in view of (e). Implication (d) gives us $x_1 x_2 \cdots x_n \in L$, again. Thus we have proved that L is precisely the language recognized by the $\mathrm{Atm}_t(D)$.

(i)⇒(iii): Suppose that L is recognized by the automaton $\mathrm{Atm}_t(D)$ induced by a graph $D = (V, E)$ with functions ℓ and f. Let us introduce the sets

$$\begin{aligned} X_N &= \{x \in X \mid f(x) \notin T\} \\ X_+ &= \{x \in X \mid \ell(x) = +\} \\ X_- &= \{x \in X \mid \ell(x) = -\} \end{aligned}$$

and the relations

$$\begin{aligned} G_1 &= \{(x,y) \in X_+ \times X_+ \mid (f(x), f(y)) \notin E\}, \\ G_2 &= \{(x,y) \in X_- \times X_- \mid (f(y), f(x)) \notin E\}, \\ G_3 &= \{(x,y) \in X_- \times X_+ \mid (f(x), f(y)) \notin E\}. \end{aligned}$$

We claim that $X^+ \setminus L$ is equal to the language defined by the regular expression (7.2). To prove one inclusion, take any element $u \in X^+ \setminus L$, say $u = x_1 \cdots x_n$, where $n \geq 1$ and $x_1, \ldots, x_n \in X$, and consider two possible cases.

Case 1: $1 \cdot u$ is defined. If $u \in X_-^*$, then $1 \cdot u = f(x_1) \notin T$. We have $u \in x_1 X_-^* \subseteq (X_- \cap X_N) X_-^*$, and so u belongs to the language (7.2), as required. Further, we may assume that some letters in u are labeled by $+$. Let x_k be the last letter in u of this kind. We get $1 \cdot u = f(x_k) \notin T$, and hence $u \in X^* x_k X_-^* \subseteq X^*(X_+ \cap X_N) X_-^*$. Therefore u lies in the language defined by (7.2), again.

Case 2: $1 \cdot u$ is undefined. Let j be the largest integer such that $1 \leq i \leq n$ and $1 \cdot x_1 \cdots x_j$ is defined. This index j always exists because $1 \cdot x_1 = f(x_1)$.

Subcase 2.1: $x_1 \cdots x_j \in X_-^*$. Then $1 \cdot x_1 \cdots x_j = f(x_1)$. Since $1 \cdot u$ is undefined, it follows that $(f(x_k), f(x_1)) \notin E$ with $\ell(x_k) = -$, or $(f(x_1), f(x_k)) \notin E$ with $\ell(x_k) = +$ for some $j < k \leq n$. Therefore $u \in x_1 X_-^* x_k X^*$ with $(x_1, x_k) \in G_2 \cup G_3$.

Subcase 2.2: Some letters in $x_1 \cdots x_j$ are labeled by $+$. Let x_i be the last letter in $x_1 \cdots x_j$ labeled by $+$. We get $1 \cdot x_1 \cdots x_j = 1 \cdot x_1 \cdots x_i = f(x_i)$. Since $1 \cdot u$ is undefined, it follows that there exists $j < k \leq n$ such that $(f(x_k), f(x_i)) \notin E$ with $\ell(x_k) = -$, or $(f(x_i), f(x_k)) \notin E$ with $\ell(x_k) = +$. Hence $u \in X_* x_i X_-^* x_k X^*$ with $(x_i, x_k) \in G_3^{-1} \cup G_1$.

Thus in all the cases u belongs to the language (7.2). This completes the proof of one inclusion.

To prove the reverse inclusion, consider an arbitrary element $u =$

$x_1 \cdots x_n$ of the language defined by (7.2). Obviously, $u \neq 1$. If $1 \cdot u$ is undefined, then $u \in X^+ \setminus L$, and we are done. Further, we may assume that $1 \cdot u$ is defined.

If $u \in xX_-^* yX^*$, for some $(x, y) \in G_2 \cup G_3$, then $1 \cdot u$ is undefined. Similarly, if $u \in X^* xX_-^* yX^*$, for $(x, y) \in G_1 \cup G_3^{-1}$, then $1 \cdot u$ is undefined, too. Therefore we may exclude these two summands of (7.2) from further consideration: u does not belong to them.

If $u \in (X_- \cap X_N)X_-^*$, or $u \in (X_- \cap X_N)X_-^*$, or $u \in X^*(X_+ \cap X_N)X_-^*$, then $1 \cdot u \notin T$, by the definition of X_-. Therefore $u \in X^+ \setminus L$, again. Thus $X^+ \setminus L$ coincides with the language given by the regular expression (7.2).

(iii)$\Rightarrow$(i): Let L be a language with the complement $X^+ \setminus L$ defined by the regular expression (7.2). Introduce a graph $G = (V, E)$ with the set $V = X$ of vertices, and the set E of edges consisting of all pairs (x, y) which are not in $G_1 \cup G_2^{-1} \cup G_3$. Let f be the mapping from X to V defined by $f(x) = x$. Put $\ell(x) = +$ for all $x \in X_+$, and $\ell(x) = -$ for all $x \in X_+$. Set $T = V \setminus X_N$. Denote by R the language recognized by $\mathrm{Atm}_t(G, T, f, \ell)$. Since $1 \notin T$, we get $1 \notin R$, and so further we have to deal only with words in X^+. In order to show that $L = R$, we verify the equality $X^+ \setminus L = X^+ \setminus R$.

Take an arbitrary element $u = x_1 \ldots x_n \in X^+ \setminus R$, where $n \geq 1$ and $x_1, \ldots, x_n \in X$. We are going to show that $u \in X^+ \setminus L$.

First, suppose that $1 \cdot u$ is defined. If $u \in X_-^*$, then $1 \cdot u = x_1$, and $x_1 \notin T$ because $u \notin R$. We have $u \in x_1 X_-^* \subseteq (X_- \cap X_N)X_-^*$, and so u belongs to the language (7.2), as required. Further, we may assume that some letters in u are labeled by $+$. Let x_k be $+$-labeled letter in u of this kind. We get $1 \cdot u = x_k \notin T$, and so $u \in X^* x_k X_-^* \subseteq X^*(X_+ \cap X_N)X_-^*$. Thus we see that $u \in X^+ \setminus L$.

Second, suppose that $1 \cdot u$ is undefined. Let i be the last index such that $1 \cdot (x_1 \cdots x_i) = x_i$. If some letters on the right to x_i in u are labeled by $+$, then we consider the nearest letter x_{i+k} be of this kind. Since $1 \cdot (x_1 \cdots x_{i+k})$ is undefined, we get $(x_i, x_{i+k}) \notin E$ or $(x_{i+j}, x_i) \notin E$ for some $j < k$. If, however, all letters on the right of x_i in u are labeled by $-$, we get $(x_{i+j}, x_i) \notin E$, for some $j > 0$, because $1 \cdot u$ is undefined. Therefore, we always have $(x_i, x_{i'}) \in G_1 \cup G_3$, for some $i' > i$, or $(x_{i''}, x_i) \in G_2^{-1} \cup G_3$ for some $i'' > i$.

Assume that $\ell(x_i) = -$. Let j be the index of the first occurrence of x_i in u. We have $x_j = x_i$ and all letters between x_j and x_i are labeled by $-$, because $1 \cdot x_1 \cdots x_i = x_i$. It easily follows that $j = 1$, and so $(x_1, x_{i'}) \in G_3$, for some $i' > 1$ or $(x_{i''}, x_1) \in G_2^{-1}$ for some $i'' > 1$. Hence $u \in x_1 X_-^* y X^*$ for some $(x_1, y) \in G_2 \cup G_3$.

Assume now that $\ell(x_i) = +$, then $(x_i, x_{i'}) \in G_1$, for some $i' > i$ or $(x_{i''}, x_i) \in G_3$ for some $i'' > i$. Therefore we get $u \in X^* x_i X_-^* y X^*$, for some $(x, y) \in G_1 \cup G_3^{-1}$. Thus u is given by the regular expression (7.2) and so $u \in X^+ \setminus L$. Thus we see that $u \in X^+ \setminus L$ in any case.

We have proved that $X^+ \setminus L \supseteq X^+ \setminus R$. To verify the reverse inclusion, consider an arbitrary element $u = x_1 \cdots x_n$ of the language $X^+ \setminus L$ defined by (7.2).

First, suppose that $u \in x X_-^* y X^*$ with $(x, y) \in G_2 \cup G_3$. If $(x, y) \in G_2$, then $\ell(y) = +$ and there is no edge (x, y). If $(x, y) \in G_3$, then $\ell(y) = -$ and there is no edge (y, x). In both cases $1 \cdot u$ is undefined.

Second, suppose that $u \in X^* x X_-^* y X^*$ with $(x, y) \in G_1 \cup G_3^{-1}$. If $(x, y) \in G_1$, then $\ell(y) = +$ and there is no edge (x, y). If $(x, y) \in G_3^{-1}$, then $\ell(y) = -$ and there is no edge (y, x). It follows that $1 \cdot u$ is undefined in any case, and so u is not accepted by $\mathrm{Atm}_t(D, T)$.

Third, suppose that $u \in (X_- \cap X_N) X_-^*$. Then $1 \cdot u = f(x_1) \notin T$, and therefore $u \notin R$.

Finally, assume that $u \in X^*(X_+ \cap X_N) X_-^*$. If $1 \cdot u$ is defined, then $1 \cdot u = f(x_k) \notin T$, and so u is not accepted by $\mathrm{Atm}_t(D, T)$. If, however, $1 \cdot u$ is undefined, then u does not belong to R, either.

Thus $X^+ \setminus L \subseteq X^+ \setminus R$. This completes the proof. □

Remark 7.1. ([147]) Condition (iii) of Theorem 7.3 can be expressed in the equivalent form using a grammar generating the language $X^+ \setminus L$. Indeed, the standard method gives us a grammar which generates the language described by the regular expression of the form (7.2) (see, for example, the proof of Proposition 6.2.3 in [180]).

Remark 7.2. ([147]) If one of the sets X_+ or X_- is empty, then the proof shows that condition (iii) remains in force, and all summands of (7.2) involving empty sets vanish.

Corollary 7.2. ([147]) *It is decidable whether a regular language is recognizable by two-sided automata of graphs.*

Proof follows from condition (iii) of our main theorem. Indeed, given a regular language L, we can find a finite automaton accepting it, and then find a regular expression for the language $X^+ \setminus L$. After that, it remains to consider all subsets X_T of X, partitions $X = X_- \dot{\cup} X_+$, and relations $G_1 \subseteq X_+ \times X$, $G_2 \subseteq X_- \times X$, and use well known algorithms to verify whether $X^+ \setminus L$ is equal to the regular language defined by the regular expression (7.2). □

Corollary 7.3. ([146]) *The class $\mathcal{G}$ is closed under intersection.*

Proof of Corollary 7.3 follows from Theorem 7.3. Indeed, it is routine to verify that $L_1 \cap L_2$ satisfies all the properties (a) through (e) of condition (ii) in Theorem 7.3 provided that L_1 and L_2 satisfy them. □

Corollary 7.4. ([146]) *The class $\mathcal{G}(X_+, \emptyset)$ is closed under the Kleene $*$-operation.*

Proof of Corollary 7.4. Fix any language $L \in \mathcal{G}(X_+, \emptyset)$. Theorem 7.3 shows that $L \in \mathcal{G}(X_+, \emptyset)$ if and only if $(L \setminus \{1\}) \in \mathcal{G}(X_+, \emptyset)$. Since $L^* \setminus \{1\} = L^+$, it suffices to verify condition (ii) of Theorem 7.3 for the language L^+. Given that $X_- = \emptyset$, we only need to check properties (a) and (b) of (ii).

First, pick any word xyu in L^+, where $x, y \in X, u \in X^*$. We have $xyu \in L^n$ for some $n \geq 1$. Consider the leftmost prefix of this word which is in L. If $x \in L$, then $yu \in L^{n-1} \subseteq L^+$. Further, assume that $xyu_1 \in L$ and $u_2 \in L^{n-1}$ for some factorization $u = u_1u_2$, where $u_1, u_2 \in X^*$. Then $yu_1 \in L$, because $L \in \mathcal{G}$. Therefore $yu = yu_1u_2 \in L^+$, again. This means that (a) holds.

Second, take $xv, yxu \in L^+$. If $y \in L$, then clearly $yxv \in L^+$. Hence we may assume that $yxu_1 \in L$ and $u_2 \in L^+$, where $u = u_1u_2$. We get $xv_1 \in L$ and $v_2 \in L^+$, where $v = v_1v_2$. Since $L \in \mathcal{G}$, we get $yxv_1 \in L$, and therefore $yxv = yxv_1v_2 \in L^+$, again. Thus condition (ii) of Theorem 7.3 is satisfied for L^+, and therefore $L^+ \in \mathcal{G}(X_+, \emptyset)$. □

Lemma 7.3. ([146]) *A language L belongs to the class $\mathcal{G}(\emptyset, X_-)$ if and only if there exists a subset X_T of X, and for each $x \in X$ there exists a subset $I_x \subseteq X$ such that the language L has the following regular expression:*

$$\sum_{x \in X_T} x I_x^*. \tag{7.3}$$

Proof. Suppose that $L \in \mathcal{G}(\emptyset, X_-)$, and let L be recognized by an automaton $\mathrm{Atm}_t(D, T, f, \ell)$ of some graph $D \in \mathcal{G}(\emptyset, X_-)$. Let $X_T = \{x \in X \mid f(x) \in T\}$. For any $x \in X$, put

$$I_x = \{y \in X \mid f(y) \in \mathrm{In}(f(x))\}.$$

One can easily verify that L is given by the regular expression (7.3).

On the other hand, let L be a language defined by (7.3). Introduce a graph $G = (V, E)$ with the set $V = X$ of vertices, and the set $E = I_x \times X_T$ of edges. Let f be the mapping from X to V defined by $f(x) = x$. Put $\ell(x) = -$ for all $x \in X_+$. Set $T = X_T$. Straightforward verification shows that L coincides with the language recognized by $\mathrm{Atm}_t(D, T, f, \ell)$. □

One of the main questions concerning classes of languages is that of which operations they are closed for. The next theorems deal with the closure properties of several classes of languages recognised by two-sided automata of graphs.

Let $\mathcal{D}$ be a class of graphs and let $X = X_+ \dot\cup X_-$ be an alphabet.

Denote by $\mathcal{G}(\mathcal{D}, X_+, X_-)$ the class of languages recognized by two-sided automata of graphs of class $\mathcal{D}$ over the alphabet X with the given labeling $\ell(X_+) = \{+\}$ and $\ell(X_-) = \{-\}$. In the case when $\mathcal{D}$ is the class of all graphs we denote this class of automata by $\mathcal{G} = \mathcal{G}(X_+, X_-)$.

Theorem 7.4. ([146]) *For every labeling $X = X_+ \cup X_-$ of the alphabet, the class $\mathcal{G} = \mathcal{G}(X_+, X_-)$ is closed under intersection.*

Theorem 7.5. ([146]) *The class $\mathcal{G}(X_+, \emptyset)$ is closed under the Kleene $*$-operation.*

The following example shows that the class $\mathcal{G}(\emptyset, X_-)$ is not closed under the Kleene $*$-operation.

Example 7.2. ([146]) Consider the graph with the set $V = X = \{x, y, z\}$ of vertices and the set $E = \{(x, y)\}$ of edges. Let $f : X \to V$ be the identity mapping. We label all letters by $-$, and take $T = \{y, z\}$. The language recognized by the resulting automaton is $L = \{z, yx^*\}$. Therefore $zyx \in L^+$, but $zx \notin L^+$, and so property (c) in (ii) of Theorem 7.3 fails for L^+.

The resuts above show that in general it is quite difficult to obtain a regular expression describing a language recognized by a two-sided automaton of a graph. Nevertheless, we show that this problem can be solved in the case of complete graphs.

Theorem 7.6. ([146]) *Let $\mathcal{D}$ be any class consisting of complete graphs. Then the following conditions are equivalent:*

(i) $L \in \mathcal{G}(\mathcal{D}, X_+, X_-)$;

(ii) *there exist subsets $Y_1 \subseteq X_+$ and $Y_2 \subseteq X_-$ such that $X^+ \setminus L$ has the following regular expression:*

$$Y_2 X_-^* + X^* Y_1 X_-^* \tag{7.4}$$

(iii) *there exist subsets* $Y_1 \subseteq X_+$ *and* $Y_2 \subseteq X_-$ *such that* L *has a regular expression of the form* (7.4) *or* (7.5)*:*

$$1 + Y_2 X_-^* + X^* Y_1 X_-^* \tag{7.5}$$

Theorem 7.7. ([146]) *Let* $\mathcal{D}$ *be any class consisting of complete graphs. Then the class* $\mathcal{G}(\mathcal{D}, X_+, X_-)$ *is closed under intersection, union and the Kleene* $*$*-operation.*

Let $K'_{1,1}$ be the graph with the set $\{a, b\}$ of vertices and the set $\{(a, a), (b, a)\}$ of edges. Put $K_{1,1} = K'_{1,1} + (b, b)$. Let $\mathcal{K}$ be the set of the four graphs on the same set $\{a, b, c\}$ of vertices such that each graph of $\mathcal{K}$ has the edges $(a, a), (b, b), (b, a), (b, c)$, it may contain the loop (c, c), and it may simultaneously contain the edges (a, b) and (a, c).

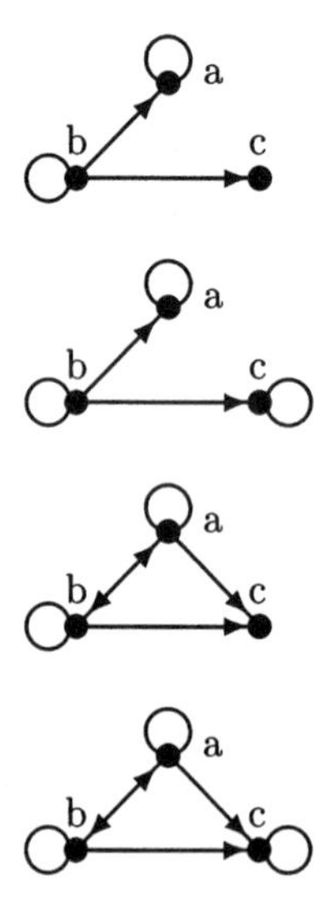

Figure 7.4: The graphs in Theorem 7.8

Theorem 7.8. ([146]) *Let* $|X| \geq 3$. *The class* $\mathcal{G}(\mathcal{D}, \emptyset, X_-)$ *is closed under the Kleene* $*$*-operation if and only if one of the following conditions is satisfied:*

(1) *not all graphs in $\mathcal{D}$ are null, and each graph of $\mathcal{D}$ either is isomorphic to one of the graphs K_2, $K_{1,1}$, $K'_{1,1}$, or is a disjoint union of some copies of K_1 and N_1;*

(2) *the following two conditions hold:*

(i) *for every graph $D = (V, E)$ of $\mathcal{D}$, the out-neighbourhood $Out(v)$ of each vertex $v \in V$, is equal to one of the sets $V, \{v\}$, or $\emptyset$;*

(ii) *the class $\mathcal{D}$ contains a graph with a subgraph isomorphic to one of graphs of $\mathcal{K}$.*

Proof of Theorem 7.4 follows from Theorem 7.3 immediately. □

Proof of Theorem 7.5. Fix any language $L \in \mathcal{G}(X_+, \emptyset)$. Theorem 7.3 shows that $L \in \mathcal{G}(X_+, \emptyset)$ if and only if $(L \setminus \{1\}) \in \mathcal{G}(X_+, \emptyset)$. Since $L^* \setminus \{1\} = L^+$, it suffices to verify condition (ii) of Theorem 7.3 for the language L^+. Given that $X_- = \emptyset$, we only need to check properties (a) and (b) of (ii).

First, pick any word xyu in L^+, where $x, y \in X, u \in X^*$. We have $xyu \in L^n$ for some $n \geq 1$. Consider the leftmost prefix of this word which is in L. If $x \in L$, then $yu \in L^{n-1} \subseteq L^+$. Further, assume that $xyu_1 \in L$ and $u_2 \in L^{n-1}$ for some factorization $u = u_1u_2$, where $u_1, u_2 \in X^*$. Then $yu_1 \in L$, because $L \in \mathcal{G}$. Therefore $yu = yu_1u_2 \in L^+$, again. This means that (a) holds.

Second, take $xv, yxu \in L^+$. If $y \in L$, then clearly $yxv \in L^+$. Hence we may assume that $yxu_1 \in L$ and $u_2 \in L^+$, where $u = u_1u_2$. We get $xv_1 \in L$ and $v_2 \in L^+$, where $v = v_1v_2$. Since $L \in \mathcal{G}$, we get $yxv_1 \in L$, and therefore $yxv = yxv_1v_2 \in L^+$, again. Thus condition (ii) of Theorem 7.3 is satisfied for L^+, and therefore $L^+ \in \mathcal{G}(X_+, \emptyset)$. □

Proof of Theorem 7.6. The equivalence of (i) and (ii) follows from the proof of Theorem 7.3, if we put $Y_1 = X_+ \cap X_N$ and $Y_2 = X_- \cap X_N$. Note that in the process of constructing an automaton $\mathrm{Atm}_t(G, V \setminus T, f, \ell)$ recognizing a language defined by the regular expression (7.4), it is enough to use one complete graph, because we can map several letters of X to the same vertex of the graph.

(ii)⇔(iii): Let D be a complete graph. It is easily seen that a language L is recognized by an automaton $\mathrm{Atm}_t(G,T,f,\ell)$ if and only if the language $X^+ \backslash L$ is recognized by the automaton $\mathrm{Atm}_t(G,V\backslash T,f,\ell)$. Therefore L belongs to $\mathcal{G}(\mathcal{D},X_+,X_-)$ if and only if $X^+ \setminus L$ lies in this class. It $1 \notin L$, then L has a regular expression of the form (7.4). Otherwise, it has a regular expression (7.5). □

Proof of Theorem 7.7. Fix two languages L, L' in $\mathcal{G}(\mathcal{D},X_+,X_-)$. By condition (iii) of Theorem 7.6, there exist subsets $Y_1, Y_1' \subseteq X_+$ and $Y_2, Y_2' \subseteq X_-$ such that $L = Y_2X_-^* + X^*Y_1X_-^*$ and $L' = Y_2'X_-^* + X^*Y_1'X_-^*$. Taking into account that the sets X_+ and X_- are disjoint, we get

$$Y_2X_-^* \cap X^*Y_1'X_-^* = \emptyset \text{ and } Y_2'X_-^* \cap X^*Y_1X_-^* = \emptyset.$$

Hence it follows that

$$L \cap L' = (Y_2 \cap Y_2')X_-^* + X^*(Y_1 \cap Y_1')X_-^* \in \mathcal{G}(\mathcal{D},X_+,X_-).$$

Similarly, we obtain

$$L \cup L' = (Y_2 + Y_2')X_-^* + X^*(Y_1 + Y_1')X_-^* \in \mathcal{G}(\mathcal{D},X_+,X_-),$$

$$\text{and } L^+ = Y_2X_-^* + X^*Y_1X_-^* = L \in \mathcal{G}(\mathcal{D},X_+,X_-).$$

This means that $\mathcal{G}(\mathcal{D},X_+,X_-)$ is closed for intersections, unions, and the Kleene $*$-operation. □

Proof of Theorem 7.8. The 'if' part: First, suppose that (1) holds. Then each language L of the class $\mathcal{G}(\mathcal{D},\emptyset,X_-)$ is equal either to X_1 or to X_1^* for some subset X_1 of X. In any case, we get $L^+ = X_1^*$, and this language is recognized by the automaton induced by a vertex with a loop.

Next, suppose that (2) is satisfied. Condition (i) implies that in every graph $D = (V,E)$ of $\mathcal{D}$ the equality $\mathrm{In}(a) \setminus \{a\} = \mathrm{In}(b) \setminus \{b\}$ holds for any two vertices $a, b \in V$.

Consider an arbitrary nontrivial language L of the class $\mathcal{G}(\mathcal{D}, \emptyset, X_-)$. Let L be recognized by an automaton $\mathrm{Atm}_t(D, T, f, \ell)$ of some graph $D = (V, E) \in \mathcal{D}$.

First, we prove that the language L is recognized by $\mathrm{Atm}_t(D', T, f, \ell)$, where $D' = (V, E \cup \{(u, v) \mid u, v \in T\}$.

The proof of Lemma 7.3 shows that the regular expression of L is given by $\sum_{x \in X_T} x I_x^*$, where $I_x = f^{-1}(\mathrm{In}(f(x))$. Our assumption yields $\mathrm{In}(a) \setminus \{a\} = \mathrm{In}(b) \setminus \{b\}$ for all $a, b \in V$. Therefore the set $I_x \setminus f^{-1}f(x)$ is the same for all $x \in X$. Denote this set by I. Let R be the language recognized by $\mathrm{Atm}_t(D', T, f, \ell)$. We are going to show that $R = L^+$. By the proof of Lemma 7.3, the regular expression of R is given by

$$\sum_{x \in X_T} x J_x^*, \text{ where } J_x = f^{-1}(\mathrm{In}_{D'}(f(x)). \tag{7.6}$$

The definition of D' implies that $\mathrm{In}_{D'}(v) = \mathrm{In}_D(v) \cup T$ for every $v \in T$. Take an arbitrary u in R. It follows that $u = xv$, where $x \in X_T, v \in J_x^*$. We may rewrite v as $v = v_1 z_2 v_2 z_3 \cdots z_n v_n$, where $z_i \in X_T$ and the words v_i do not contain letters from X_T. We have $v_1 \in I_x^*$ and $v_i \in I_{z_i}^*$ for all $i = 2, \ldots, n$. Therefore $u = xv \in L^n \subseteq L^+$. On the other hand, take an arbitrary u in L^+. It follows that $u = x_1 u_1 \cdots x_n u_n$, where $x_i \in X_T, u_i \in I_{x_i}^*$. Since D' contains all edges between vertices in T, we get $x_2, \ldots, x_n \in J_{x_1}$. Besides, we see that the set J_x is identical for all $x \in X_T$. Hence $u_i \in J_{x_1}^*$ for all $i = 2, \ldots, n$, and so $u \in R$.

Further, let a be the vertex with a loop in $K'_{1,1}$, let b be another vertex of $K'_{1,1}$, and let c be an isolated vertex in $K'_{1,1} \cup N_0$, i.e., a vertex without edges incident to it. Consider the automaton $\mathrm{Atm}_t(K'_{1,1} \cup N_0, \{a\}, f', \ell)$, where the mapping $f' : X \to \{a, b\}$ is defined by

$$f(y) = \begin{cases} a & \text{if } y \in X_T, \\ b & \text{if } y \in J \setminus X_T, \\ c & \text{if } y \in X \setminus (X_T \cup J). \end{cases}$$

It is easily seen that the automaton $\mathrm{Atm}_t(K'_{1,1} \cup N_0, \{a\}, f', \ell)$ recognizes the language (7.6).

A routine verification shows that, for any automaton of the graph $K'_{1,1} \cup N_1$, each graph with a subgraph isomorphic to one of the graphs in the set $\mathcal{K}$ has a two-sided automaton recognizing exactly the same language as the given automaton of $K'_{1,1} \cup N_1$ does. Therefore L^+ belongs to $\mathcal{G}(\mathcal{D}, \emptyset, X_-)$, in view of (ii).

The 'only if' part: First, suppose all graphs in $\mathcal{D}$ are null. Then any language L of the class $\mathcal{G}(\mathcal{D}, \emptyset, X_-)$ is equal to X_1 for some subset X_1 of X. Hence $L^+ = X_1^+$, a contradiction.

Now, suppose that condition (1) does hot hold. We shall prove that condition (2) is satisfied in this case.

Let us prove condition (i). Suppose to the contrary that there exists a graph $D(V, E) \in \mathcal{D}$ which does not satisfy condition (i). Then $\mathrm{In}(a) \setminus \{a\} \neq \mathrm{In}(b) \setminus \{b\}$ for some vertices $a, b \in V$. We may assume that $c \in \mathrm{In}(a) \setminus \mathrm{In}(b)$ for some $c \in V$.

Choose any two distinct letters x, y in X, and define a mapping $f : X \to V$ so that $f(x) = a, f(y) = b, f(X \setminus \{x, y\}) = \{c\}$. Further, we label all letters by $-$. The following two cases are possible:

Case 1: $c \notin \{a, b\}$. Then we take $T = \{a, b\}$. The language recognized by the automaton $A(D, T, f, \ell)$is $L = xI_x^* + yI_y^*$. Then we get $yxz \in L^+$ but $yz \notin L^+$, because $f(z) \in \mathrm{In}(x) \setminus \mathrm{In}(y)$, where $z \in X \setminus \{x, y\}$.

Case 2: $c = b$. Then we take $T = \{a, b\}$. The language recognized by the resulting automaton is $L = xI_x^* + yI_y^*$. Hence we get $yxy \in L^+$ but $yy \notin L^+$, because $f(y) \in \mathrm{In}(x) \setminus \mathrm{In}(y)$.

Therefore condition (c) in (ii) of Theorem 7.3 fails for L^+, in both the cases. Hence L^+ does not belong to the class $\mathcal{G}(\emptyset, X_-)$, and moreover, to the class $\mathcal{G}(\mathcal{D}, \emptyset, X_-)$, a contradiction.

It remains to verify condition (ii). Since we have assumed that (1) fails, there exists a graph $D(V, E) \in \mathcal{D}$ with edge (b, a) but without edge (c, a), where a, b, c are pairwise distinct vertices in V. Condition (i) implies $(b, b), (b, c) \in E$.

If $(a,a) \in E$, then the subgraph of D induced by the vertices a, b, and c contains the edges $(a,a), (b,a), (b,b)$, and (b,c) and does not contain (c,a). Thus we are done because, each graph of this kind belongs to $\mathcal{K}$ by the definition.

Therefore we may assume that $(a,a) \notin E$. Choose arbitrarily two different letters x, y in X and denote the set $X \setminus \{x,y\}$ by I. Define a mapping $f : X \to V$ so that $f(x) = a, f(y) = b$, and $f(I) = \{c\}$. Further, we label all letters by $-$ and take $T = \{a\}$. The language recognized by the obtained automaton is $L = xy^*$. Our assumption shows that the language

$$\begin{aligned} L^+ &= xy^* + xy^*xy^* + xy^*xy^*xy^* + \cdots \\ &= x(y^* + y^*xy^* + y^*xy^*xy^* + \cdots). \end{aligned} \tag{7.7}$$

belongs to $\mathcal{G}(\mathcal{D}, \emptyset, X_-)$, and so it is recognized by $\mathrm{Atm}_t(D', T', f', \ell)$ for some graph $D' = (V', E') \in \mathcal{D}$. Without loss of generality we may assume that $V' = f'(X)$. Put $u = f'(x)$ and $v = f'(y)$. The regular expression (7.7) yields that $T' = \{u\}$, $(f')^{-1}(u) = x$, and $(f')^{-1}(v) = y$. By Lemma 7.3, we get $y^* + y^*xy^* + y^*xy^*xy^* + \cdots = A^*$, where $A = (f')^{-1}(\mathrm{In}(u))$. Hence it is clear that $A = \{x,y\}$, and therefore $(u,u), (v,u) \in E'$.

Take an arbitrary $z \in I$ and denote $f'(z)$ by w. Since $(f')^{-1}(\mathrm{In}(u)) = \{x,y\}$, the set E' does not contain the edge (w,u). Condition (i) implies $(v,v), (v,w) \in E'$. Thus we have proved that the subgraph of the graph D' induced by the vertices u, v and w is isomorphic to one of the graphs of $\mathcal{K}$. $\square$

Example 7.3. ([146]) The languages $L_1 = \{x_1x_2, x_2\}$ and $L_2 = \{x_2x_3, x_3\}$ belong to $\mathcal{G}(X_+, \emptyset)$, because they satisfy condition (ii) of Theorem 7.3. However, their union $L_1 \cup L_2$ contains the words x_1x_2, x_2x_3, but does not contain $x_1x_2x_3$. Hence $L_1 \cup L_2$ does not satisfy condition (b). Therefore $L_1 \cup L_2 \notin \mathcal{G}(X_+, \emptyset)$.

The languages $L_3 = \{x_1\}$ and $L_2 = \{x_2x_3, x_3\}$ satisfy condition (ii) of Theorem 7.3, and so they are recognized by two-sided automata of graphs. However, (ii) does not hold for $L_3 \cdot L_2 = \{x_1x_2x_3, x_1x_3\}$.

Chapter 8

Tree Languages

Theorem 8.1. ([164]) *The graph algebra of a graph D is the syntactic algebra of a tree language if and only if the following conditions hold:*

(i) *all vertices of in-degree zero have pairwise distinct out-neighbourhoods in D;*

(ii) *D does not have three vertices with equal in-neighbourhoods and equal out-neighbourhoods.*

Lemma 8.1. *Let $D = (V, E)$ be a graph, τ a translation of $Alg(D)$, and let $x \in V$. Then*

$$\tau(x) = \begin{cases} \ell(\tau(x)) & \text{if } G[\tau(x)] \text{ is a subgraph of } D, \\ 0 & \text{otherwise.} \end{cases}$$

Proof. The proof uses induction. If $\tau = 1$, then the graph $G[\tau(x)]$ consists of the only isolated vertex x, and $\tau(x) = x = \ell(\tau(x))$. Assume that the assertion has been proven for some translation δ, and let $\tau = u^f\delta$, where $f \in \{l, r\}$. Denote the root $\ell(\delta(x))$ of $G[\delta(x)]$ by v. Then $V(\tau(x)) = V(\delta(x)) \cup \{u\}$, and

$$E(\tau(x)) = E(\delta(x)) \cup \begin{cases} \{(u, v)\} & \text{if } f = l, \\ \{(v, u)\} & \text{if } f = r. \end{cases}$$

If $G[\tau(x)]$ is a subgraph of D, then $G[\delta(x)]$ is a subgraph of D, too. Hence

$$\tau(x) = u^f \delta(x) = \begin{cases} u^l \delta(x) = uv = u = \ell(\tau(x)) & \text{if } f = l, \\ u^r \delta(x) = vu = v = \ell(\tau(x)) & \text{if } f = r. \end{cases}$$

Suppose that $G[\tau(x)]$ is not a subgraph of D. Then the following two cases are the only possible ones.

Case 1: $G[\delta(x)]$ is not a subgraph of D, and so both for $f = l$ and for $f = r$ we get

$$\tau(x) = u^f \delta(x) = u^f 0 = 0.$$

Case 2: $(u, v) \notin E$ for $f = l$, or $(v, u) \notin E$ for $f = r$, and therefore

$$\tau(x) = u^f \delta(x) = \begin{cases} u^l \delta(x) = uv = 0 & \text{if } f = l, \\ u^r \delta(x) = vu = 0 & \text{if } f = r \end{cases}$$

□

In particular, all trivial translations in every graph algebra evaluate to 0.

Corollary 8.1. ([164]) *Let Q be a subset of Alg(D) such that $0 \notin Q$, and let $x \in V$. Then the context $Cont_Q(x)$ consists of all translations $\tau \in Tr(Alg(D))$ such that $G[\tau(x)]$ is a subgraph of D with the root $\ell(\tau(x))$ in Q.*

Proof follows from the definition of the context and Lemma 8.1. □

Proof of Theorem 8.1. The 'if' part: Suppose that the graph $D = (V, E)$ satisfies conditions (i) and (ii). By the axiom of choice we can select vertices of D and put them into a subset Q of Alg(D) so that, for each $u \in V$, there exists precisely one $\tilde{u} \in Q$ satisfying $\text{Out}(u) = \text{Out}(\tilde{u})$ and $\text{In}(u) = \text{In}(\tilde{u})$.

We claim that the set Q is disjunctive. Take any two elements x, y of Alg(D) such that $\text{Cont}_Q(x) = \text{Cont}_Q(y)$. We shall verify that $x = y$.

First, let us suppose that $\text{Cont}_Q(x) = \emptyset$ and $x \in V$. If $x \in Q$, then $1 \in \text{Cont}_Q(x)$. Therefore $x \notin Q$. By the definition of Q and (i), we see that

Q contains all vertices of in-degree zero, and so $\text{In}(x) \neq \emptyset$. Take $v \in \text{In}(x)$. By the definition of Q there exists $\tilde{v} \in Q$ such that $x \in \text{Out}(v) = \text{Out}(\tilde{v})$. Hence $\tilde{v}^l \in \text{Cont}_Q(x)$, a contradiction. Therefore if $\text{Cont}_Q(x) = \emptyset$, then $x = y = 0$. Obviously, $\text{Cont}_Q(0) = \emptyset$.

Next, suppose that $\text{Cont}_Q(x) \neq \emptyset$, and therefore $x, y \in V$. The following three cases are possible:

Case 1: $\text{In}(x) \neq \text{In}(y)$. Assume that there exists z in $\text{In}(x) \setminus \text{In}(y)$. By the definition of Q there exists $\tilde{z}$ in Q such that $\text{Out}(z) = \text{Out}(\tilde{z})$. We have $x \in \text{Out}(\tilde{z})$ and $z \notin \text{Out}(\tilde{z})$. Therefore $z \in Q$, and so $z^l \in \text{Cont}_Q(x) \setminus \text{Cont}_Q(y)$, a contradiction.

Case 2: $\text{Out}(x) \neq \text{Out}(y)$. Choose an element z in, say, $\text{Out}(x) \setminus \text{Out}(y)$. Take any $\tau \in \text{Cont}_Q(x)$. Then $\tau z^r \in \text{Cont}_Q(x) \setminus \text{Cont}_Q(y)$. The contradiction shows that this case is impossible.

Case 3: $\text{In}(x) = \text{In}(y)$ and $\text{Out}(x) = \text{Out}(y)$. Clearly, any vertex is in Q if and only if the trivial translation 1 belongs to its context w.r.t. Q. It follows that x and y are either both in Q, or both not in Q. If $x, y \in Q$, then the definition of Q implies $x = y$. If $x, y \notin Q$, then there exists precisely one $\tilde{x} \in Q$ satisfying $\text{Out}(x) = \text{Out}(\tilde{x})$ and $\text{In}(x) = \text{In}(\tilde{x})$. Condition (ii) yields $x = y$, again.

Thus we have verified that Q is a disjunctive subset. Lemma 3.5 implies that $\text{Alg}(D)$ is syntactic.

The 'only if' part: Suppose that the graph algebra $\text{Alg}(D)$ is a syntactic algebra of a tree language. By Lemma 3.5 there exists a disjunctive subset Q of $\text{Alg}(D)$. Note that the set $\text{Alg}(D) \setminus Q$ is disjunctive. Therefore we may always assume that $0 \notin Q$.

Let x be a vertex of in-degree zero. If $x \notin Q$, then $\text{Cont}_Q(x) = \text{Cont}_Q(0) = \emptyset$, which contradicts the definition of a disjunctive set. Hence $x \in Q$, and Corollary 8.1 yields

$$\text{Cont}_Q(x) = \{1, u_{i_1}^r \cdots u_{i_k}^r \mid u_i \in \text{Out}(x)\}.$$

Therefore all vertices of in-degree zero have different out-neighbourhoods, that is, condition (i) holds.

Suppose to the contrary that (ii) does not hold, i.e., there exist three vertices in V with the same in-neighbourhood and the same out-neighbourhood. We can choose two of them, say x and y, which both belong either to Q, or to $\mathrm{Alg}(D) \setminus Q$. We prove that $\mathrm{Cont}_Q(x) = \mathrm{Cont}_Q(y)$. To this end, consider any $\tau \in \mathrm{Cont}_Q(x)$. We are going to verify that $\tau \in \mathrm{Cont}_Q(y)$. If $\tau = 1$, then $x \in Q$. By our assumption $y \in Q$, and so $\tau \in \mathrm{Cont}_Q(y)$.

Assume now that $\tau = u_1^{f_1} \cdots u_n^{f_n}$, where $f_i \in \{r, l\}$ for some $n \geq 1$. By Corollary 8.1, $G[\tau(x)]$ is a subraph of D and $\ell(\tau(x)) \in Q$. Since $\mathrm{In}(x) = \mathrm{In}(y)$ and $\mathrm{Out}(x) = \mathrm{Out}(y)$, we have $(y, u_i) \in E$ if and only if $(x, u_i) \in E$ and, similarly, $(u_i, y) \in E$ if and only if $(u_i, x) \in E$. Therefore $G[\tau(y)]$ is a subgraph of D. Further, we see that either $\ell(\tau(y))$ equals $\ell(\tau(x))$, or $\ell(\tau(y)) = y$ and $\ell(\tau(x)) = x$. In the former case $\ell(\tau(y)) \in Q$, and Corollary 8.1 implies that $\tau \in \mathrm{Cont}_Q(y)$. In the latter case we get $x \in Q$, and so $y \in Q$ by the the choice of x and y. Therefore $\tau \in \mathrm{Cont}_Q(y)$, again. Thus $\mathrm{Cont}_Q(x) = \mathrm{Cont}_Q(y)$. This contradicts the definition of a disjunctive set and completes the proof. □

Lemma 8.2. *A* 0*-direct union of syntactic groupoids is a syntactic groupoid.*

Proof. Let $S = \bigcup_{i \in I} S_i$ be a 0-direct union of groupoids S_i, where $i \in I$. Suppose that all the groupoids S_i are syntactic. Then by Lemma 3.5 every S_i contains a disjunctive subset Q_i. Note that we can always assume that $0 \notin Q$ for all $i \in I$. Indeed, if $0 \in Q_i$ for some $i \in I$, we can take the subset $S_i \setminus Q_i$ instead of Q. Put $Q = \bigcup_{i \in I} S_i$. The context of 0 with respect to Q is empty, as well as with respect to each Q_i, and $\mathrm{Cont}_{Q_i}(x) = \mathrm{Cont}_Q(x)$, for any element $x \in S_i, x \neq 0$. Therefore $\mathrm{Cont}_Q(x) \neq \mathrm{Cont}_Q(y)$ whenever $x \neq y$. □

Proposition 8.1. ([164]) *The graph algebra of an undirected graph is syntactic if and only if it is isomorphic to a* 0*-direct union of subdirectly irre-*

ducible groupoids.

Proof. It is easily seen that Alg(D) is a 0-direct union of the graph algebras of all connected components of the graph D.

Suppose that Alg(D) is a syntactic algebra. Theorem 8.1 implies that D has at most one isolated vertex, and at most one pair of distinct vertices have the same neighbourhoods. By Lemma 4.2 all connected components of D are subdirectly irreducible. Therefore Alg(D) is isomorphic to a 0-direct union of subdirectly irreducible groupoids.

Conversely, suppose that Alg(D) is isomorphic to a 0-direct union of subdirectly irreducible groupoids. Lemma 3.6 shows that all these groupoids are syntactic. By Lemma 8.2, a 0-direct union of syntactic groupoids is syntactic, as well. □

Proposition 8.1 does not generalize from undirected to arbitrary graphs.

Example 8.1. ([164]) The graph algebra of the subgraph D' induced by the vertices u_4, u_5, and u_6 of the graph in Figure 3.56. By Theorem 8.1, Alg(D') is a syntactic algebra. However, it cannot be represented as a 0-direct union of subdirectly irreducible groupoids. Moreover, Alg(D') is indecomposable into 0-direct unions of proper subgroupoids, and it is not subdirectly irreducible itself. Indeed, the meet of the following two congruences

$$\begin{aligned}\delta_1 &= \{\{u_5, u_6\}, \{u_4\}, \{0\}\}, \\ \delta_2 &= \{\{u_4, u_5, 0\}, \{u_6\}\}\end{aligned}$$

is the equality relation, and so Alg(D') is subdirectly reducible.

Chapter 9

Equational Theories

Let $D_1 = (V_1, E_1), \ldots, D_n = (V_n, E_n)$ be graphs. The direct product $D = D_1 \times \cdots D_n$ is the graph $D = (V, E)$, where $V = V_1 \times \cdots \times V_n$ and

$$E = \{((x_1, \ldots, x_n)(y_1, \ldots, y_n)) \mid (x_i, y_i) \in E_i \text{ for } i = 1, \ldots, n\}.$$

More generally, given a family of graphs $D_i = (V_i, E_i)$ indexed by the elements of a set I, the direct product $\prod_{i \in I} D_i$ is the graph with vertex set $\prod_{i \in I} V_i$ such that two vertices in the product are adjacent if and only if so are all their components.

Let $D = (V_1, E_1)$ and $D_2 = (V_2, E_2)$ be graphs. A mapping $f : V_1 \to V_2$ is called a *homomorphism* if, for every edge $(x, y) \in E_1$, the image $(f(x), f(y))$ belongs to E_2.

Let $D = (V, E)$ be a graph. A graph $D' = (V', E')$ is called a *subgraph* of D if $V' \subseteq V$. A subgraph $D' = (V', E')$ of D is called an *induced subgaph* if and only if $E' = E \cap (V' \times V')$, that is the edges of D' are precisely the edges of D connecting elements of V'.

Let G_0 be the graph in Figure 9.1. The graph algebra $\mathrm{Alg}(G_0)$ is called the Murski's groupoid.

A variety or equational class is *inherently nonfinitely based* if it is

Figure 9.1: The graph of Murski's groupoid.

is not finitely based, is locally finite and is not contained in any finitely based locally finite variety. An algebra is *inherently nonfinitely based* if it generates an inherently nonfinitely based variety.

Theorem 9.1. ([197]) *Let D be an undirected graph. The graph algebra $Alg(D)$ is simple if and only if D is connected and any two distinct vertices of D have distinct neighbourhoods.*

A graph $D = (V, E)$ is said to be *looped* if E contains all loops (v, v) for $v \in V$.

Theorem 9.2. ([197]) *Let $\mathcal{K}$ be a variety of groupoids containing either all graph algebras of loopless undirected graphs, or all graph algebras of looped undirected graphs. Then*

(i) *if $\mathcal{K}$ is locally finite, then it is inherently nonfinitely based;*

(ii) *$\mathcal{K}$ has simple groupoids of every cardinality larger than 4.*

Theorem 9.3. ([197],[206]) *Let $\mathcal{K}$ be a variety of groupoids containing all graph algebras of loopless undirected graphs. Then $\mathcal{K}$ has an infinite ascending chain of subvarieties and an infinite descending chain of subvarieties.*

Recall that the congruence $\Theta(x, y)$ generated by the pair (x, y) is the smallest congruence containing (x, y).

Lemma 9.1. ([197]) *A universal algebra is subdirectly irreducible if and only if it has two congruences β and γ such that $\beta \cap \gamma = \iota$.*

Lemma 9.2. ([206]) *Let $A_1, \ldots, A_n$ be the graph algebras of undirected graphs $D_1, \ldots, D_n$, and let ϱ be the relation defined on $A = A_1 \times \cdots \times A_n$*

consisting of all pairs $((x_1, \ldots, x_n), (y_1, \ldots, y_n))$ *such that* $x_i, y_i \in A_i$ *for* $i = 1, \ldots, n$, *and either* $x_i = y_i$ *for all* $i = 1, \ldots, n$, *or there exist* i, j *with* $x_i = y_j = 0$. *Then* ϱ *is a congruence on* A *and the quotient groupoid* A/ϱ *is isomorphic to the graph algebra of the graph* $D = D_1 \times \cdots \times D_n$.

Proof. Take any pair (x, y) in ϱ, where $x = (x_1, \ldots, x_n)$, $y = (y_1, \ldots, y_n)$ and an element $z = (z_1, \ldots, z_n)$ in A. If $x = y$, then $xz = yz$ in A, and so $(xz, yz) \in \varrho$. Otherwise, if $x_i = y_j = 0$ for some i, j, then $x_i z_i = y_j z_j = 0$, whence $(xz, yz) \in \varrho$ again. Similarly, $(zx, zy) \in \varrho$ in all cases too. Thus ϱ is a congruence on A.

Let $D = D_1 \times \cdots \times D_n$ and $D = (V, E)$. Consider the mapping $f : A \to (V \cup \{0\})$ defined by $f(w) = w^\varrho$. Note that the set of all elements of A with at least one zero component is equal to the class 0^ϱ of A/ϱ. If $x = (x_1, \ldots, x_n)$ and $y = (y_1, \ldots, y_n$ have no zero entries, then the product xy has no zero entries if and only if $x_i y_i = x_i$ for all i, that is, if and only if there is an edge from x to y in D. It is routine to verify that f is a homomorphism from A onto $\mathrm{Alg}(D)$, and that ϱ is the kernel congruence of f. Hence the Homomorphism Theorem (Theorem 2.53) completes the proof. □

Theorem 9.4. ([206]) *Let* $D = (V, E)$ *be a loopless undirected graph. Then* *Alg*(D) *is isomorphic to a quotient of a subalgebra of a cartesian product of copies of Murski's groupoid* *Alg*(G_0).

Proof. For each $v \in V$, define a function $f_v : V \times V \to \mathrm{Alg}(G_0)$ by setting

$$
\begin{aligned}
f_v(x, y) &= 2 \text{ if } x, y \neq v \\
f_v(x, v) &= \begin{cases} 2 & \text{if } v \text{ is adjacent to } x \text{ in } D, \\ 1 & \text{otherwise.} \end{cases}
\end{aligned}
$$

Denote by G the subgroupoid of $\mathrm{Alg}(G_0)^{V \times V}$ generated by all functions f_v for $v \in V$. Let ϱ be the congruence on A introduced in Lemma 9.2, then it is easy to verify that A/ϱ is isomorphic to $\mathrm{Alg}(D)$. □

The symbol A^V, as before, denotes the set of all functions from V to A. If A is a universal algebra, then A^V can be regarded as a universal algebra of the same type since all operations of A carry over to A^V defined componentwise. The smallest variety containing the class $\mathcal{K}$ is denoted by var$(\mathcal{K})$.

Theorem 9.5. ([175]) *Let $\mathcal{C}$ be a class of undirected graphs, and let $\mathcal{K}$ be the class of graph algebras of all graphs in $\mathcal{C}$. The graph algebra of a finite connected undirected graph H belongs to the variety var$(\mathcal{K})$ if and only if H is an induced subgraph of a product of members of $\mathcal{C}$. If D is an arbitrary undirected graph, then Alg(D) is in var$(\mathcal{K})$ if and only if var$(\mathcal{K})$ contains the graph algebras of all connected components of the finite induced subgraphs of D.*

A *rooted graph* is a graph with a distinguished vertex. For every groupoid term t, define a rooted undirected graph $G(t)$ by induction on the length of t. The vertex set $V(t)$ of $G(t)$ is the set of variables occurring in t, and the root of $G(t)$ is the leftmost variable of t. If t is a variable, then $G(t)$ has no edges. If $t = t_1t_2$ for two terms t_1, t_2, then the set $E(t)$ of edges of $G(t)$ is equal to the union of $G(t_1)$, $G(t_2)$ with the new edge connecting the root of $G(t_1)$ with the root of $G(t_2)$.

For example, the graph G_0 defining the Murski's groupoid coincides with the graph of the term $(x_2x_2)(x_1x_2)$.

Lemma 9.3. ([175]) *Let $D = (V, E)$ be an undirected graph, $t(x_1, \ldots, x_n)$ a groupoid term with leftmost variable x_1, and let f be a mapping from $V(t)$ to V. Then the value of $t(f(x_1), \ldots, f(x_n))$ in Alg(D) is $f(x_1)$ if f is a graph homomorphism from $G(t)$ to D, and is 0 otherwise.*

This lemma does not generalize from undirected to arbitrary graphs.

Lemma 9.4. ([175]) *For each finite connected rooted undirected graph D, there exists a term t such that $D = D(t)$.*

Proof. We proceed by induction on the number of edges in D. The induction basis is obvious, since if D has no edges, then it corresponds to a variable. For the induction step, denote the root of a finite undirected graph $D = (V, E)$ by r. Since $|E| > 0$, it follows that the set E contains an edge (r, x). Remove this edge from D, and denote by D_r and D_x the connected components of r and x in the resulting graph with roots r and x. By the induction hypothesis, $D_r = D(t_r)$ and $D_x = D(t_x)$ for some terms t_r and t_x. Then $D = G(t_r t_x)$. □

Proof of Theorem 9.5. Suppose that H is a finite connected undirected graph such that $\mathrm{Alg}(H)$ belongs to the variety $\mathrm{var}(\mathcal{K})$. Denote the vertices of H be $x_1, \ldots, x_n$. One can see immediately that H is an induced subgraph of a product of members of $\mathcal{C}$ if and only if the following two conditions hold.

(1) For every two distinct vertices x and y of H there is a homomorphism f from H to a member of G such that $f(x) \neq f(y)$.

(2) For every two non-adjacent vertices x and y of H there exists a homomorphism f from H to a member of G such that $f(x)$ and $f(y)$ are not adjacent.

It remains to verify these conditions to complete the proof.

Take a vertex x in H and consider it a the root of H. By Lemma 9.4, the rooted graph H coincides with $G(t_x)$ for some term t_x. To prove condition (1) choose two distinct vertices x and y and consider the equation $t_x = t_y$. Since H is connected, we see that the set $V(t_x)$ of variables of the terms t_x is equal to V, and likewise $V(t_y) = V$. Let f be the identical map on the set $V = \{x_1, \ldots, x_n\}$. Lemma 9.3 implies that the equation $t_x = t_y$ fails in the variety $\mathrm{Alg}(H)$. Therefore it fails in some groupoid $\mathrm{Alg}(D)$, where $D = (W, E)$ is a member of $\mathcal{C}$. This means that there are elements $w_1, \ldots, w_n$ of W such that

$$f_x(w_1, \ldots, w_n) \neq f_y(w_1, \ldots, w_n).$$

Now consider a new mapping f sending x_i to w_i. By Lemma 9.3, this mapping is a homomorphism required in condition (1).

The statement (2) can be verified in the same way by considering the equation $t_x t_y = t_x$. This completes the proof. □

Corollary 9.1. ([197], Theorem 1) *Let D be a finite connected undirected graph. The variety var(Alg(D)) contains the graph algebras of all loopless undirected graphs if and only if G_0 is an induced subgraph of D.*

Proof. Theorem 9.5 allows us to deal with finite connected undirected graphs only. We may use conditions (1) and (2) given in the proof of Theorem 9.5. Clearly, every proper homomorphic image of a complete graph contains a loop. Hence it follows that condition (2) with $x = y$ is necessary. Conversely, it is obvious that every loopless graph has all homomorphisms into G_0 as required in (1) and (2), and so satisfies these conditions. □

Corollary 9.2. ([208]) *The lattice of subvarieties of the variety generated by Murskii's groupoid contains a sublattice isomorphic to the lattice of all subsets of a countable set and the lattice isomorphic to the lattice of real numbers.*

Proof. It has been proved in [83] and [84] that there exists a countable family of finite loopless connected undirected graphs $D_1, D_2, \ldots$ such that D_i has no homomorphism into D_j if $i \neq j$. By Theorem 9.5, for every i the variety var(Alg(D_i)) is not contained in the variety generated by the set $\{ \text{Alg}(D_j) \mid j \neq i\}$. Therefore all subsets of the set $\{ \text{Alg}(D_i) \mid i = 1, 2, \ldots\}$ generate paiwise distinct varieties. □

Theorem 9.6. ([10]) *An undirected graph has a finitely bases graph algebra if and only if it has no induced subgraph isomorphic to one of the four graphs of Figure 9.2.*

Corollary 9.3. ([10]) *If a graph algebra of an undirected graph is not finitely based, then it is inherently nonfinitely based.*

Theorem 9.7. ([10]) *For any undirected graph G, either the variety of groupoids with as unary operation generated by the graph algebra of G is*

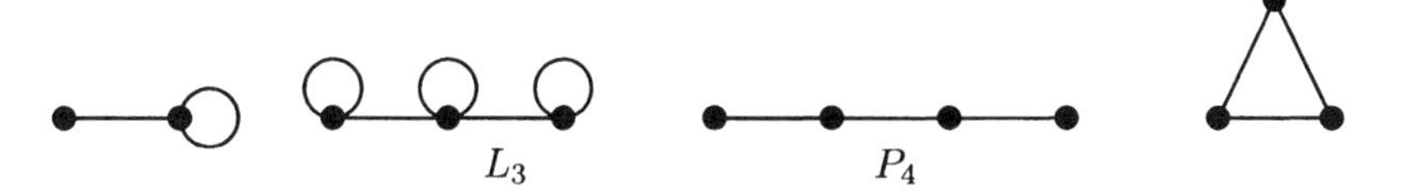

Figure 9.2: The graphs of Theorem 9.6.

inherently nonfinitely based, or the graph algebra of G generates one of the following eleven finitely based varieties in the class of groupoids with zero as a unary operation:

(i) *the variety defined by ten equations*

$$x0 = 0x = 0, \tag{9.1}$$

$$xy = (xy)y, \tag{9.2}$$

$$x(yz) = (xy)(yz), \tag{9.3}$$

$$(xy)z = (xz)y, \tag{9.4}$$

$$xy = x(yx), \tag{9.5}$$

$$x((yz)u) = (x(yz))(yu), \tag{9.6}$$

$$x(y(zu)) = (x(yz))(uz), \tag{9.7}$$

$$(x(yz))(uv) = (x(yz))(uz), \tag{9.8}$$

$$x(xy) = x(yy), \tag{9.9}$$

$$(xx)(yz) = (x(yy))(zz); \tag{9.10}$$

(ii) *the variety defined by equations (9.1) to (9.10) and* $x(yy) = y(xx)$;

(iii) *the variety defined by equations (9.1) to (9.10) and* $xx = yy$;

(iv) *the variety defined by equations (9.1) to (9.10),* $xx = yy$ *and* $x(yz) = z(yx)$;

(v) *the variety defined by equations (9.1) and* $xy = uz$;

(vi) *the variety defined by equation* $x = y$;

(vii) *the variety defined by equations (9.1) to (9.10) and* $x(yz) = z(yx)$;

(viii) *the variety defined by equations (9.1) to (9.10) and* $x(yz) = (xy)z$;

(ix) *the variety defined by equations (9.1)* $x(yz) = (xy)z$, $xy = yx$ *and* $xy = x(yy)$*;*

(x) *the variety defined by equations (9.1)* $x(yz) = (xy)z$, $xx = x$ *and* $x(yz) = x(zy)$*;*

(xi) *the variety defined by equations (9.1)* $x(yz) = (xy)z$, $xx = x$ *and* $xy = yx$*.*

For a positive integer n, let L_n and P_n denote the undirected graphs with n vertices in the form of one path with or without loops on all vertices, as illustrated in Figure 9.2 for the graphs L_3 and P_4. For graphs D and G, the disjoint union of D and G without edges between them is denoted by $D + G$.

Theorem 9.8. ([10]) *For any undirected graph* G*, either the variety of groupoids generated by the graph algebra of* G *is inherently nonfinitely based, or the graph algebra of* G *generates one of the following eleven finitely based varieties:*

(i) *the variety generated by* $P_2 + L_2$*;*

(ii) *the variety generated by* $P_3 + L_1$*;*

(iii) *the variety generated by* P_3*;*

(iv) *the variety generated by* $P_2 + L_1$*;*

(v) *the variety generated by* $P_1 + L_2$*;*

(vi) *the variety generated by* P_2*;*

(vii) *the variety generated by* $P_1 + L_1$*;*

(viii) *the variety generated by* L_2*;*

(ix) *the variety generated by* P_1*;*

(x) *the variety generated by* L_1*;*

(xi) *the variety defined by the equation* $x = xy$;

(xii) *the variety defined by the equation* $xx = xy$;

(xiii) *the variety defined by equations (9.2) to (9.10) and* $xx = x(yy)$;

(xiv) *the variety defined by the equation* $x = y$.

For a class $\mathcal{G}$ of graphs, let $V(\mathcal{G})$ be the class of all graphs H such that $\text{Alg}(H) \in \text{var}\{ \text{Alg}(D) \mid D \in \mathcal{G}\}$. Following [176], we say that the class $\mathcal{G}$ is a *graph variety* or an *equational class* if $V(\mathcal{G}) = \mathcal{G}$.

Theorem 9.9. ([175], [225]) *A class of undirected graphs is a graph variety if and only if it is closed under direct products, induced subgraphs, disjoint unions and unions of all finite induced subgraphs.*

Theorem 9.10. ([176]) *Let* $\mathcal{G}$ *be a class of graphs. Then the lattice of subvarieties of var*$\{$ *Alg*$(D) \mid D \in \mathcal{G}\}$ *is isomorphic to the lattice of subvarieties of* $V(\mathcal{G})$.

Theorem 9.11. ([176]) *Every subvariety of a variety generated by graph algebras is also generated by graph algebras.*

Theorem 9.12. ([176]) *Let* $\mathcal{V}$ *be a variety generated by a class of undirected graph algebras. The lattice of subvarieties of* $\mathcal{V}$ *satisfies the descending chain condition if and only if* $\mathcal{V}$ *is one of the seventeen subvarieties of the variety generated by Alg*(L_3) *and Alg*(P_4). *Therefore, for varieties generated by graph algebras, the descending chain condition implies teh ascending chain condition.*

Theorem 9.13. ([176]) *Let* $\mathcal{V}_0$ *be the variety generated by Alg*(L_3) *and Alg*(P_4). *Then* $\mathcal{V}_0$ *has seventeen subvarieties, and inclusions among these subvarieties are given by the following assertions.*

(1) $V(P_1) \subset V(P_2) \subset V(P_3) \subset V(P_4)$.

(2) $V(L_1) \subset V(L_2) \subset V(L_3)$.

(3) *If* $L_i \in V(L_j, P_k)$, *then* $L_i \in V(L_j)$.

(4) *If* $P_i \in V(L_1, P_k)$, *then* $P_i \in V(P_k)$.

(5) $P_3 \notin V(L_2, P_2)$, $P_4 \notin V(L_2, P_3)$, $P_2 \notin V(L_3, P_1)$, $P_3 \in V(L_3, P_2)$.

Theorem 9.14. ([185]) *For every undirected graph* G, *the following conditions are equivalent:*

(i) G *is complete without loops;*

(ii) *every subalgebra of* Alg(G) *has at most two congruences.*

Let $T(X)$ be the set of all terms over the alphabet $X = \{x_1, x_2, \dots\}$ using the binary operation $\cdot$ and unary operation 0. Consider equations of the form $t_1 = t_2$, where $t_1, t_2 \in T(X)$. For a graph D, we write $D \models t_1 = t_2$ or $\mathrm{Alg}(D) \models t_1 = t_2$ if the equality $t_1 = t_2$ holds for every assignment of elements of $\mathrm{Alg}(D)$ to the variables in t_1, t_2. If Σ is a set of equations, then we write $D \models \Sigma$ or $\mathrm{Alg}(D) \models \Sigma$ if $D \models t_1 = t_2$ for all $t_1 = t_2 \in \Sigma$.

Let $\mathcal{L}$ be a class of graphs, and let $\mathcal{K} \subseteq \mathcal{L}$. The class $\mathcal{K}$ is called *equational* in $\mathcal{L}$, if there exists a set of equations such that

$$\mathcal{K} = \{D \in \mathcal{L} \mid D \models \Sigma\}.$$

We use notation

$$\Sigma(\mathcal{K}) = \{t_1 = t_2 \mid t_1, t_2 \in T(X),\ (\forall D \in \mathcal{K})\ D \models t_1 = t_2\},$$

$$V_{\mathcal{L}}(\mathcal{K}) = V_{\mathcal{L}}(\Sigma(\mathcal{K})) = \{D \in \mathcal{L} \mid D \models \Sigma(\mathcal{K})\}.$$

The class $V_{\mathcal{L}}(\mathcal{K})$ is called the *graph variety* enerated by $\mathcal{K}$ in $\mathcal{L}$. Evidently,

$$V_{\mathcal{L}}(\mathcal{K}) = \{D \in \mathcal{L} \mid \mathrm{Alg}(D) \in \mathrm{var}(\mathrm{Alg}(G), G \in \mathcal{K})\},$$

and $V_{\mathcal{L}}(\mathcal{K}) = \mathcal{K}$ if and only if $\mathcal{K}$ is an equational class in $\mathcal{L}$.

For example, the class of all undirected graphs is equational in the class of all graphs, because a graph D is undirected if and only if $D \models x_1(x_2x_1) = x_1x_2$. Similarly, a graphs D does not contain any undirected edges or loops if and only $D \models x_1(x_2x_1) = 0$.

Let us introduce a few closure operations used in characterizing graph varieties. Let $\mathcal{K}$ be a class of graphs. The following symbols will be used to denote new classes derived from $\mathcal{K}$:

$$
\begin{aligned}
I\mathcal{K} &= \text{all isomorphic copies of the graphs in } \mathcal{K};\\
S\mathcal{K} &= \text{all induced subgraphs of the graphs in } \mathcal{K};\\
\overline{S}\mathcal{K} &= \text{all subgraphs of the graphs in } \mathcal{K};\\
P\mathcal{K} &= \text{all direct products of the graphs in } \mathcal{K};\\
P_f\mathcal{K} &= \text{all finite direct products of the graphs in } \mathcal{K};\\
U\mathcal{K} &= \text{all disjoint unions of the graphs in } \mathcal{K};\\
U_f\mathcal{K} &= \text{all finite disjoint unions of the graphs in } \mathcal{K}.
\end{aligned}
$$

If $D = (V, E)$ is a graph, then by $D^1 = (V^1, E')$ we denote the graph obtained by adding a new vertex 1 to V with a loop and all edges from 1 to all vertices in V added to E'. The direct product $\prod_{i\in I} D_i^1$ is called a *(full) pointed direct product* of D_i, $i \in I$. Every induced subgraph of $\prod_{i\in I} D_i^1$ is called a *pointed subproduct* of D_i, $i \in I$. It is said to be finite if I is finite. For a vertex x of the prointed product, let

$$Y(x) = \{i \in I \mid x(i) \neq 1\}.$$

Evidently, $y \in \text{Out}^*(x)$ implies $Y(x) \subseteq Y(y)$. A pointed subproduct $W = (V(W), E(W))$ of D_i, $i \in I$, is said to be *restricted* if the following two conditions hold:

- $Y(x) \neq \emptyset$ for every $x \in V(W)$; (9.11)
- for all $x \in V(W)$ and $y, y' \in \text{Out}_W^*(x)$, if $(y, y') \notin E(W)$, then $(y(i), y'(i)) \notin E(D_k)$ for some $i \in Y(x)$. (9.12)

Denote by $P^{rp}(\mathcal{K})$ (and $P_f^{rp}(\mathcal{K})$) the class of all (resp., all finite) restricted pointed subproducts of the graphs in $\mathcal{K}$.

Theorem 9.15. ([224]) *A class of finite graphs is equational if and only if it is closed with respect to finite restricted pointed subproducts and isomorphic copies.*

Proposition 9.1. ([224]) *For every class $\mathcal{K}$ of graphs,*

$$USP\mathcal{K} \subseteq IP^{rp}\mathcal{K}.$$

Proof. Conditions (9.11) and (9.12) are inherited by induced subgraphs, it follows that every induced subgraph of a restricted pointed subproduct of D_i, $i \in I$, is a restricted pointed subproduct of the same graphs too.

For each subset $J \subseteq I$, the direct product $\prod_{j\in J} D_j$ is isomorphic to the restricted pointed subproduct of D_i, $i \in I$, induced by all x with $Y(x) = J$.

Every disjoint union W of graphs D_i, $i \in I$, is isomorphic to a restricted pointed subproduct of D_i, $i \in I$. Indeed, we can embed each D_i into $\prod_{i\in I} D_i^1$ by identifying its vertices with all x such that $Y(x) = i$, and consider the disjoint union of these embedded copies of the D_i. Conditions (9.11) and (9.12) are satisfied, and so we get a restricted pointed subproduct. □

Let $t = t(x_1, \ldots, x_n) \in T(X)$ be a term with variables $x_1, \ldots, x_n$. Denote by LEFT(t) the leftmost variable in t. We associate a graph $D(t) = (V(t), E(t))$ to t, where $V(t)$ is the set of variables that occur in t, and $E(t)$ is defined inductively by setting

$$E(t) = \emptyset \text{ if } t \text{ is a variable or } t = 1,$$

$$E(t_1t_2) = E(t_1) \cup E(t_2) \cup \{\text{LEFT}(t_1), \text{LEFT}(t_2)\} \text{ for } t = t_1t_2 \in T(X).$$

A (*relational*) *homomorphism* from a graph $D_1 = (V_1, E_1)$ to $D_2 = (V_2, E_2)$ is a mapping $f : V_1 \to V_2$ such that $(f(x), f(y)) \in E_2$ for all $(x, y) \in E_1$. If $t = t(x_1, \ldots, x_n)$ is a term, and $D = (V, E)$ is a graph, then we put

$$f(t) = t(f(x_1), \ldots, f(x_n)),$$

where the right hand side is calculated in Alg(D). The following two lemmas immediately follow from the definitions, because in every graph algebra $x_ix_j = x_i$ if and only if $(x_i, x_) \in E$, or $x_ix_j = 0$ otherwise.

Lemma 9.5. ([224]) *Let $D = (V, E)$ be a graph, $t = t(x_1, \ldots, x_n) \in T(X)$, and let $h : V(t) \to V$ be an assignment of vertices of D to the variables of t. Then the equality*

$$t(h(x_1), \ldots, h(x_n)) = h(x_1)$$

holds in Alg(D) if and only if h is a homomorphism from $D(t)$ to D. Otherwise, $t(h(x_1), \ldots, h(x_n)) = 0$.

Lemma 9.6. *Two terms t_1, t_2 are equivalent, i.e., $D \models t_1 = t_2$ for all graphs D, if and only if $D(t_1) = D(t_2)$ and $\mathrm{LEFT}(t_1) = \mathrm{LEFT}(t_2)$.*

Lemma 9.7. *A graph $D = (V, E)$ with $V = \{x_{i_1}, \ldots, x_{i_m}\}$ coincides with the term graph $D(t)$ of a term $t(x_{i_1}, \ldots, x_{i_m})$ with $\mathrm{LEFT}(t) = x_{i_1}$ if and only if either $m = 1$ and $E \in \{\emptyset, \{(x_{i_1}, x_{i_1})\}\}$, or $V = Out^*(x_{i_1})$.*

Proof. Necessity is clear. To prove sufficiency, we proceed by induction on the number of edges in D. The induction basis is obvious: if D is a null graph, then it is the term graph of a variable. For the induction step, choose an edge $(x_{i_1}, x_j) \in E$, delete it, and denote by D_1 and D_2 the subgraphs of $(V, E \setminus \{(x_{i_1}, x_j)\})$ induced by the sets $\mathrm{Out}^*(x_{i_1})$ and $\mathrm{Out}^*(x_j)$, respectively. By the induction hypothesis there exist terms t_1 and t_2 such that $D_1 = D(t_1)$ and $D_2 = D(t_2)$, where $x_{i_1} = \mathrm{LEFT} t_1$ and $x_j = \mathrm{LEFT} t_2$. Since $V = \mathrm{Out}^*(x_{i_1})$ in D, it follows that $E = E(t_1) \cup E(t_2) \cup \{(x_{i_1}, x_j\}$. Therefore $D = D(t_1 t_2)$. □

Lemma 9.8. *Let D be a graph, and let t_1, t_2 be terms with variables $V(t_1) = V(t_2) = \{x_1, \ldots, x_m\}$ and $\mathrm{LEFT}(t_1) = \mathrm{LEFT}(t_2) = x_1$. Then $G \models t_1 = t_2$ if and only if the set of all homomorphisms from $V(t_1)$ to $V(D)$ coincides with the set of all homomorphisms from $V(t_2)$ to $V(D)$.*

Proof follows from Lemmas 9.5 and 9.6. □

Lemma 9.9. *Let D be a graph, and let t_1 and t_2 be terms with different sets of variables. Then $D \models t_1 = t_2$ if and only if $D \models t_1 = 0$ and $D \models t_2 = 0$.*

Proof. Suppose that $x_i \in V(t_1) \setminus V(t_2)$. For every

$$h : V(t_1) \cup V(t_2) \to V(D) \cup \{0\}$$

with $h(x_i)$ we get $h(t_1) = 0$ by Lemma 9.5. Hence $h(t_2) = 0$, because $D \models t_1 = t_2$. It follows that $h(t_1) = 0$ and $h(t_2) = 0$ for all assignments h. □

Theorem 9.16. ([224]) *Let $\mathcal{G}_{df}$ be the class of all finite graphs, and let $\mathcal{K} \subseteq \mathcal{G}_{df}$. Then*

$$V_{\mathcal{G}_{df}}(\mathcal{K}) = IP_f^{rp}\mathcal{K}.$$

In particular, a class is a graph variety if and only if it is closed for finite restricted pointed subproducts and isomorphic copies.

Corollary 9.4. ([224]) *Let $\mathcal{G}_{uf}$ be the class of all finite undirected graphs, and let $\mathcal{K} \subseteq \mathcal{G}_{uf}$. Then*

$$V_{\mathcal{G}_{uf}}(\mathcal{K}) = IU_fSP_f\mathcal{K}.$$

In particular, a class is a graph variety if and only if it is closed for finite restricted pointed subproducts and isomorphic copies.

The following remark shows that the question of which infinite graphs belong to a graph varieties reduces to finite induced subgraphs of those graphs.

Remark 9.1. ([224]) *Let $\mathcal{G}_d$ be the class of all graphs, $\mathcal{K} \subseteq \mathcal{G}_{df}$, and let D be a (possibly infinite) graph. Then D belongs to $V_{\mathcal{G}_d}(\mathcal{K})$ if and only if all finite induced subgraphs of D are in $V_{\mathcal{G}_d}(\mathcal{K})$.*

Lemma 9.10. ([224]) *A graph D is isomorphic to a restricted pointed subproduct of members of a class $\mathcal{K}$ of graphs if and only if the following conditions hold:*

(1) *for each $x \in V(D)$, there exist $G \in \mathcal{K}$ and a homomorphism $f : D \to G^1$ such that $f(x) \neq 1$;*

(2) *for each pair of distinct vertices x_1, x_2 in $V(D)$, there exist $G \in \mathcal{K}$ and a homomorphism $f : D \to G^1$ such that $f(x_1) \neq f(x_2)$;*

(3) *for each $x \in V(D)$ and every $y_1, y_2 \in Out^*(x)$ with $(y_1, y_2) \notin E(D)$, there exist $G \in \mathcal{K}$ and a homomorphism $f : D \to G^1$ such that $f(x) \neq 1$ and $(f(y_1), f(y_2)) \notin E(G)$.*

Assertions (1) and (3) of Lemma 9.10 yield the following condition

(4) for each pair of vertices $y_1, y_2 \in V(D)$ with $(y_1, y_2) \notin E(D)$, there exist $G \in \mathcal{K}$ and a homomorphism $f : D \to G^1$ such that $f(x) \neq 1$ and $(f(y_1), f(y_2)) \notin E(G^1)$.

Indeed, if $y_2 \in \mathrm{Out}^*_D(y_1)$, then the required f exists by (3). Otherwise, if $y_2 \in \mathrm{Out}^*_D(y_1)$, then by (1) there exists a homomorphism f_{y_1} from D to G^1 such that $f_{y_1}(y_1) \neq 1$. Define a mapping f from D to G^1 by setting, for $x \in V(D)$,

$$f(x) = \begin{cases} 1 & \text{if } y_1 \in \mathrm{Out}^*_D(x) \\ f_{y_1}(x) & \text{otherwise.} \end{cases}$$

It is readily verified that f is a homomorphism and

$$(f(y_1), f(y_1)) = (f_{y_1}(y_1), 1) \notin E(G^1),$$

because $y_2 \notin \mathrm{Out}^*_D(y_1)$.

If D is an undirected graph, then condition (3) can be replaced by (4), and D is a pointed subproduct of memebers of $\mathcal{K}$ if and only if conditions (2) and (4) hold. Indeed, $y_1, y_2 \in \mathrm{Out}^*(x)$ implies $x \in \mathrm{Out}^*(y_1)$. Hence by (4) there exists f with $(f(y_1), f(y_2) \notin E(G^1)$. Therefore $f(y_1) \neq 1$ and $f(x) \neq 1$.

The information on the automorphism group of a graph D turns out useful in verifing conditions (1) through (4). For example, if D is a vertex transitive graph, then it suffices to verify these conditions for one vertex of D.

Proof of Lemma 9.10. The 'only if' part. Let W be a restricted pointed subproduct of graph $D_i \in \mathcal{K}$, where $i \in I$. For $i \in I$, denote by p_i the projection of W onto D_i such that $p_i(a) = a(i)$ when $a \in \prod_{i\in I} D_i^1$. Evidently, all these projections are homomorphisms. Condition (1) follows from (9.11), because $p_i(a) \neq 1$ whenever $i \in Y(a)$. Next, suppose that $a \neq b$. Then there exists $i \in I$ such that $p_i(a) \neq p_i(b)$. Hence (2) follows. Finally, consider any $b, c \in \text{Out}^*(a)$ with $(b, c) \notin E(W)$. By (9.12), there exists $i \in Y(a)$ with $(p_i(b), p_(c)) \notin E(D_i)$ and $p_i(a) \neq 1$. Therefore we get (3) too.

The 'if' part. Let $\varphi_i : W \to D_i$, $i \in I$, be the set of homomorphisms satisfying (1) to (4). Consider the subgraph W' induced in $\prod_{i\in I} D_i^1$ by the set

$$V(W') = \{(\varphi_i(a))_{i\in I} \mid a \in V(W)\}$$

of vertices. Clearly, $h : a \mapsto (\varphi_i(a))_{i\in I}$ is a homomorphism from W into W'. Condition (2) shows that h is injective. It follows from (4) that h is invertible, and so it is an isomorphism. Further, (1) yields us (9.11). Finally, take any $b, c \in \text{Out}^*(a)$ with $(b, c) \in E(W)$. It follows from (3) that there exists $i \in Y(a)$ such that $\varphi_i(a) = a(i) \neq 1$ and $(b(i), c(i)) = (\varphi_i(b), \varphi_i(c)) \notin E(D_i)$. This completes the proof. □

Several important classes of graphs can be defined by identities of the form $t = 0$.

Definition 9.1. Let $G = (V, E)$ be a graph, and let $x \in V$. The graph $G' = (V', E')$ is obtained from G by *doubling the vertex* x if $V' = V \cup x'$, where $x' \notin V$, and

$$E' = E \cup \{(x', y) \mid (x, y) \in E\} \cup \{(y, x') \mid (y, x) \in E\}.$$

If G_{i+1} is obtained from G_i by doubling a vertex, for $i = 0, 1, \ldots, n-1$, then we say that G_n is obtained from G_0 by vertex-doubling, and we write $G_n \in D\{G_0\}$

Proof of Theorem 9.15. In order to prove the inclusion $V_{\mathcal{G}_{df}}(\mathcal{K}) \supseteq IP_f^{rp}\mathcal{K}$ let us take a class $\mathcal{K}$ of graphs and a set Σ of identities. We are going to verify the stronger assertion that $\mathcal{K} \models \Sigma$ implies $P^{rp}\mathcal{K} \models \Sigma$.

Take any $W \in P^{rp}\mathcal{K}$ and any equation $t = t'$ in Σ. Let $V(t) = \{x_1, \ldots, x_n\}$ and $\mathrm{LEFT}(t) = x_1$. By Lemma 9.9, we may assume that $V(t') = V(t)$ ot $t' = 0$. Suppose to the contrary that $t = t'$ does not hold in W. This means that there exists an assignment $h : V(t) \to \mathrm{Alg}(W)$ such that $h(t) \neq h(t')$. Put $h(x_i) = a_i$ for $i = 1, \ldots, n$. It suffices to consider the case where $h(t) \neq 0$, that is $h(t) = h(\mathrm{LEFT}(t)) = a_0$. By Lemma 9.5, we may assume in addition that $a_i \neq 0$ for all $i = 1, \ldots, n$. Thus $h : V(t) \to V(W)$. Consider three possible cases.

Case 1: $h(t') \neq 0$. Then $t' \neq 0$, and setting $\mathrm{LEFT}(t') = x_j$, we get $h(t') = a_j \neq a_0$. Lemma 9.5 shows that h is a homomorphism from $D(t)$ to W, and a homomorphism from $D(t')$ to W. By Lemma 9.10, condition (2), there exists G in $\mathcal{K}$ and a homomorphism $\varphi : W \to G^1$ such that $\varphi(a_1) \neq \varphi(a_j)$ and the composition h' of h and φ is a homomorphism from $D(t)$ and $D(t')$ to G^1 satisfying $h'(t) = \varphi(a_1) \neq \varphi(a_j) = h'(t')$. For every $a \in V(G)$, we get $1 \notin \mathrm{Out}^*_{G^1}(a)$, and in view of Lemma 9.7 $\mathrm{Out}^*_{D(t)}(x_1) = \mathrm{Out}^*_{D(t')}(x_j) = V(t)$ implies $1 \notin \mathrm{Out}^*_{G^1}(h'(x_1)) = \mathrm{Out}^*_{G^1}(h'(x_j)) = h'(V(t))$, because at least one of two elements $h'(x_1) \neq h'(x_j)$ is distinct from 1. Thus, 1 does not occur in the image of h', and so h' is a homomorphism into G. Therefore $t = t'$ does not hold in G under the assignment $h' : V(t) \to V(G)$, contrary to $G \in \mathcal{K} \models t = t'$. This contradition show that the first case is impossible.

Case 2: $h(t') = 0$ and $V(t) = V(t')$. Then, by Lemma 9.5, $h : V(t) \to V(W)$ is a homomorphism from $D(t)$ into W, but it is not a homomorphism from $D(t')$ into W. This means that there exists an edge $(x_r, x_s) \in E(t')$ such that $(a_r, a_s) \notin E(W)$. By Lemma 9.7, $x_r, x_s \in \mathrm{Out}^*_{D(t)}(x_1)$, and so $a_r, a_s \in \mathrm{Out}^*_W(a_1)$. Condition (3) of Lemma 9.10 implies that there exist $G \in \mathcal{K}$ and a homomorphism $\varphi : W \to G^1$ such that $\varphi(x_1) \neq 1$ and $(\varphi(a_r), \varphi(a_s)) \notin E(G)$. Since $V(t) = \mathrm{Out}^*_{D(t)}(x_1)$ and $1 \notin h'(V(t)) = \mathrm{Out}^*_{G^1}(\varphi(a_1))$, it follows that the composition h' of h and φ is a mapping from $V(t)$ to $V(G)$, which is a homomorphism from $D(t)$ into G, but is not a homomorphism from $D(t')$ to G. By Lemma 9.5, $h'(t) = h'(x_1) \neq 0 = h'(t')$. This contradicts $G \models t = t'$, and shows that the second case is impossible either.

Case 3: $t' = 0$. A contradiction follows again like in Case 2 if we use a mapping $\varphi : W \to G^1$ such that $\varphi(a_1) \neq 1$ that exists by condition (1) of Lemma 9.10. Thus, we get contraditions in all three cases. 1

In the second half of the proof we verify the reversed inclusion

$$V_{\mathcal{G}_{df}}(\mathcal{K}) \subseteq IP_f^{rp}\mathcal{K}.$$

Take any finite graph $W in V_{\mathcal{G}_{df}}(\mathcal{K})$. Let $V(W) = \{x_1, \ldots, x_n\}$. We need to verify the conditions (1) through (4) in Lemma 9.10.

Let us begin with (2). Choose any $x, y \in V(W)$, where $x \neq y$. Denote by W_x and W_y the subgraphs induced in W by the sets $\mathrm{Out}_W^*(x)$ and $\mathrm{Out}_W^*(y)$, respectively. Lemma 9.7 tells us that there exist terms t_x and t_y such that $D(t_x) = W_x$, $D(t_y) = W_y$, $\mathrm{LEFT}(t_x) = x$, and $\mathrm{LEFT}(t_y) = y$. The identical assignment $\iota : x_i \mapsto x_i$, $i = 1, \ldots, n$, yields that $\iota(t_x) = x \neq y = \iota(t_y)$ in $\mathrm{Alg}(W)$. Therefore $W \not\models t_x = t_y$. Given that $W in V_{\mathcal{G}_{df}}(\mathcal{K})$, there exists $G \in \mathcal{K}$ such that $t_x = t_y$ does not hold in G, i.e., there is an assignment $\varphi : V(t_x) \cup V(t_y) \to V(G) \cup \{0\}$ with $\varphi(t_x) \neq \varphi(t_y)$. Clearly, we may assume that $\varphi(t_x) \neq 0$. Hence $\varphi(t_x) = x$, because $x = \mathrm{LEFT}(t_x)$. The restriction $\varphi' : V(t_x) \to V(G)$ of φ is a homomorphism from $G(t_x)$ in G by Lemma 9.5. We extend φ' to $\varphi'' : V(W) \to V(G^1)$ by setting $\varphi''(x) = 1$ for all $z \in V(W) \setminus V(t_x)$. Evidently, φ'' is a homomorphism from W into G^1, because there does not exist any edge from a vertex in $V(t_x) = \mathrm{Out}_W^*(x)$ to a vertex in $V(W) \setminus V(t_x)$. Since $\varphi''(y) \in \{1, \varphi(t_y)\}$ by definition, we get $\varphi''(x) = \varphi''(t_x) = \varphi(t_x) \neq \varphi''(y)$. Hence condition (2) of Lemma 9.10 follows.

Second, we verify condition (3). Take any $x, y, y' \in V(W)$ where $y, y' \in \mathrm{Out}_W^*(x)$ and $(y, y') \notin E(W)$. Lemma 9.7 gives us a term $t \in T(X)$ with $E(t) = E(t_x) \cup \{(y, y')\}$ and $\mathrm{LEFT}(t) = x$. For example, we can obtain t from t_x by substituting yy' for y, because $V(t) = V(t_x)$ and $y, y' \in V(t_x) = \mathrm{Out}_W^*(x)$. As in the preceding paragraph, consider the terms t, t_x and the equation $t = t_x$. Since $yy' = 0$ and $\iota(t) = 0 \neq x = \iota(t_x)$ for the identical assignment ι, the equation $t = t_x$ does not hold in W.

Therefore it does not hold in some graph G in $\mathcal{K}$. This means that there exists an assignment $\varphi : V(t) \to V(G)$ such that $\varphi(t) \neq \varphi(t_x)$. Since $E(t) \supseteq E(t_x)$, we get $\varphi(t) = 0$ and $\varphi(t_x) = \varphi(t_x) = \varphi(\mathrm{LEFT}(t_x)) = \varphi(x)$. Therefore $(\varphi(y), \varphi(y')) \notin E(G)$, since otherwise $\varphi(t) =]varphi(t_x)$. Again, we can extend φ to $\varphi'' : V(W) \to V(G^1)$ by putting $\varphi''(z) = 1$ for all $z \in V(W) \setminus V(t_x)$. The mapping φ'' is a homomorphism from W to G^1, and we get $\varphi''(x) = \varphi(x) \neq 1$ and $(\varphi''(y), \varphi''(y')) = (\varphi(y), \varphi(y')) \notin E(G^1)$. Thus condition (3) is established.

It remains to verify condition (1). Take any $a \in V(W)$. If $(b, b') \notin E(W)$ for some $b, b' \in \mathrm{Out}^*_W(a)$, then (3) implies that there exists a homomorphism $\varphi : W \to G^1$ such that $\varphi(a) \neq 1$. Indeed, otherwise it would follow that the subgraph K induced in W by the set $V(K) = \mathrm{Out}^*_W(a)$ was a complete graph. If $|V(K)| \geq 2$, then by (2) there exists $b \in V(K)$ and a homomorphism φ with $\varphi(b) \neq 1$. Clearly, a homomorphism φ like this exists for all $b \in V(K)$. If, however, $|V(K)| = 1$, then K is an isolated looped vertex. At least one graph with a looped vertex must exist in $\mathcal{K}$, because otherwise $\mathcal{K} \models x_1 x_1 = 0$ would imply $W \models x_1 x_1 = 0$, a contradition to $aa = a$. Hence we can choose G in $\mathcal{K}$ with a looped vertex $z \in V(G)$. Evidently, the mapping $\varphi : W \to G^1$ defined by $\varphi(a) = z$ and $\varphi(b) = 1$, for all $b \neq a$, is the desired homomorphism. Hence (1) follows.

The two inclusions show that the exact equality holds, completing the proof. □

Theorem 9.17. *Let $\mathcal{K}$ be a set of finite loopless undirected graphs. Then $\mathcal{K} = IU_f\overline{S}D\mathcal{K}$ if and only if there exists $F \subseteq T(X)$ such that*

$$\mathcal{K} = \{D \in \mathcal{G}_{uf} \mid \forall t \in F\; G \models t = 0\}.$$

Proof. Let F be a set of terms in $T(X)$. Denote by $\mathcal{K}$ the class of all finite graphs G such that $G \models t = 0$ for all $t \in F$. We have to verify that $\overline{S}D\mathcal{K} \subseteq \mathcal{K}$.

To this end suppose that $G \models t = 0$. By Lemma 9.5, the deletion of an edge or a vertex cannot change the validity of the identity $t = 0$, and hence $\overline{S}\mathcal{K} \subseteq \mathcal{K}$. Let G' be the graph obtained grom G by doubling a vertex a. Suppose that G' does not satisfy $t = 0$. Then there exists a homomorphism $h : G(t) \to G'$ such that $h(t) = h(\mathrm{LEFT}(t))$. The vertex a and its double a' must occur among the images $h(x_i)$, for $x_i \in V(t)$, because otherwise h turns out to be a homomorphism from $G(t)$ to G contradicting to $G \models t = 0$. Assume that $h(x_1) = a$ and $h(x_2) = a'$. Denote by t' the term obtained from t by substituting x_1 for x_2 throughout. Then $G \models t' = 0$ because $G \models t = 0$. If $(x_1, x_2) \in E(t)$, then $(h(x_1), h(x_2)) \in E(G')$ contradicts $(a, a') \notin E(G')$. Therefore $(x_1, x_2) \notin E(t)$. Hence the definition of G' shows that, for every $x \in V(t) \setminus \{x_1, x_2\}$, $(h(x), h(x_1)) \in E(G')$ if and only if $(h(x), h(x_2)) \in E(G')$. A similar equivalence holds for $(h(x_1), h(x)) \in E(G')$. Therefore the mapping h' from $G(t')$ to G, defined by $h'(x_i) = h(x_i)$ for $x_i \in V(t')$, is also a homomorphism. Hence $h'(t') = h'(\mathrm{LEFT}(t'))$ and G does not satisfy $t' = 0$. This contradiction shows that our assumption was wrong and $G' \models t = 0$. Thus $D\mathcal{K} \subseteq \mathcal{K}$.

Theorem 9.15 and Proposition 9.1 imply that $U_f \subseteq \mathcal{K}$. Therefore $IU_f\overline{S}D\mathcal{K} \subseteq \mathcal{K}$.

Conversely, suppose that $IU_f\overline{S}D\mathcal{K} \subseteq \mathcal{K}$ for some set $\mathcal{K}$ of finite loopless undirected graphs. Denote by $\mathcal{L}$ the class of all finite undirected graphs H such that $H \notin \mathcal{K}$, but $\mathcal{K}$ contains all proper subgraphs of H. Given that $\overline{S}\mathcal{K} \subseteq \mathcal{K}$, we get

$$\mathcal{K} = \{G \in \mathcal{G}_{uf} \mid H \notin I\overline{S}\{G\} \forall H \in \mathcal{L}\}.$$

Obviously, every $H \in \mathcal{L}$ is connected, and so there is a term t_H with $G(t_H) \cong H$. Since H does not belong to $I\overline{S}D\mathcal{K}$, Lemmas 9.5 and 9.11 show that $G \models t_H = 0$ for all $G \in \mathcal{K}$, and

$$\mathcal{K} = \{G \mid G \models t_H = 0 \forall H \in \mathcal{K}\}.$$

This completes the proof. □

Lemma 9.11. *Let D and G be finite loopless graphs. Then there exists a homomorphism from D into G if and only if $D \in I\overline{S}D\{G\}$.*

Proof. The 'if' part. Consider any H in $\overline{S}D\{G\}$. Define a mapping h from $V(H)$ to $V(G)$ by setting $h(a') = a$ for all a' in $V(H)$ that are equal to a or arise from a by the doubling operation. Since $(a', b') \in E(H)$ implies $(a, b) \in E(G)$ whenever $(a', b') \in V(H)$ are doubles of $a, b \in V(G)$, it follows that h is a homomorphism.

The 'only if' part. Consider a homomorphism h from H to G. For all a in $V(h)$ we double the vertex a

$$|\{b \in V(H) \mid h(b) = h(a)\}|$$

times. This produces a graph $G' \in D\{G\}$ which is the full preimage of G with respect to h. Therefore H is a subgraph of G' and $H \in I\overline{S}D\{G\}$, as required. □

The formal deduction rule $\vdash$ is defined by the following four rules:

(DE1) $p = q \vdash q = p$;

(DE2) $p = q, q = r \vdash q = r$;

(DE3) (replacement rule) if $p, q, r \in T(X)$ and p occurs as a subterm of r, and if s is obtained by replacing p by q in r, then $p = q \vdash r = s$;

(DE4) (substitution rule) if $p, q, r \in T(X)$, and p', q' are obtained from p, q by replacing every occurrennce of a variable x by r, then $p = q \vdash p' = q'$.

We write $\Sigma \vdash p = q$ if there exists a sequence $p_1 = q_1, \ldots, p_n = q_n$ of equalities such that each $p_i = q_i$ either belongs to Σ, or coincides with $p = q$, or is a result of applying one of the rules (DE1) to (DE4) to preceding equalities in the sequence, and if the last equality $p_n = q_n$ is $p = q$.

Denote by Σ_0 the set of all identities that hold in all graph algebras. We write $\Sigma \vdash_g p = q$ if and only if $\Sigma \cup \Sigma_0 \vdash p = q$. The notation $\Sigma \models_g p = g$ means that the equation $p = g$ holds in all graph algebras where all equations of the set Σ hold.

Theorem 9.18. ([222]) (Completeness Theorem for Equational Logic of Graph Algebras)

$$\Sigma \models_g p = q \iff \Sigma \vdash_g p = q.$$

Chapter 10

Groupoid Rings

Let G be a groupoid, and let R be a not necessarily associative ring. The *groupoid ring* $R[G]$ consists of all finite sums $\sum_{i=1}^{n} r_i g_i$ where $r_i \in R$, $g_i \in G$, with addition and multiplication defined by the rules

$$\sum_{g\in G} r_g g + \sum_{g\in G} r'_g g = \sum_{g\in G} (r_g + r'_g) g,$$

$$\left(\sum_{g\in G} r_g g\right)\left(\sum_{h\in G} r'_h h\right) = \sum_{g,h\in G} (r_g r'_h) gh.$$

Theorem 10.1. ([186]) *Let $G = (V, E)$ be an undirected graph without loops and multiple edges. The graph algebra Alg(G) is simple if and only if, for each simple ring R, the groupoid ring $R[$Alg$(G)]$ is simple.*

The proof uses the following easy lemma.

Lemma 10.1. *If R is a simple ring, s, t are nonzero elements of R, and $k \geq 2$, then there exist $r_1, \ldots, r_k$ in R such that $s = r_1(r_2(\ldots(r_k t)\ldots))$.*

Proof of Theorem 10.1. First, suppose that Alg(G) is simple. Theorem 9.1 tells us that the graph G is connected and $\mathrm{Out}_G(x) \neq \mathrm{Out}_G(y)$ for any distinct x, y in V. In order to prove that the ring $R[G]$ is simple, it suffices to verify that the principal ideal $\langle u\rangle$ generated by any nonzero element u of $R[\,\mathrm{Alg}(G)]$ is equal to the whole ring. We proceed by induction on the number$L(u)$ of summands in the expression of u.

For the induction basis, assume that $L(u) = 1$. Then $u = rx$ for some $0 \neq r \in R$ and $x \in V$. Since $0 \neq s \in R$ and $y \in V$, it follows that there exists a walk $y, y_1, \ldots, y_{k-1}, x$ of length k in G. Lemma 10.1 says that there exist $r_1, \ldots, r_k$ such that

$$s = r_1(r_2(\ldots(r_k t)\ldots)).$$

Hence $sy = r_1 y \cdot (r_2 y_1 \cdot (r_3 y_2(\ldots(r_k y_{k-1} \cdot tx))\ldots)))$. Therefore $sy \in \langle u\rangle$, and so $\langle u\rangle$ contains all the generators of $R[G]$. Thus $R[G] = \langle u\rangle$.

For the induction step, assume that it has already been shown that $\langle u\rangle = R[G]$ for all u with $L(u) < n$. Take any $u = r_1x_1 + \cdots + r_nx_n$. Given that x_{n-1} and x_n are distinct, we may assume that there exists y in $\mathrm{Out}_G(x_{n-1}) \setminus \mathrm{Out}_G(x_{n-1})$. Hence $\langle 1y\rangle u = \langle 1y\rangle(r_1x_1 + \cdots + r_{n-1}x_{n-1})$ and $1 \leq L(\langle 1y\rangle u) \leq n-1$. Therefore $\langle u\rangle = R[\,\mathrm{Alg}(G)]$, because $\langle u\rangle \supseteq \langle\langle 1y\rangle u\rangle$.

Second, suppose that $\mathrm{Alg}(G)$ is not simple. Then G is not connected or there exist vertices $x \neq y$ of G with $\mathrm{Out}_G(x) = \mathrm{Out}_G(y)$.

If G is not connected, then denote by H a connected component of G. Clearly, $R[\,\mathrm{Alg}(H)]$ is an ideal of $R[\,\mathrm{Alg}(G)]$, and so $R[\,\mathrm{Alg}(G)]$ is not simple either.

If, however, there exist vertices $x \neq y$ of G with $\mathrm{Out}_G(x) = \mathrm{Out}_G(y)$, then then consider the set $I = \{rx - ry \mid r \in R\}$. Evidently, I is closed under addition, and for every $s \in R \setminus \{0\}$, $z \in V$, we have

$$(sz) \cdot (rx = ry) = 0;$$

$$(rx = ry) \cdot (sz) = \begin{cases} rsx - rxy & \text{if } x \in \mathrm{Out}_G(x) \\ 0 & \text{otherwise.} \end{cases}$$

Therefore I is an ideal of $R[\,\mathrm{Alg}(G)]$. Thus, the groupoid ring $R[\,\mathrm{Alg}(G)]$ is not simple again. This completes the proof. □

Theorem 10.2. ([186]) *Let $G = (V, E)$ be an undirected graph without loops and multiple edges. Then the graph algebra Alg(G) is a subalgebra of a simple graph algebra.*

Proof. Consider two possible cases.

Case 1. The graph G is connected. Put $S(G) = (V \cap V', F)$, where $V' = \{x' \mid x \in V\}$ and $F = E \cup \{(x, x') \mid x \in V, x' \in V'\}$. Then the graph G is an induced subgraph of $S(G)$. Clearly, $\text{Out}_{S(G)}(x) = \text{Out}_G(x) \cup \{x'\}$ and $\text{Out}_{S(G)}(x') = \{x\}$. Therefore $S(G)$ is a connected graph where distinct vertices have distinct neighbourhoods. Therefore $\text{Alg}(G)$ is a subgroupoid of the simple groupoid $\text{Alg}(S(G))$.

Case 2. The graph G is not connected. Let us introduce a new connected graph $S(G)$ containing G as an induced subgraph. Put $S(G) = (V \cup V' \cup \{t\}, T)$, where $V' = \{x' \mid x \in V\}$, $t \notin V \cup V'$, and $T = E \cup \{(x, x') \mid x \in V\} \cup \{(t, x) \mid x \in V\}$. Eidently, distinct vertices of $S(G)$ have distinct neighbourhoods. Hence $\text{Alg}(G)$ is a subgroupoid of the simple groupoid $\text{Alg}(S(G))$. □

Corollary 10.1. ([186]) *If R is a simple ring and G is an undirected graph without loops and multiple edges, then the groupoid ring $R[\,Alg(G)]$ can be embedded into a simple groupoid ring $R[\,Alg(S(G))]$.*

Chapter 11

Dualities, Topologies, Flatness

We refer to [28] for preliminaries on topological spaces. Let B be a finite algebra. The *alter ago* of B is a structured topological space $\mathbb{B}$, where the topology is the discrete topology on B and the additional structure consists of a system of (possibly infinitely many) operations, partial operations and relations on B each of which must be a subuniverse of the appropriate finite direct power of the algebra B. We write $A \in SP(B)$ if A is isomorphic to a subalgebra of a direct power of B. Then the set $\mathrm{Hom}(A, B)$ of all homomorphisms from A to B is a topologically closed subuniverse of $\mathbb{B}^A$. The structured topological space $\mathbb{D}(A)$ of A is called the *dual* of A with respect to $\mathbb{B}$.

Similarly, for a structured topological space $\mathbb{X}$ isomorphic to a topologically closed substructure of some nontrivial power of $\mathbb{B}$, the set of continuous structure preserving maps from $\mathbb{X}$ into $\mathbb{B}$ is denoted by $\mathrm{Hom}(\mathbb{X}, \mathbb{B})$. It is a nonempty subuniverse of the algebra B^X. The corresponding subalgebra is called the *dual* of $\mathbb{X}$ and is denoted by $E(\mathbb{X})$. There exists a natural embedding e of the algebra A into its double dual $E(\mathbb{D}(A))$. In order to define e, we associate to each $a \in A$ the evaluation map e_a defined by

$$e_a(\alpha) = \alpha(a) \text{ for all } \alpha \in \mathbb{D}(A).$$

Generally speaking the set $E(\mathbb{D}(A))$ may have elements that are not evaluation maps of this sort, and the image of e may be properly contained in $E(\mathbb{D}(A))$. If, however, e is a may onto $E(\mathbb{D}(A))$, then we say that $\mathbb{B}$ *yields a duality on* A. The algebra B is said to be *dualizable* if it has an elter ego $\mathbb{B}$ such that $\mathbb{B}$ yields a duality on A for every algebra $A \in SP(B)$.

If an algebra B is dualizable, then it follows that it is dualizable by the largest alter ego equipped with all the appropriate operations, partial operations, and relations. This alter ego is called the *brute force* alter ego of B. The algebra B fails to be dualizable if and only if there exists $A \in SP(B)$ such that the brute force alter ego of B does not yield a duality on A.

Let $A \in SP(B)$. The *dual* of A with respect to $\mathbb{B}$ is the structured topological space corresponding to $\text{Hom}(A, B)$, which is a topologically closed subspace of $\mathbb{B}^A$.

Likewise, suppose that $\mathbb{X}$ is a structured topological space isomorphic to a topologically closed substructure of some nontrivial power of $\mathbb{B}$. Then the set $\text{Hom}(\mathbb{X}, \mathbb{B})$ of continuous structure preserving mps from $\mathbb{X}$ to $\mathbb{B}$ is a nonempty subset of the algebra B^X. The corresponding subalgebra is denoted by $E(\mathbb{X})$ and is called the *dual* of $\mathbb{X}$. The mapping e assigning each a to the evaluation map at a is an embedding of the algebra A into its double dual $E(\mathbb{D}(A))$. If this mapping is onto, then we say that $\mathbb{B}$ yields a κ-duality on A. The algebra B is said to be *κ-dualizable* if it has a κ-alter ego $\mathbb{B}$ which yields a κ-duality on A for every algebra $A \in SP(B)$.

Theorem 11.1. ([48]) *For every finite undirected graph D, the following conditions are equivalent:*

(i) *the graph algebra $Alg(D)$ is dualizable;*

(ii) *each connected component of D is either complete, or bipartite complete, or an isolated vertex;*

(iii) *D is 4-transitive;*

(iv) *the equation $(x \cdot z) \cdot (y \cdot w) = (x \cdot z) \cdot (y \cdot w)$ holds in $Alg(D)$;*

(v) *the variety generated by $Alg(D)$ is finitely based.*

Let B be a finite algebra, and let κ be a cardinal. A κ-alter ego of B is a structured topological space $\mathbb{B}$ where the topology is the discrete topology on B, and the additional structure consists of a system of (possibly infinitely many) operations, partial operations, and relations on B each of which is a subset of some direct power B^λ of the algebra B, where λ is a cardinal smaller than κ. These operations, partial operations and relations are said to be κ-algebraic on B. Notice that an ordinary alter ego is an ω-alter ego.

A graph $D = (V, E)$ is 4-*transitive* if

$$(u, v), (v, w), (w, z) \in E \Rightarrow (z, u)$$

for all $u, v, w, z \in V$.

The following four small graphs occur in the next theorem.

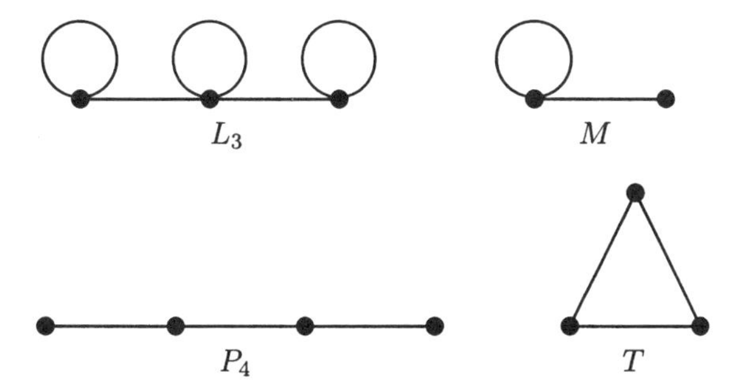

Theorem 11.2. ([48]) *For every finite undirected graph D, the following conditions are equivalent:*

(i) *the graph algebra $Alg(D)$ is inherently non-κ-dualizable for every cardinal κ;*

(ii) *Alg(D) is not dualizable;*

(iii) *at least one of the graphs M, L_3, T, or P_4 is an induced subgraph of D;*

(iv) *Alg(D) is inherently nonfinitely based;*

(v) *Alg(D) is not finitely based.*

A triple (A, Ω, T) is called a *topological algebra* if (A, Ω) is a universal algebra, T is a topology on A, and every operation of Ω is continuous with respect to T. A universal algebra (A, Ω) is called a *DT-algebra* if the discrete topology is the only topology T that can be defined on A so that (A, Ω, T) becomes a topological algebra.

An undirected graph $D = (V, E)$ is said to be *locally finite* if the neighbourhoods $\mathrm{Out}(x)$ are finite for all $x \in V$.

Theorem 11.3. ([184]) *If D is a locally finite undirected graph, then Alg(D) is not DT.*

An undirected graph $D = (V, E)$ is said to be *locally cofinite* if the complements $V \setminus \mathrm{Out}(x)$ are finite for all $x \in V$.

Theorem 11.4. ([184]) *If D is a loopless locally cofinite connected undirected graph, then Alg(D) is DT.*

Theorem 11.5. ([184]) *Let $D = (V, E)$ be an undirected graph with a loop at every vertex and with a finite nonempty subset U of V such that $|\,Out(x)| < \infty$ for all $x \in U$ and $|V \setminus Out(x)| < \infty$ for all $x \notin U$. Then Alg(D) is DT.*

Theorem 11.6. ([184]) *If D is a loopless connected undirected graph, then Alg(D) is a subalgebra of a simple DT graph algebra.*

Definition 11.1. Given a nonempty set S, a binary operation $\wedge$ is called a *meet-semilattice operation* on S if it is associative, idempotent and commutative. Thus $\wedge$ is a meet-semilattice operation on S if and only if S is a semilattice with respect to $\wedge$.

Definition 11.2. A *flat algebra* is an algebra (A, Ω) whose type Ω includes a binary meet-semilattice operation $\wedge$ and a constant 0 such that

(i) $(A, \wedge)$ is a semilattice with zero and has height 1.

(ii) 0 is the zero of the universal algebra A, that is, for every n-ary operation f of A, if $0 \in \{a_1, \ldots, a_n\}$, then $f(a_1, \ldots, a_n) = 0$.

For every graph $D = (V, E)$, the graph algebra $\mathrm{Alg}(D)$ becomes a flat graph algebra if we define the meet- semilattice operation by setting

$$x \wedge y = \begin{cases} x & \text{if } x = y, \\ 0 & \text{otherwise.} \end{cases}$$

Theorem 11.7. ([49]) *An undirected graph has a finitely based flat graph algebra if and only if it has no induced subgraph isomorphic to one of the four graphs of Figure 9.2.*

Chapter 12

Open Problems

Problem 1. ([146]) *How do properties of graph labelings determine properties of classes of languages defined on these graphs?*

Problem 2. ([146]) *When is the class of languages recognised by two-sided automata of bipartite graphs closed for intersection, union, complement, concatenation and the Kleene star operation?*

Problem 3. *When is the language recognized by a two-sided automaton star-free?*

Problem 4. *Find the group complexity of the language recognized by two-sided automaton* $Atm(D, T, f, \ell)$.

Problem 5. *Introduce nondeterministic automata defined by labeled graphs via their algebras. Investigate properties of languages recognized by these automata.*

Problem 6. *Introduce pushdown automata defined by labeled graphs via their algebras. Investigate properties of languages recognized by these automata.*

Problem 7. *For various important classes* $\mathcal{K}$ *of graphs, find conditions on a language* L *necessary and sufficient for there to be a graph in* $\mathcal{K}$ *with automaton* $Atm(D, T, f, \ell)$ *accepting* L.

Problem 8. *For various important classes $\mathcal{K}$ of graphs, describe group complexities of the languages recognized by automata* $Atm\langle D, T, f, \ell\rangle$ *of graphs in $\mathcal{K}$.*

Problem 9. *For various important classes $\mathcal{K}$ of graphs, describe all 2D and 3D picture languages or images encoded by automata* $Atm\langle D, T, f, \ell\rangle$ *of graphs in $\mathcal{K}$.*

Problem 10. *For various important classes $\mathcal{K}$ of graphs, investigate closure properties of the class of all 2D or 3D picture languages or images encoded by automata* $Atm\langle D, T, f, \ell\rangle$ *of graphs in $\mathcal{K}$ with respect to standard operations on images.*

Appendix A. Glossary of Notation

Symbol	Concept	Page
$\mid$	Divisibility relation	16
$\emptyset$	Empty set	3
$=$	Equality relation	31
$=_S$	The equality relation on S	31
$\subseteq,\supseteq$	subset	3
$\subset,\supset$	proper subset	3
$\circ$	Composition of functions	9
$\leq, \geq$	Partial order	33, 69
$\{\ldots\}$	List notation for sets	2
η_L	Nerode equivalence	181
η_T	Nerode equivalence	175
φ^{-1}	Converse relation	31
ι	Equality relation	31
ι	Equality relation on FSA	175
ι_S	The equality relation on S	31
μ_L	Myhill congruence	180, 211
ρ^{∞}	Transitive closure of ρ	32
$\rho^{\natural}$	Natural map of congruence ρ	59
ρ^{e}	Equivalence generated by ρ	32
ρ^{n}	Power of relation	32
ρ^{r}	Reflexive closure of ρ	32
$\rho \circ \eta$	Composition of relations	31
1_X	Identical function on X	10
2^X	Power set of X	3
$[1:n]$	The set $\{1,\ldots,n\}$	10
$[a]$	The equivalence class of a	32

Symbol	**Concept**	**Page**
$a \in X$	a belongs to X	1
$a \notin X$	a is not in X	1
A/ϱ	Quotient algebra	132
$a\varrho b$	a is in relation ϱ to b	31
(a, b)	Ordered pair	7
$(a, b) \in \varrho$	a is in relation ϱ to b	31
AB	Product of sets A and B in groupoid	49
AB	Product of subrings	89
AB	Product in a semiring	91
AB	Product of matrices	108
$A + B$	Sum of matrices	108
$A + B$	Sum of sets in a ring	89
$A + B$	Sum in a semiring	91
$\mathrm{Alg}(D)$	Graph algebra	50
$\mathrm{alph}(w)$	The set of letters in w	40
$\mathrm{Ann}(I)$	Annihilator	126
$\mathrm{Atm}_r(G, T)$	Right automaton	227
$\mathrm{Atm}_\ell(D, T)$	Left automaton	234
$\mathrm{Atm}_t(D)$	Two-sided automaton	249
$\mathrm{Atm}_t(D, T)$	Two-sided automaton	249
$\mathrm{Atm}_t(D, T, f, \ell)$	Two-sided automaton	249
$AX^t = B^t$	System of linear equations	109
$\mathrm{Cay}(G, T)$	Cayley graph	60
$\mathrm{Con}(\mathcal{A})$	Lattice of congruences of FSA	175
$\mathrm{Cont}_T(x)$	Context	179
$\mathrm{Cont}_{T,S}(x)$	Context	179
$\mathcal{D}$	Green's relation	81
$\deg(f)$	Degree of polynomial	99
$D_1 + D_2$	The sum of graphs D_1 and D_2	35
$D_1 \oplus D_2$	The direct sum of graphs D_1 and D_2	35
$d_H(u, v)$	Hamming distance	98
$E(D)$	The set of edges of the graph D	34
E/S	Rees quotient	70
$\mathrm{ET}(A)$	The set of elementary translations	134
$f : X \to Y$	Mapping from X to Y	9
$f(x)$	Image of x under f	9
$f(V)$	Image of V	9
f_A	Transition function	198
$\mathbb{F}_2$	The field of order two	91

Symbol	Concept	Page
$\mathbb{F}_3$	The field of order three	92
F_p	The field of order p	92
F^n	Standard vector space	93
$\mathbb{F}_{p^r}$	The field of p^r elements	97
FSA	Finite state automaton	146
$F[x]$	Polynomial ring	99
$F[x]/(f(x))$	The ring of polynomials modulo $f(x)$	100
$\lvert g\rvert$	The order of the element g	51
$\gcd(a_1, \ldots, a_n)$	Greatest common divisor	19
G^1T	Left ideal generated by T in G	57
$GF(2)$	The field of order two	91
$GF(3)$	The field of order three	92
$GF(4)$	The field of 4 elements	98
$GF(p)$	The field of order p	92
$GF(p^r)$	The field of p^r elements	97
gH	Left coset	65
$\lvert G : H\rvert$	The index of H in G	65
$(g; i, \lambda)$	Element of Rees matrix semigroup	74
G/N	Quotient group	66
(Q, X, δ, q_0, T)	Finite state automaton	144
$(Q, X, Y, \delta, \lambda, q_0)$	Mealy machine	186
$(Q, X, Y, \delta, q_0, T)$	Finite transducer	188
$\mathcal{H}$	Green's relation	81
H_x	Green's $\mathcal{H}$-class	81
Hg	Right coset gH	65
$h^{\perp}(x)$	Reciprocal polynomial	126
$\mathrm{id}(T)$	Ideal generated by T	55
$\mathrm{id}(T)$	Ideal in a ring	89
$\mathrm{id}_r(T)$	Right ideal in a ring	89
$\mathrm{id}_l(T)$	Left ideal in a ring	89
$\mathcal{J}$	Green's relation	81
J_x	Green's $\mathcal{J}$-class	81
$\mathcal{K}_E$	Variety defined by E	135
K_n	Complete graph of order n	36
K_n^m	The direct sum of N_m and K_n	36
$\mathcal{L}$	Green's relation	81
$\mathrm{lcm}(a_1, \ldots, a_n)$	Least common multiple	21
L_x	Green's $\mathcal{L}$-class	81
$M_{m,n}(R)$	The set of $m \times n$ matrices	107
$M_n(R)$	The set of $n \times n$ matrices	107
$M(G; I, \Lambda; P)$	Rees matrix semigroup	74

Symbol	**Concept**	**Page**
$M^0(G; I, \Lambda; P)$	Rees matrix semigroup	74
$m \mod n$	Remainder on division of m by n	16
$\mathbb{N}$	The set of all positive integers	2
$[n]$	The set $\{1, \ldots, n\}$	10
$n!$	n factorial	10
N_m	Null graph of order m	35
$n \mid m$	n divides m	16
$\mathbb{N}_0$	The set of nonnegative integers	2
$(\mathbb{N}, +)$	The additive semigroup of positive integers	52
$(\mathbb{N}, \cdot)$	Multiplicative semigroup of positive integers	52
$\mathcal{P}_m(S)$	The set of all m-element subsets of S	4
$\mathcal{P}(X)$	The power set of X	3
$P = [p_{\lambda i}]$	Sandwich-matrix	74
$\mathbb{Q}$	The set of all rational numbers	2
$\mathbb{Q}_+$	Set of positive rational numbers	64
$\mathbb{R}$	The set of all real numbers	2
$\mathcal{R}$	Green's relation	81
R_x	Green's $\mathcal{R}$-class	81
$\mathbb{R}_+$	Set of positive real numbers	64
$\mathrm{RCont}_T(x)$	Right context	181
$\mathrm{RCont}_{T,S}(x)$	Right context	181
R^1T	Left ideal in a ring	89
R^1TR^1	Ideal in a ring	89
$\mathrm{sign}(f)$	Sign of permutation f	14
$\sim_{\mathcal{V}}$	Equational congruence	135
S_y	Component of semilattice decomposition	70
S^1TS^1	Ideal generated by T	55
$\mathrm{Syn}(L)$	Syntactic semigroup	180
$\mathrm{Syn}(L)$	Syntactic algebra	211
$\langle T \rangle$	Subsemigroup generated by T	55
T^+	Subsemigroup generated by T	55
(T)	Ideal generated by T	55
$\langle T \rangle$	Subring generated by T	89
TR^1	Right ideal in a ring	89
TG^1	Right ideal generated by T in G	57
$T_X(\Omega)$	Term algebra	133
S^0	Semigroup with zero adjoined	73
S^1	Semigroup with identity adjoined	73
S_n	Symmetric group	10
$\lvert S_n \rvert$	Order of symmetric group	10
$S \times T$	Direct product of groupoids	52

Symbol	**Concept**	**Page**
$\mathrm{Th}(\mathcal{K})$	Equational theory of $\mathcal{K}$	135
$\cup_{y \in Y} S_y$	Semilattice Y of subsemigroups S_y	70
$V^{\perp}$	Orthogonal complement	113
$V(D)$	The set of vertices of the graph D	34
$\mathcal{V}_{\mathrm{Th}(\mathcal{K})}$	Variety generated by $\mathcal{K}$	135
$\lvert w \rvert$	The length of word w	39
$\lvert w \rvert_x$	The number of occurrences of x in w	40
$\lvert w \rvert_Y$	The number of letters of Y in w	40
$\langle X \rangle$	Subgroupoid generated by X	51
$w_H(u)$	Hamming weight	98
$\lfloor x \rfloor$	The floor of x	66
$\langle x_1, \cdots \mid r_1, \ldots \rangle$	Finitely presented semigroup.	60
x^{-1}	The inverse of x	58
$\langle x \rangle$	Cyclic semigroup	53
$x * y$	Infix notation for binary operation $*$	15
$x \equiv y \mod n$	x is congruent to y modulo n	17
X^+	Subgroupoid generated by X	51
$\overline{X}$	The complement of X	5
X^c	Complement of X	5
$\lvert X \rvert$	Cardinality of X	3
$X = Y$	Equal sets	1
$X \cup Y$	Union of X and Y	3
$X \cap Y$	Intersection of X and Y	4
$x \mapsto f(x)$	f maps x to $f(x)$	9
$X \times Y$	Direct product of sets X and Y	8
$X \neq Y$	Unequal sets	1
Y^X	The set of all functions $f : X \to Y$	9
$\mathbb{Z}_2$	The field of order two	91
$\mathbb{Z}_3$	The field of order three	92
$\mathbb{Z}_n = \mathbb{Z}/n\mathbb{Z}$	Cyclic group of order n	66
$\mathbb{Z}_p$	The field of order p	92
$\mathbb{Z}_{p^r}$	The field of p^r elements	97
$\mathbb{Z}_{p^\infty}$	Quasicyclic group	66

Bibliography

[1] Aho, A., Hopcroft, J. and Ullman, J. (1974) "The Design and Analysis of Computer Algorithms". Addison-Wesley, New York.

[2] Almeida, J. (1990) "On pseudovarietiess, varieties of languages, filters of congruences, pseudoidentities and related topics", *Algebra Universalis* **27**, 333–350.

[3] Almeida, J., Silva, P. V. and Gomes, G. M. S. (Eds) (1996) "Semigroups, Automata and Languages", World Scientific, New York.

[4] Anton, H. and Rorres, C. (2000) "Elementary Linear Algebra: Applications Version", Wiley, New York, Brisbane.

[5] Arbib, M. A. (1968) "Algebraic Theory of Machines, Languages, and Semigroups", New York, Academic Press.

[6] Babai, L. (1995) "Automorphism groups, isomorphism, reconstruction", *Handbook of Combinatorics*, Elsevier Sci., 1447–1540.

[7] Baeten, J. C. M. (1991) "Applications of Process Algebra", Cambridge University Press, Cambridge.

[8] Baeten, J. C. M. and Weijland, W. P. (1990) "Process Algebra", Cambridge University Press, Cambridge.

[9] Banchoff, W. (1995) "Linear Algebra Through Geometry", Springer, Undergraduate Texts in Mathematics, New York.

[10] Baker, K. A., McNulty, G. F. and Werner, H. (1987) "The finitely based varieties of graph algebras", *Acta Sci. Math. (Szeged)* **51** (1-2), 3–15.

[11] Batten, L.M. and Beutelspacher, A. (1993) "The Theory of Finite Linear Spaces", Cambridge University Press, Cambridge.

[12] Beachy, J. A. (1999) "Introductory Lectures on Rings and Modules", London Mathematical Society Student Texts **47**, Cambridge University Press, Cambridge.

[13] Bergstra, J. A., Ponse, A. and Smolka, S. A. (2001) "Handbook of Process Algebra", Elsevier, North-Holland.

[14] K.N. Berk, P. Carey, "Data Analysis With Microsoft Excel", Duxbury Press, 2003.

[15] Berstel, J. (1989) "Finite automata and rational languages: an introduction", *Lect. Notes Computer Science* **386**, New York, 2–14.

[16] Bespamyatnikh, S. N. and Kelarev, A. V.(2002) "An algorithm for analysis of data in geographic information systems", In: *"Australasian Workshop on Combinatorial Algorithms"*, AWOCA 2002 (7–10 July 2002, Kingfisher Bay Resort, Fraser Island), Queensland, Brisbane, 1–10.

[17] Bespamyatnikh, S. N. and Kelarev, A. V. (2003) "Algorithms for shortest paths and d-cycle problems", *J. Discrete Algorithms*, in press.

[18] Biggs, N. (1994) "Algebraic Graph Theory", Cambridge University Press.

[19] Blyth, R. (1998) "Basic Linear Algebra", Springer, Undergraduate Mathematics Series, Berlin.

[20] Blyth, R. (2001) "Further Linear Algebra", Springer, Undergraduate Mathematics Series, Berlin.

[21] Bremaud, P. (1999) "Markov Chains", Springer, Texts in Applied Mathematics, Berlin.

[22] Brualdi, R. A. and Ryser, H. J. (1991) "Combinatorial Matrix Theory", Cambridge University Press, London.

[23] Calude, C. (1998) "People and Ideas in Theoretical Computer Science", Springer, Berlin.

[24] Cazaran, J. and A.V. Kelarev (1997) "Generators and weights of polynomial codes", *Archiv Math. (Basel)* **69**, 479–486.

[25] Cazaran, J. and Kelarev, A. V. (1999) "On finite principal ideal rings", *Acta Math. Univ. Comeniae* **68**, (1), 77–84.

[26] Cazaran, J. and Kelarev, A. V. (1999) "Semisimple Artinian graded rings", *Comm. Algebra* **27** (8), 3863–3874.

[27] Choffrut, C. and Karhumaki, J. (1997) "Combinatorics of Words", In: Rozenberg, G. and Salomaa, A. (Eds.) *"Handbook of Formal Languages"*, Vol. 1, Springer, New York, 329–438.

[28] Clark, D. M. and Davey, B. A. (1998) "Natural Dualities for the Working Algebraist", Cambridge University Press, Cambridge.

[29] Clase, M. V., Jespers, E., Kelarev, A. V. and Okniński, J. (1995) "Artinian semigroup-graded rings", *Bull. London Math. Soc.* **27**, 441–446.

[30] Clase, M. V., Jespers, E., and del Rio, A. (1996) "Semigroup graded rings with finite supports", *Glasgow Math. J.* **38**, 11–18.

[31] Clase, M. V., and A. V. Kelarev, (1994) "Homogeneity of the Jacobson radical of semigroup graded rings", *Comm. Algebra* **22**, 4963–4975.

[32] Clifford, A. H. and Preston, G. B. (1963) "The Algebraic Theory of Semigroups", Amer. Math. Soc., Providence, Vols. 1 and 2.

[33] Cohen, J. S. (2002) "Computer Algebra and Symbolic Computation. Elementary Algorithms", A.K. Peters, Natick, MA.

[34] Cohen, J. S. (2002) "Computer Algebra and Symbolic Computation: Mathematical Methods", A.K. Peters, Natick, MA.

[35] Cohen, A. M. and Cuypers, H. S. H. (1999) "Algebra Interactive! Learning Algebra in an Exciting Way", Springer-Verlag, Berlin.

[36] Cohn, P. M. (2000) "Introduction to Ring Theory", Springer, Undergraduate Mathematics Series, London.

[37] Comprehensive R Archive Network, http://cran.r-project.org/

[38] Crochemore, M. and Rytter, W. (1994) "Text Algorithms", Oxford University Press, New York.

[39] Crochemore, M. and Rytter, W. (1997) Automata for matching patterns, In: Rozenberg, G. and Salomaa, A. (Eds.) *"Handbook of Formal Languages"*, Vol. 2. Springer-Verlag, Berlin, 1997, 399–462.

[40] Dăscălescu, S., Jarvis, P.D., Kelarev, A.V. and Năstăsescu, C., "On associative superalgebras of matrices", Rocky Mountain J. Math, in press.

[41] Dăscălescu, S. and Kelarev, A. V. (1999) "Finiteness conditions for semigroup-graded modules", *Revue Roumaine Math. Pures Appl.* **44** (1), 37–50.

[42] Dăscălescu, S., Kelarev, A.V. and Năstăsescu, C. (2001) "Semigroup gradings of upper triangular matrix rings", *Romanian J. Pure Applied Math.* **46**, (5), 611–615.

[43] Dăscălescu, S., Kelarev, A. V. and Torrecillas, B. (1997) "FBN Hopf module algebras", *Comm. Algebra* **25**, 3521–3529.

[44] Dăscălescu, S., Kelarev, A.V. and van Wyk, L. (2001) "Semigroup gradings of full matrix rings", *Comm. Algebra* **29**, (11), 5023–5031.

[45] Dăscălescu, S., Năstăsescu, C. and Raianu, S. (2001) "Hopf Algebras. An Introduction", *Monographs and Textbooks in Pure and Applied Mathematics* **235**, Marcel Dekker, New York.

[46] Dassow, J. and Păun, Gh. (1990) "Regulated Rewriting in Formal Language Theory", Springer, Berlin.

[47] Davenport, J. H., Siret, Y. and Tournier, E. (1993) "Computer Algebra. Systems and Algorithms for Algebraic Computation", Academic Press, London.

[48] Davey, B. A., Idziak, P. M., Lampe, W. A. and McNulty, G. F. (2000) "Dualizability and graph algebras", *Discrete Math.* **214** (1-3), 145–172.

[49] Delić, D. (2001) "Finite bases for flat graph algebras", *J. Algebra* **246**, 453–469.

[50] de Luca, A. and Varriccio, S. (1997) "Regularity and Finiteness Conditions", In: Rozenberg, G. and Salomaa, A. (Eds.) *"Handbook of Formal Languages"*, Vol. 1, Springer, New York, 747–810.

[51] de Luca, A. and Varriccio, S. (1998) "Finiteness and Regularity in Semigroups and Formal Languages", Monographs in Theoretical Computer Science, Springer-Verlag, Berlin.

[52] Denecke, K. (2002) "Universal Algebra and Applications in Theoretical Computer Science", CRC, Chapman & Hall.

[53] Drensky, V. and Lakatos, P. (1989) "Monomial ideals, group algebras and error-correcting codes", *Lecture Notes in Computer Science* **357**, 181–188.

[54] Dretzke, B. J. and Heilman, K. A. (2001) "Statistics with Excel", Prentice Hall.

[55] Durbin, J. R. (2000) "Modern Algebra: An Introduction", Wiley, New York.

[56] Ebbinghaus, H.-D., Flum, J. and Thomas, W. (1994) "Mathematical Logic", Springer, Undergraduate Texts in Mathematics, Berlin.

[57] Ehrig, H., Engels, G., Kreowski, H.-J., and Rozenberg, G. (1999) "Handbook of Graph Grammars and Computing by Graph Transformation", Vol. 1, 2, World Sci. Publishing, River Edge, NJ, 1999.

[58] Eilenberg, S. (1976) "Automata, Languages and Machines", Vol. A, B, Academic Press, New York.

[59] Fan, Y., Xiong, Q. Y. and Zheng, Y. L. (2000) "A Course in Algebra", World Scientific, River Edge, NJ.

[60] Fearnley-Sander, D., Kelarev, A. V. and Stokes, T. (2002) "Noncommutative modal rings and internalized equality", *Algebra Colloquium* **9**, (1), 65–80.

[61] The GAP Group, *GAP – Groups, Algorithms, and Programming, Version 4.3*; 2002, (`http://www.gap-system.org`).

[62] Gardner, B.J. (1989) "Radical Theory", Longman, Pitman.

[63] Gardner, B. J. and Kelarev, A. V. (1997) "Invariant radicals", *Proc. Roy. Soc. Edinburgh Sect. A* **127**, 773–780.

[64] Gardner, B. J. and Kelarev, A. V. (1998) "Two generalizations of T-nilpotence", *Algebra Colloquium* **5**, 449–458.

[65] Gardner, B.J. and Wiegandt, R. (2003) "Radical Theory of Rings", Marcel Dekker.

[66] Gecseg, F. (1986) "Products of Automata", Springer, Monographs in Theoretical Computer Science, Berlin.

[67] Gécseg, F. and Steinby, M. (1984) "Tree Languages", Académiai Kiadó, Budapest.

[68] Gécseg, F. and Steinby, M. (1997) "Tree Languages", In: Rozenberg, G. and Salomaa, A. (Eds) *"Handbook of Formal Languages"*. Vol. 3, Springer, New York, 1–68.

[69] Gentili, F., Menini, L., Tornambè, A. and Zaccarian, A. (1998) "Mathematical Methods for System Theory", World Scientific, New York.

[70] Godsil, C. and Royle, G. F. (2001) "Algebraic Graph Theory", Springer, New York.

[71] Gomes, G. M. S., Pin, J.-E. and Silva, P. V. (Eds) (2002) "Semigroups, Algorithms, Automata and Languages", (Proc. Thematic Term, Coimbra, Portugal, May - July 2001) World Scientific, New York.

[72] Graham, R. L., Grötschel, M. and Lovász, L. (1995) "Handbook of Combinatorics", Vol. 1, 2, Elsevier, Amsterdam.

[73] Graham, R. L., Rothschild, B. L. and Spencer, J. H. (1990) "Ramsey Theory", Wiley, Discrete Math. & Optimization.

[74] Grillet, P.-A. (1996) "Computing finite commutative semigroups", *Semigroup Forum* **53** (2), 140–154.

[75] Grillet, P.-A. (1995b) "Semigroups. An Introduction to the Structure Theory", Marcel Dekker, New York.

[76] Grillet, P.-A. (1999) "Algebra", Wiley, New York.

[77] Gross, J. and Yellen, J. (1998) "Graph Theory and its Aplications", Barnes & Noble.

[78] Gusfield, D. (1997) "Algorithms on Strings, Trees, and Sequences", Cambridge University Press.

[79] Halanay, A. and Ionescu, V. (1994) "Time-Varying Discrete Linear Systems", Springer, Operator Theory: Advances & Applications, Berlin.

[80] Harris, J. M., Hirst, J. and Mossinghoff, M. J. (2000) "Combinatorics and Graph Theory", Springer, Undergraduate Texts in Mathematics, Berlin.

[81] Hazewinkel, M. (Ed.) (1996) "Handbook of Algebra, Vol. 1", Amsterdam, North-Holland.

[82] Hebisch, U. and Weinert, H. J. (1990) "Semirings – Algebraic Theory and Applications in Computer Science", World Scientific, Singapore.

[83] Hederlin, Z. and Pulter, A. (1965) "Symmetric relations (undirected graphs) with given semigroups", *Monatsch. Math.* **69**, 318–322.

[84] Hederlin, Z., Vopenka, P. and Pulter, A. (1965) "A rigid relation exists on any set", *Comment. Math. Univ. Carolinae* **6**, 149–155.

[85] Heydemann, M.-C. (1997) "Cayley graphs and interconnection networks", *Graph Symmetry: Algebraic Methods and Applications* (Montreal, Canada, July 1–12, 1996), Kluwer, Dordrecht, 167–224.

[86] Herstein, I. N., (1994) "Noncommutative Rings", *Carus Mathematical Monographs* **15**, Mathematical Association of America, Washington.

[87] Holmgren, R. A. (1996) "A First Course in Discrete Dynamical Systems", Springer, Universitext, Berlin.

[88] Holub, J., Iliopoulos, C. S., Melichar, B. and Mouchard, L. (1999) "Distributed string matching using finite automata", *"Australasian Workshop on Combinatorial Algorithms"*, AWOCA 99, Perth, 114–127.

[89] Howie, J. M. (1995) "Fundamentals of Semigroup Theory", Clarendon Press, Oxford.

[90] Howie, J. M. (1991) "Automata and Languages", Clarendon Press, Oxford.

[91] Hungerford, T. W. (1990) "Abstract Algebra. An Introduction", Saunders, San Francisco – Sydney – Tokyo.

[92] Hungerford, T. W. (1997) "Algebra", *Graduate Texts in Mathematics* **73**, Springer-Verlag, New York–Berlin.

[93] Istrail, S., Pevzner, P. and Shamir, R. (Eds.) (2003) "Computational Molecular Biology", Topics in Discrete Mathematics, Elsevier.

[94] Ito, M. (2001) "Algebraic Theory of Automata and Languages", World Scientific, New York.

[95] Ito, M. and Imaoka, T. (2002) "Words, Languages and Combinatorics III", Proc. Third International Colloquium, Kyoto, Japan (14-18 March 2000).

[96] Java^{TM}, http://java.sun.com/

[97] Jespers, E., Kelarev, A. V. and Okniński J. (2001) "On the Jacobson radical of graded rings", *Comm. Algebra* **29**, (5), 2185–2191.

[98] Kargapolov, M. I. and Merzljakov, Ju. M. (1979) "Fundamentals of the Theory of Groups", Springer-Verlag, New York.

[99] Kelarev, A. V. (1990a) "On the Jacobson radical of semigroup rings of commutative semigroups", *Math. Proc. Cambridge Philos. Soc.* **108**, 429–433.

[100] Kelarev, A. V. (1990c) "When is the radical of a band sum of rings homogeneous?", *Comm. Algebra* **18**, 585–603.

[101] Kelarev, A. V. (1991a) "Hereditary radicals and bands of associative rings", *J. Austral. Math. Soc. Ser. A* **51**, 62–72.

[102] Kelarev, A. V. (1992a) "The regular radical of semigroup rings of commutative semigroups", *Glasgow Math. J.* **34**, 133–141.

[103] Kelarev, A. V. (1992b) "Radicals and semilattice sums of rings revisited", *Comm. Algebra* **20**, 701–709.

[104] Kelarev, A. V. (1992c) "The Jacobson radical of commutative semigroup rings", *J. Algebra* **150**, 378–387.

[105] Kelarev, A. V. (1992d) "Radicals of graded rings and applications to semigroup rings", *Comm. Algebra* **20**, 681–700.

[106] Kelarev, A. V. (1992e) "On the Jacobson radical of graded rings", *Comment. Math. Univ. Carolinae* **33**, 21–24.

[107] Kelarev, A. V. (1993a) "A sum of two locally nilpotent rings may be not nil", *Arch. Math. (Basel)* **60**, 431–435.

[108] Kelarev, A. V. (1993b) "On the Jacobson radical of strongly group graded rings", *Comment. Math. Univ. Carolinae* **35**, 575–580.

[109] Kelarev, A. V. (1993c) "A general approach to the structure of radicals in some ring constructions", "Theory of Radicals", Szekszárd, 1991, *Coll. Math. Soc. János Bolyai* **61**, 131–144.

[110] Kelarev, A. V. (1993d) "Strongly semigroup graded rings with nilpotent radicals", *Simon Stevin* **67**, 323–331.

[111] Kelarev, A. V. (1993e) "On semigroup graded PI-algebras", *Semigroup Forum* **47**, 294–298.

[112] Kelarev, A. V. (1994a) "Nil properties for rings which are sums of their additive subgroups", *Comm. Algebra* **22**, (13), 5437–5446.

[113] Kelarev, A. V. (1994b) "Finiteness conditions for special band graded rings", *Demonstratio Math.* **27**(1994), 171–178.

[114] Kelarev, A. V. (1994c) "Radicals of semigroup rings of commutative semigroups", *Semigroup Forum* **48**, 1–17.

[115] Kelarev, A. V. (1994d) "The group of units of a commutative semigroup ring", *J. Algebra* **170**, 902–912.

[116] Kelarev, A. V. (1995a) "Artinian band sums of rings", *J. Austral. Math. Soc. Ser. A* **57**, 66–72.

[117] Kelarev, A. V. (1995b) "Radicals of contracted graded rings", *J. Algebra* **176**, 561–568.

[118] Kelarev, A. V. (1995c) "On groupoid graded rings", *J. Algebra* **178**, 391–399.

[119] Kelarev, A. V. (1995d) "Applications of epigroups to graded ring theory", *Semigroup Forum* **50**, 327–350.

[120] Kelarev, A. V. (1996a) "On the structure of the Jacobson radical of graded rings", *Quaestiones Math.* **19**, 331–340.

[121] Kelarev, A. V. (1996b) "Radicals of algebras graded by cancellative linear semigroups", *Proc. Amer. Math. Soc.* **124**, 61–65.

[122] Kelarev, A. V. (1997a) "A primitive ring which is a sum of two Wedderburn radical subrings", *Proc. Amer. Math. Soc.* **125**, (7), 2191–2193.

[123] Kelarev, A. V. (1997b) "Generalized radical semigroup rings", *Southeast Asian Bulletin Math.* **21**, 85–90.

[124] Kelarev, A. V. (1997c) "On classical Krull dimension of group-graded rings", *Bull. Austral. Math. Soc.* **55**, 255–259.

[125] Kelarev, A. V. (1998a) "An answer to a question of Kegel on sums of rings", *Canad. Math. Bull.* **41**, (1), 79–80.

[126] Kelarev, A. V. (1998b) "Semigroup rings in semisimple varieties", *Bull. Austral. Math. Soc.* **57**, 387–391.

[127] Kelarev, A. V. (1998c) "Semisimple rings graded by inverse semigroups", *J. Algebra* **205**, 451–459.

[128] Kelarev, A. V. (1998d) "Recent results and open questions on radicals of semigroup-graded rings", *Fund. Appl. Math.* **4**, (4), 1115–1139.

[129] Kelarev, A. V. (1998e) "Directed graphs and nilpotent rings", *J. Austral. Math. Soc. Ser. A* **65**, 326–332.

[130] Kelarev, A. V. (1999a) "Rings which are sums of finite fields", *Finite Fields and their Applications* **5**, 89–95.

[131] Kelarev, A. V. (1999b) "Band-graded rings", *Math. Japonica* **49**, (3), 467–479.

[132] Kelarev, A. V. (1999c) "Two ring constructions and sums of fields", *New Zealand J. Math.* **28**, 43–46.

[133] Kelarev, A. V. (1999d) "On rings with inverse adjoint semigroup", *Southeast Asian Bull. Math.* **23**, (3), 431–436.

[134] Kelarev, A. V. (1999e) "Semigroup varieties and semigroup algebras", *Semigroup Forum* **59**, 461–466.

[135] Kelarev, A. V. (1999f) "Semigroup rings which are products of Galois rings", *Semigroup Forum* **59**, 467–469.

[136] Kelarev, A. V. (2000a) "Varieties and sums of rings", *Contrib. Algebra & Geometry* **41** (2), 325–328.

[137] Kelarev, A. V. (2000b) "Finiteness conditions and sums of rings", *Publ. Math. (Debrecen)* **57** (1-2), 115–119.

[138] Kelarev, A. V. (2001) "On a theorem of Cohen and Montgomery for graded rings", *Proc. Royal Soc. Edinburgh A* **131**, 1163–1166.

[139] Kelarev, A. V. (2002) "Ring Constructions and Applications", World Scientific, New York.

[140] Kelarev, A. V. (2002) "An algorithm for repeated convex regions in geographic information systems", *Far East J. Applied Math.* **8**, (1), 75–79.

[141] Kelarev, A. V. (2002) "On undirected Cayley graphs", *Australasian J. Combinatorics* **25**, 73–78.

[142] Kelarev, A. V. (2003) "An algorithm for 3D image analysis", *Far East J. Applied Math.*, in press.

[143] Kelarev, A. V. "Directed graphs and Lie superalgebras of matrices", to appear.

[144] Kelarev, A. V., Göbel, R., Rangaswamy, K.M., Schultz, P. and Vinsonhaler, C. (2001) "Abelian Groups Rings and Modules", *Contemporary Mathematics* **273**, American Mathematical Society.

[145] Kelarev, A. V. and McConnell, N. R. (1995) "Two versions of graded rings", *Publ. Math. (Debrecen)* **47**, (3-4), 219–227.

[146] Kelarev, A. V., Miller, M. and Sokratova, O. V. (2001) "Directed graphs and closure properties for languages", *12 Australasian Workshop on Combinatorial Algorithms, AWOCA 2001,* (ed. E.T. Baskoro), Putri Gunung Hotel, Lembang, Bandung, Indonesia, July 14-17, 118–125.

[147] Kelarev, A. V., Miller, M. and Sokratova, O. V. Languages recognized by two-sided automata of graphs, *Proc. Estonian Academy of Science*, in press.

[148] Kelarev, A. V. and Okniński, J. (1995) "Group graded rings satisfying polynomial identities", *Glasgow Math. J.* **37**, 205–210.

[149] Kelarev, A. V. and Okniński, J. (1996) "The Jacobson radical of graded PI-rings and related classes of rings", *J. Algebra* **186** (3), 818–830.

[150] Kelarev, A. V. and Plant, A. (1995) "Bergman's lemma for graded rings", *Comm. Algebra* **23** (12), 4613–4624.

[151] Kelarev, A. V. and Plant, A. (1996) "Graded rings with invariant radicals", *Bull. Austral. Math. Soc.* **53**, 143–147.

[152] Kelarev, A. V. and Praeger, C. E. (2003) "On transitive Cayley graphs of groups and semigroups", *European J. Combinatorics* **24**, (1), 59–72.

[153] Kelarev, A. V. and Quinn, S. J. (2000) "A combinatorial property and power graphs of groups", *Contrib. General Algebra* **12**, "58. Arbeitstagung Allgemeine Algebra" (Vienna University of Technology, June 3-6, 1999) Springer-Verlag, 2000, 229–235.

[154] Kelarev, A. V. and Quinn, S. J. (2001) "Directed graphs and combinatorial properties of semigroups", *J. Algebra* **251**,(1), 16–26.

[155] Kelarev, A. V. and Quinn, S. J. (2003) "A combinatorial property and Cayley graphs of semigroups", *Semigroup Forum* **66**, 89–96.

[156] Kelarev, A. V., Quinn, S. J. and Smolikova R. (2000) "On fuzzy regular languages", *Artificial Intelligence in Science and Technology 2000*, Hobart, Engineering, 291–295.

[157] Kelarev, A. V., Quinn, S. J. and Smolikova R. (2001) "Power graphs and semigroups of matrices", *Bull. Austral. Math. Soc.* **63**, 341–344.

[158] Kelarev, A. V., Quinn, S. J. and Sokratova, O. V. (2000) "Automata with languages recognized by graph algebras", *Artificial Intelligence in Science and Technology 2000*, Hobart, Engineering, 296–301.

[159] Kelarev, A. V. and Sokratova, O. V. (2000) "Syntactic semigroups and graph algebras", *Bull. Aust. Math. Soc.* **62**, 471–477.

[160] Kelarev, A. V. and Sokratova, O. V. (2000) "An algorithm for languages recognized by graph algebras", *11 Australasian Workshop on Combinatorial Algorithms, AWOCA 2000,* Information Technology, University of Newcastle, 43–52.

[161] Kelarev, A. V. and Sokratova, O. V. (2001) "A class of semisimple automata", *Korean J. Comput. Applied Math.* **8** (1), 1–8.

[162] Kelarev, A. V. and Sokratova, O. V. (2001) "Languages recognized by a class of finite automata", *Acta Cybernetica* **15**, 45–52.

[163] Kelarev, A. V. and Sokratova, O. V. (2001) "Information rates and weights of codes in structural matrix rings", *Lect. Notes Computer Science* **2227**, 151–158.

[164] "Directed graphs and syntactic algebras of tree languages", *J. Automata, Languages & Combinatorics* **6** (2001)(3), 305–311.

[165] Kelarev, A. V. and Sokratova, O. V. (2002) "Two algorithms for languages recognized by graph algebras", *International Journal of Computer Mathematics* **79**, (12), 1317–1327.

[166] Kelarev, A. V. and Sokratova, O. V. (2003) "On congruences of automata defined by directed graphs", *Theoretical Computer Science*, in press.

[167] Kelarev, A. V. and Solé, P. (2001) "Error-correcting codes as ideals in group ring", *Contemporary Mathematics* **273**, 11–18.

[168] Kelarev, A. V. and Stokes, T. (1999) "Central internal algebras and varieties", *Comm. Algebra* **27** (8), 3851–3862.

[169] Kelarev, A. V. and Stokes, T. (1999) "Interior algebras and varieties", *J. Algebra* **221**, 50–59.

[170] Kelarev, A. V. and Stokes, T. (2000) "Equality algebras and varieties", *Fund. Appl. Math.* **6** (3), 1–5.

[171] Kelarev, A. V. and Stokes, T. (2001) "Rings used in modal logic and their radicals", *Vietnam Math. J.* **29** (2001)(2), 179–190.

[172] Kelarev, A. V. and Trotter, P. G. (2001) "A combinatorial property of automata, languages and their syntactic monoids", *"Words, Languages, and Combinatorics"* (Kyoto, 14-19 March 2000).

[173] Kelarev, A. V., van der Merwe, B. and van Wyk, L. (1998) "The minimum number of idempotent generators of an upper triangular matrix algebra", *J. Algebra* **205**, 605–616.

[174] Khoussainov, B. and Nerode, A. (2001) "Automata Theory and its Applications", Birkhauser, Berlin.

[175] Kiss, E. W. (1985) "A note on varieties of graph algebras", *Universal algebra and lattice theory (Charleston, S.C., 1984)*, Springer, Berlin, 163–166.

[176] Kiss, E. W., Pöschel, R. and Pröhle, P. (1990) "Subvarieties of varieties generated by graph algebras", *Acta Sci. Math. (Szeged)* **54** (1-2), 57–75.

[177] Kozen, D. C. (1997) "Automata and Computability", Springer, Undergraduate Texts in Computer Science, Berlin.

[178] Kuich, W. (1997) "Semirings and formal power series", In: Rozenberg, G. and Salomaa, A. (Eds.) *"Handbook of Formal Languages"*, Vol. 1, Springer, New York, 609–677.

[179] Kuich, W. and Salomaa, A. (1985) "Semirings, Automata, Languages", Springer, Monographs in Theoretical Computer Science, Berlin.

[180] Lallement, G. (1979) "Semigroups and Combinatorial Applications", Wiley, New York.

[181] Lampe, W. A. (1991) "Further properties of lattices of equational theories", *Algebra Universalis* **28** (4), 459–486.

[182] Lang, S. (1997) "introduction to Linear Algebra", Springer, Undergraduate Texts in Mathematics, Berlin.

[183] Lang, S. (1996) "Linear Algebra", Springer, Undergraduate Texts in Mathematics, Berlin.

[184] Lee, S.-M. (1988) "Graph algebras which admit only discrete topologies", *Congr. Numer.* **64**, 147–156.

[185] Lee, S.-M. (1989) "On hereditarily simple graph algebras", *Bull. Malaysian Math. Soc. (2)* **12**, (1), 29–31.

[186] Lee, S.-M. (1991) "Simple graph algebras and simple rings", *Southeast Asian Bull. Math.* **15**, (2), 117–121.

[187] D.M. Levine, P.P. Ramsey, R.K. Smidt, "Applied Statistics For Engineers and Scientists Using Microsoft Excel and MINITAB", Prentice Hall, 2001.

[188] Lidl, R. and Niederreiter, H. (1983) "Finite Fields", Addison-Wesley, New York.

[189] Lidl, R. and Niederreiter, H. (1994) "Introduction to Finite Fields and their Applications", Cambridge University Press, Cambridge.

[190] Lidl, R. and Pilz, G. (1998) "Applied Abstract Algebra", Springer-Verlag, New York.

[191] Lidl, R. and Wiesenbauer, J. (1980) "Ring Theory and Applications" (German), Wiesbaden.

[192] Lima, P. U. and Saridis, G. N. (1996) "Design of Intelligent Control Systems Based on Hierarchical Stochastic Automata", World Scientific, New York.

[193] Magma Computational Algebra System, http://magma.maths.usyd.edu.au/magma/

[194] Mahmoud, M. S. and Singh, M. G. (1984) "Discrete Systems: Analysis, Control, and Optimization", Springer, Communication & Control Engineering, Berlin.

[195] Matescu, A. and Salomaa, A. (1997) "Formal languages: an introduction and a synopsis", In: Rozenberg, G. and Salomaa, A. (Eds.) *"Handbook of Formal Languages"*, Vol. 1, Springer, New York, 329–438.

[196] McKenzie, R. N., McNulty, G. F. and Taylor, W. F. (1987) "Algebras, Lattices, Varieties, Vol. 1", Wadworth & Brooks, Monterey, CA.

[197] McNulty, G. F. and Shallon, C. R. (1983) "Inherently nonfinitely based finite algebras", *Universal algebra and lattice theory (Puebla, 1982)*, Springer, Berlin, 206–231.

[198] McPherson, G. (1990) "Statistics in Scientific Investigation: its Basis, Application, and Interpretation", Springer-Verlag, New York.

[199] McPherson, G. (2001) "Applying and Interpreting Statistics: a Comprehensive Guide", New York, Springer.

[200] Meduna, A. (2000) "Automata and languages", Springer, London.

[201] Menezes, A.J., van Oorschot, P. C. and Vanstone, S. A. (1996) "Handbook of Applied Cryptography", CRC Press, New York–Tokyo.

[202] Mordeson, J. N. and Malik, D. S. (2002) "Fuzzy Automata and Languages", CRC Press, Chapman & Hall.

[203] Murota, K. (2000) "Matrices and Matroids for Systems Analysis", Springer, Algorithms and Combinatorics, Berlin.

[204] Murskiĭ, V. L. (1965) "The existence in three-valued logic of a closed class with finite basis, not having a finite complete system of identities", *Soviet Math. Dokl.* 6 (1965), 1020–1024.

[205] Nehaniv C. L. and Ito, M. (Eds) (1999) "Algebraic Engineering", (Proc. Int. Workshop Formal Languages and Computer Systems, Kyoto, Japan 18 – 21 March 1997, and First Int. Conf. Semigroups and Algebraic Engineering Aizu, Japan 24 – 28 March 1997), World Scientific, New York.

[206] Oates-Williams, S. (1980) "Murski's algebra does not satisfy min", *Bull. Austral. Math.* **22**, 199–203.

[207] Oates-Williams, S. (1981) "Graphs and universal algebras", *Combinatorial mathematics, VIII (Geelong, 1980)*, Springer, Berlin, 351–354.

[208] Oates-Williams, S. (1984) "On the variety generated by Murskiĭ's algebra", *Algebra Universalis* **18**, (2), 175–177.

[209] Oehmke, R. H. (1990) "On semisimplicity for automata", *"Words, Languages and Combinatorics"* (Kyoto), World Scientific, Singapore, 373–384.

[210] Oehmke, R. H. (1992) "Automata with complemented lattices of congruences", *"Words, Languages and Combinatorics"* (Kyoto), World Scientific, Singapore, 294–300.

[211] Oehmke, R. H. (1998) "The direct product of right congruences", *Korean J. Comput. Appl. Math.* **5**, 475–480.

[212] Passman, D. S. (1991) "A Course in Ring Theory", Wadsworth & Brooks, Pacific Grove, CA.

[213] Paun, G., Rozenberg, G. and Salomaa, A. (1998) "DNA Computing: New Computing Paradigms", Springer, Berlin, New York.

[214] Paun, G. and Salomaa, A. (1997) "New Trends in Formal Languages", *Lect. Notes Comp. Sci.* **1218**, Springer-Verlag, Berlin.

[215] Penner, R. C. (1999) "Discrete Mathemtics. Proof Techniques and Mathematical Structures", World Scientific, New York.

[216] Pin, J. E. (1989) "Formal Properties of Finite Automata and Applications", *Lect. Notes Computer Science* **386**, Springer, New York.

[217] Pless, V. S., Huffman, W. C. and Brualdi, R. A. (1998) "Handbook of Coding Theory", Vol. I, II, North-Holland, Amsterdam.

[218] Pless, V. S., Huffman, W. C. and Brualdi, R. A. (1998) "An introduction to algebraic codes", In: *"Handbook of Coding Theory"*, Vol. I, II, North-Holland, Amsterdam, 3–139.

[219] Plotkin, B. I., Greenglaz, L. Ja. and Gvaramija, A. A. (1992) "Algebraic Structures in Automata and Databases Theory", World Scientific, New York.

[220] Poli, A. and Huguet, L. (1992) "Error-Correcting Codes: Theory and Applications", Prentice-Hall.

[221] Pöschel, R. (1986) "Shallon algebras and varieties for graphs and relational systems", *Algebra und Graphentheorie.* Siebenleht, Bergakademie Freiberg, 53–56.

[222] Pöschel, R. (1989) "The equational logic for graph algebras", *Z. Math. Logik Grundlag. Math.* **35** (3), 273–282.

[223] Pöschel, R. (1989) "Graph varieties containing Murskiĭ's groupoid", *Bull. Austral. Math. Soc.* **39** (2), 265–276.

[224] Pöschel, R. (1990) "Graph algebras and graph varieties", *Algebra Universalis* **27** (4), 559–577.

[225] Pöschel, R. and Wessel, W. (1987) "Classes of graphs definable by graph algebra identities or quasi-identities", *Comment. Math. Univ. Carolin.* **28** (3), 581–592.

[226] Roberts, F. (1989) "Applications of Combinatorics and Graph Theory to the Biological and Social Sciences", Springer, New Brunswick.

[227] Robinson, D.J.S. (1982) "A Course in the Theory of Groups", Springer, New-York, Berlin.

[228] Roman, S. (1995) "Advanced Linear Algebra", Graduate Texts in Mathematics, Springer, Berlin.

[229] Rozenberg, G. and Salomaa, A. (1993) "Current trends in theoretical computer science: essays and tutorials," World Scientific, New York.

[230] Rozenberg, G. and Salomaa, A. (Eds.) (1997), "Handbook of Formal Languages", Vol. 1, 2, 3, Springer, New York.

[231] Salomaa, A. and Soittola, M. (1978) "Automata-Theoretic Aspects of Formal Power Series", Springer, Texts and Monographs in Computer Science, Berlin.

[232] Salomaa, A., Wood, D. and Yu, S. (2001) "A Half-Century of Automata Theory", World Scientific, New York.

[233] Scherk, J. (2000) "Algebra. A Computational Introduction", *Studies in Advanced Mathematics*, Chapman & Hall, Boca Raton, FL.

[234] Shevrin, L. N. (1961) "Nilsemigroups with certain finiteness conditions", *Mat. Sb.* (N.S.) **55** (97), 473–480.

[235] Shevrin, L. N. and Ovsyannikov, A. J. (1996) "Semigroups and their Subsemigroup Lattices", Kluwer, Dordrecht.

[236] Shparlinski, I. (1992) "Computational and Algorithmic Problems in Finite Fields", Kluwer.

[237] Shparlinski, I. (1999) "Finite Fields: Theory and Computation", Kluwer.

[238] Shparlinski, I. (1999) "Number Theoretic Methods in Cryptography: Complexity Lower Bounds", Birkhauser.

[239] Simon, M. (1999) "Automata Theory", World Scientific, New York.

[240] Stanton, D. and White, D. (1986) "Constructive Combinatorics", Springer, Undergraduate Texts in Mathematics, Berlin.

[241] Steinby, M. (1981) "Some aspects of recognazability and rationality", *Lecture Notes Comp. Sci.* **117**, 360–372.

[242] Steinby, M. (1992) "A theory of tree languages varieties", In: *"Tree automata and languages"*, North-Holland, 57–81.

[243] Steinby, M. (1994) "Classifying regular languages by their syntactic algebras", *Lecture Notes Comp. Sci.* **812**, 396–409.

[244] Straubing, H. (1994) "Finite Automata, Formal Logic and Circuit Complexity", Springer, Progress in Theoretical Computer Science, Berlin.

[245] Street, A. P. and Wallis, W. D. (1977) "Combinatorial Theory: An Introduction". *Charles Babbage Research Centre*, Brisbane.

[246] Sundararajan, D. (2001) "The Discrete Fourier Transform. Theory, Algorithms and Applications", World Scientific, New York.

[247] Tornambè, A. (1995) "Discrete-Event System Theory. An Introduction", World Scientific, New York.

[248] Triola, M. F. (2003) "Elementary Statistics Using Excel", Addison-Wesley.

[249] van Tilborg, H. C. A. (1998) *Coding theory at work in cryptography and vice versa*, In: "Handbook of Coding Theory", Elsevier, New York, 1195–1227.

[250] Van, D. L. and Ito, M. (Eds) (2002) "The Mathematical Foundations of Informatics", Proc. Conf. Hanoi, Vietnam 25–28 October 1999), World Scientific, New York.

[251] van Leeuwen, J. (1990) "Handbook of Theoretical Computer Science. Vol. A, B Algorithms and Complexity", Elsevier, Amsterdam.

[252] Vinogradov (1972) "Foundations of Number Theory", Moscow, Science.

[253] Wechler, W. (1992) "Universal Algebra for Computer Scientists" Springer, Monographs in Theoretical Computer Science,

[254] Yan, S. Y. (1998) "An Introduction to Formal Languages and Machine Computation", World Scientific, New York.

[255] Yu, S. (1997) "Regular Languages", In: Rozenberg, G. and Salomaa, A. (Eds.) *"Handbook of Formal Languages"*, Vol. 1, Springer, New York, 41–110.

Index